T0362926

MARIAN WILKINSON is a multi-award-winning journalist with a career that has spanned radio, television and print.

She has covered politics, national security, refugee issues and climate change as well as serving as a foreign correspondent in Washington DC for the *Sydney Morning Herald* and *The Age*. She was a deputy editor of the *Sydney Morning Herald*, executive producer of the ABC's *Four Corners* program and a senior reporter with *Four Corners*.

As environment editor for the *Sydney Morning Herald*, she reported on the rapid melt of Arctic sea ice for a joint *Four Corners–Sydney Morning Herald* production that won a Walkley Award for journalism and the Australian Museum's Eureka prize for environmental journalism. She also covered the UN climate conferences in Bali and Copenhagen.

As a member of the International Consortium of Investigative Journalists (ICIJ), she reported on the Panama Papers and the Paradise Papers for *Four Corners*.

She has written several books, including the political biography *The Fixer*, on former Labor powerbroker Graham Richardson, and *Dark Victory*, on Australia's response to asylum seekers, which she co-authored with David Marr.

In 2018 she was inducted into the Australian Media Hall of Fame.

OTHER BOOKS BY MARIAN WILKINSON

The Fixer
Dark Victory (co-authored with David Marr)

THE CARBON CLUB

HOW A NETWORK OF INFLUENTIAL CLIMATE SCEPTICS,
POLITICIANS AND BUSINESS LEADERS FOUGHT TO
CONTROL AUSTRALIA'S CLIMATE POLICY

MARIAN WILKINSON

ALLEN&UNWIN
SYDNEY · MELBOURNE · AUCKLAND · LONDON

First published in 2020

Copyright © Marian Wilkinson 2020

All rights reserved. No part of this book may be reproduced or transmitted in
any form or by any means, electronic or mechanical, including photocopying,
recording or by any information storage and retrieval system, without prior
permission in writing from the publisher. The Australian *Copyright Act 1968*
(the Act) allows a maximum of one chapter or 10 per cent of this book,
whichever is the greater, to be photocopied by any educational institution for
its educational purposes provided that the educational institution (or body
that administers it) has given a remuneration notice to the Copyright Agency
(Australia) under the Act.

Allen & Unwin
83 Alexander Street
Crows Nest NSW 2065
Australia
Phone: (61 2) 8425 0100
Email: info@allenandunwin.com
Web: www.allenandunwin.com

A catalogue record for this
book is available from the
National Library of Australia

ISBN 978 1 76087 599 2

Internal design by Philip Campbell Designs
Index by Garry Cousins
Set in 11.25/15.75 pt Sabon by Midland Typesetters, Australia
Printed and bound in Australia by Griffin Press, part of Ovato

10 9 8 7 6 5 4 3

The paper in this book is FSC® certified.
FSC® promotes environmentally responsible,
socially beneficial and economically viable
management of the world's forests.

*For Matthew,
with love and thanks*

CONTENTS

CONTENTS

PROLOGUE

I ALWAYS EXPECTED the damaging impacts brought on by a hotter planet would ratchet up in my lifetime. For years I'd spoken to so many scientists, reported on so many jolting changes, from the melting of the Arctic sea ice to the bleaching of the Great Barrier Reef, that this seemed inevitable. But even I was surprised by what happened on the eve of the new decade.

Bushfires are nothing new. Australia is the land of droughts, flooding rains and fire. But the bushfires we saw in 2019 and into 2020 were different. The length of the fire season, the intensity and destructiveness of the fires, fitted a pattern climate scientists had been predicting for years.

The warning signs were there that winter. My brother-in-law with the volunteer Rural Fire Service in the Northern Rivers region of New South Wales was already getting call-outs in July. The long drought, the tinder-dry conditions, the higher temperatures all meant an early start to the fire season. As the weeks went by, the conditions turned from bad to worse in small towns and communities spread across his area. As I sat at my desk in Sydney, my sister would phone in with stories of long nights waiting at home, houses gone up in flames, a tea-tree plantation destroyed, jobs lost and, tragically, lives too.

At the same time, I was also, weirdly, monitoring fires on the other side of the world. That July my husband was canoeing through the Arctic Circle on the adventure of a lifetime. He and his party were paddling down the Porcupine River from Canada into Alaska, spotting moose and bear and camping on the river-bank. But halfway through their journey I got a nervous satellite text asking me to check on wildfires coming their way. They had no internet and were surrounded by smoke. They were worried the village of Old Crow where they were heading had gone up in flames. When I finally spoke to the local police officer, I discovered the village was safe and the thick smoke was blowing in from Alaska, where huge wildfires were burning across the state. That July was the hottest month Alaska had ever recorded. Their fire season too had started early and was more severe than usual. And it struck me acutely that the warnings I was seeing from former fire chiefs in Australia were not academic; fire seasons in North America and Australia, once distinct, were now overlapping. Sharing resources in our fire season would soon be that much harder.

By the time the family all got together in early summer in Sydney, our city was blanketed by smoke from bushfires burning to the west and north. The air quality hovered between poor and hazardous. Even at the beach there was no escape, as we swam out through waves littered with burnt leaves and ash. But the worst was yet to come. New Year's Day 2020 brought news of unforgettable devastation from around the country. We spent our time glued to our phones, to the radio and TV—and trying to call friends caught up in the fires.

For many Australians, politics and science collided that day. Prime Minister Scott Morrison, who once held up a lump of coal in parliament, could no longer avoid the words 'climate change'. When he ordered the navy to evacuate thousands of summer holidaymakers sheltering on beaches from blazing fires and called out 3000 defence force reservists to help state authorities, we knew our emergency services had been stretched beyond their capacity.

Climate adaptation suddenly became the new buzz-phrase for Morrison's cabinet ministers. The treasurer spoke in billion-dollar figures as he totted up the costs of the bushfire recovery.

Finally it seemed government budgets would have to factor in a premium for climate change, recognition that natural disasters were becoming more frequent and more intense. This was a cost we never wanted to put in the ledger, hoping to leave it to future generations. That was no longer possible.

The fires brought another remarkable change. A few Liberal Party politicians rediscovered their mettle and talked about action on climate change. The party's climate sceptics, for so long encouraged, elected to leadership roles and lionised by many in the media, were being pushed into the background.

As I watched this happening I knew it had happened before and the battle was on.

The idea for this book began fifteen years ago after I had finished up a stint as a foreign correspondent in Washington DC. I had collected a stack of documents on the close working relationship between US politicians, business lobbyists and influential climate sceptics. The documents sat in bound folders on my bookshelf, but every so often I found myself re-reading them as I reported on the bitterly divisive politics of climate change in Australia.

I understood why the climate-sceptic scientists were so important to the politicians and business leaders who encouraged and cultivated them. Their scepticism, passionately believed in most cases, removed the moral imperative to act on climate change. They made the case for protecting Australia's carbon-intensive economy, its profits and its many jobs, from the scientific findings of so-called 'climate alarmists'—even though these included the overwhelming majority of climate scientists.

As a result, the climate sceptics were always the loud chorus accompanying the battle over climate change in Australia. Any plan to seriously grapple with the wickedly difficult job of transforming the country into a cleaner economy was scuppered by the carbon club of politicians, business leaders and their climate-sceptic allies. They were dug in for the long haul.

The battle is still in full swing but the stakes today are so much higher. The economic shock caused by the COVID-19 pandemic has made the task of addressing climate change that much harder, especially in Australia. The profits and the jobs from the carbon-heavy industries remain the backbone of many communities. Seriously cutting Australia's greenhouse gas emissions—climate mitigation—is again bitterly contested. Yet the risks to Australia from rising global emissions are impossible to ignore. Global temperatures are on track to hit well over 3 degrees Celsius of warming by the end of the century, with devastating consequences. By the start of this decade world leaders were supposed to get serious about trying to limit the rise to a 1.5 degree Celsius threshold.

Australia's political leaders, Liberal, National and Labor, are painfully aware the battle over climate change is not for the faint-hearted. Over the last two decades it has broken the careers of three prime ministers and many others besides. This book is the story of that long battle. We need to know it because we are still trapped in it.

THE CARBON CLUB

CHAPTER ONE

HEARTS AND MINDS

WHEN SIMON MCRAE arrived at Canberra's National Convention Centre he was pumped. Dressed in a smart suit with shoes to match, he could have passed for any other conference delegate that day, except that he had a pair of handcuffs in his pocket and a secret button on his briefcase. Once inside, he mingled discreetly with a second delegate carrying an identical black briefcase who had arrived with him. A group of high-powered Americans began to make their way into the auditorium.

The head of the US Frontiers of Freedom foundation, Malcolm Wallop, had flown in to open the Countdown to Kyoto conference. On that wintery day in August 1997 Wallop had one aim—to ignite a provocative debate on whether the world really needed a new global agreement to protect the planet from climate change.

A retired senator from Big Horn, Wyoming, Wallop was a Republican A-lister who counted the party's grandee, Dick Cheney, among his good friends. Wallop was the founder of the Frontiers of Freedom and it was supported by some of the wealthiest men in America. Its dollars were helping to bankroll this Canberra show.

Wallop landed with a who's who of American climate-change sceptics. In an extraordinary coup for the Australian conference, Republican senator Chuck Hagel had joined Wallop's party. A few weeks earlier, Hagel, with Democrat senator Robert Byrd, had successfully shepherded through the US Senate a unanimous resolution designed, quite simply, to blow a hole in plans for a new global climate agreement that had been years in the making.

The final negotiations on the agreement were due to take place in just four months in the Japanese city of Kyoto. The Countdown to Kyoto conference in Canberra wanted to deliver a bold message: Americans and Australians must join together to torpedo the planned Kyoto Protocol.

Wallop's partner and co-host that morning was one of the most divisive businessmen in Australia, Hugh Morgan. Charming, eccentric and highly influential, Morgan was a doyen of the Melbourne Liberal Party establishment and the ideological godfather of the Australian Right. He ran Western Mining, one of the biggest companies in the country, sat on the Reserve Bank board and had the ear of the prime minister, John Howard.

Officially the conference organiser was the grandly titled Asia-Pacific Economic Cooperation Study Centre, run by a former Australian diplomat and free trade advocate, Alan Oxley. But he was working with Hugh Morgan's close friend, speechwriter and alter ego Ray Evans, who was known as a vocal denier of the science of global warming.

'He thought the whole thing was "crap",' recalled the American climate sceptic Myron Ebell, who had flown into Canberra with Wallop. 'He thought it was completely bogus and there was no crisis. If there was some warming it will largely be beneficial for humankind.'

Ebell had been working for Frontiers of Freedom when Evans turned up in Washington looking for money to fund the Canberra conference. 'Ray wanted not only an American co-sponsor, he wanted fundraising, which was the problem because he wanted to get Americans to fly to Canberra to speak but, of course, that costs a lot of money,' Ebell recalled.

Frontiers of Freedom, it turned out, was happy to help. Global warming was a hot issue in Washington and plenty of lobbyists and think tanks wanted to get in on the action. Wallop arrived in Canberra with Hagel and a small group of heavy-hitting climate-science sceptics who had advised the senator. They included Dr Pat Michaels, whose blistering attacks on the world's top climatologists had just earned him a cover story in *New Scientist* magazine.

Wallop was ready to open the conference with a ripper speech written jointly with Ebell. His pitch was blunt. The science of climate change was still in dispute. There was no planetary emergency. And a Kyoto agreement would be economic suicide for the industrialised nations of the world.

But just as Wallop was about to take the stage, a large van pulled up outside the Convention Centre. The doors flew open and around twenty Greenpeace activists in bright-orange jumpsuits leapt out and ran into the auditorium blowing whistles and yelling at the top of their voices. The orange invaders sat down right in front of the stage, linked arms and refused to move. The stunned conference organisers called the Federal Police, but the racket forced the delegates to retreat to the centre's dining room.

'They were chanting something, I forget what, something very loud,' said Ebell.

Simon McRae was watching all this with his handcuffs still in his pocket but now on full alert. He and his colleague picked up their briefcases and followed the rest of the delegates into the dining area. There they split up and found tables on opposite sides of the room. They sat down, discreetly pulled their handcuffs out of their pockets and shackled their briefcases to the table leg. With a second set of cuffs hidden up their sleeves they fastened the briefcases to their wrists. McRae knew his co-conspirator was going to hit the button on his briefcase as soon as the Frontiers of Freedom boss got up to speak.

Right on cue, an almighty thunderclap rang out across the dining room followed by a wailing siren and a voice screaming at 90 decibels, 'Warning! Warning! Warning! This is a climate emergency!' A loudspeaker was going off inside the briefcase warning of the catastrophic costs of climate change, from rising sea levels to crippling droughts.

The Americans were not happy. 'It was really starting to piss them off,' McRae recalled. As his friend, another undercover Greenpeace activist, was dragged away by security, McRae sat silently watching from across the room. He waited for Senator Hagel to take the podium. A few minutes into the senator's speech, McRae hit the button on his own briefcase. This time security went for the briefcase first. 'They got hold of it pretty quickly and disconnected it from the table and I let them drag me out,' said McRae.

Released onto the street, the Greenpeace warrior had a WTF moment. 'These people are fucking mad,' he said to himself. 'They honestly think they run the world.'

———

Greenpeace had made a big noise on the first day of the Count-down to Kyoto conference and it starred in the news that night. But what happened on the second day would reverberate louder and longer in Australian politics. The Howard government's Minister for the Environment, Robert Hill, arrived at lunchtime, took the podium and slapped down the lionised climate sceptics so carefully assembled by Malcolm Wallop and Hugh Morgan.

'I have stated many times, and will do so again, that Australia accepts the balance of scientific evidence which suggests that human activity is accelerating the increase in the earth's average temperature, thus enhancing the natural greenhouse effect, and causing the climate to change,' Hill lectured them. 'No country can ignore the potential ramifications—least of all Australia.'

In Howard's government, Robert Hill was known as a 'wet', a moderate 'small-l' Liberal lawyer from South Australia. His political skills had won him the plum position of government leader in the Senate after John Howard won the 1996 election. But Hill was from the minority faction in a Liberal Party dominated by its right wing. He succeeded because he was clever, diplomatic when necessary, and, at times, wickedly stubborn. But he was also a tad naïve. The cabinet had been wrangling for over a year about what to do about Kyoto.

Hill liked to insist that in the Howard government, 'apart from a few with extreme views, there wasn't really a mainstream debate about the science of climate change'. Yet right at the top of the government, the prime minister and some of his key advisers were still sceptical about the threat from global warming almost a decade after world leaders first decided to act on it.

Back in the sweltering American summer of 1988, climate change had hit the headlines in a front-page story in the *New York Times*. 'Global Warming Has Begun, Expert Tells Senate', reported on a dramatic presentation by the renowned climate scientist James Hansen to a US Senate Committee in Washington. Hansen told the senators he could say with 99 per cent confidence the planet was warming and with high confidence it was linked to the 'greenhouse effect'.

By then scientists had been studying the greenhouse effect for years with growing unease. Since the Industrial Revolution, the burning of fossil fuels—coal, oil and gas—and the deforestation of the planet had been spewing heat-trapping greenhouse gases like carbon dioxide and methane into the atmosphere at an astonishing rate. Hansen and his fellow climate scientists at NASA's Goddard Institute for Space Studies were now convinced this was upsetting the natural greenhouse effect, which kept the planet's temperature in balance.

Hansen's testimony marked a turning point in the public debate. Human interference, he said, was already causing the earth's surface, oceans and atmosphere to warm. While he stressed the effects were still difficult to model, he stated that the warming was likely to increase extreme weather events like heatwaves and droughts, with unknown consequences for people around the world. 'In my opinion,' said Hansen, 'the greenhouse effect has been detected, and it is changing our climate now.'

Just months after Hansen's testimony, the United Nations resolved to act. It agreed to muster several thousand scientific and policy experts from around the world to gather advice. The new Intergovernmental Panel on Climate Change—the IPCC— was created by the UN Environment Programme and the World Meteorological Organization (WMO). It was asked to get the best handle on the science of climate change and its impacts, and on possible strategies to address it. Australia offered its top scientists and officials to help with the effort.

Four years later, in 1992, US president George H.W. Bush flew into Rio de Janeiro for the 'Earth Summit' and called for 'concrete action to protect the planet'. The Republican president signed a groundbreaking deal to keep greenhouse gas emissions at a level that would prevent dangerous climate change caused by human activity. Australia's then Labor government also signed up. The UN Framework Convention on Climate Change or UNFCCC was unanimously approved by the US Senate, and Australia was among the first countries to ratify it.

Almost immediately there was a ferocious backlash from the world's largest corporations, especially the big fossil fuel producers. Exxon, Texaco Oil, Peabody Coal, Ford, General Motors and a parade of other corporations, including the 'Big Australian' miner BHP, had banded together several years earlier into a lobby group called the Global Climate Coalition to fight for their interests in the lead-up to the Rio summit. The lobby promoted sceptics who questioned the climate science and politicians who questioned the cost of action. Now it took up the fight against the new UN climate convention.

The companies' most potent argument was that the UN convention was unfair. It expected the US, Australia, Japan, the Europeans and the former Eastern bloc to shoulder responsibility for most of the excess greenhouse gases already in the atmosphere because they had benefited most from the last 150 years of industrialisation.

China, India and the rest of the developing world, still struggling with crushing poverty, were asked only to 'mitigate' their growing emissions. By contrast, the rich nations, including Australia, were pushed to cut their emissions back to 1990 levels.

By August 1997 this push became a giant shove. Greenhouse gas emissions were still going up, not down, and the approaching Kyoto talks were now likely to insist on legally binding cuts to emissions from the rich nations. The corporations and their lobbyists hit the panic button, warning of dire consequences for world prosperity. In the US, the Global Climate Coalition sponsored a multimillion-dollar advertising campaign attacking Kyoto with the simple slogan 'It's Not Global and It Won't Work'.

This attack was echoed loudly by the Frontiers of Freedom leader when he came to Canberra for the Countdown to Kyoto conference. 'The question before us is whether we are going to allow the noose that was put around our necks at Rio to be drawn tighter or whether we are going to resist,' Wallop demanded. Australia's environment minister put the problem more prosaically. 'It became difficult when people began to calculate the cost,' said Hill.

Despite his pushback against the climate sceptics, Hill knew his own government was also deeply worried about a climate deal in Kyoto. For more than a year, Howard had mounted a mighty diplomatic effort to fight binding emissions cuts on the rich economies and to shift more responsibility back onto developing countries like China. He had personally lobbied European leaders and the Democrat president Bill Clinton, who was now in the White House. Howard met with a lukewarm response. Clinton's vice-president, Al Gore, had been ringing the alarm bell on global warming for years and wanted a deal in Kyoto.

In the final round of UN talks before Kyoto, Australia stood almost alone with oil-rich Saudi Arabia and Russia trying to stop the momentum for binding emissions cuts on rich countries. 'At that stage, Australia was pretty well isolated,' recalled Roger Beale, then head of the Environment Department and Hill's top adviser.

The Howard government faced an awful dilemma. The scientific advice from the IPCC was warning Australia was likely to be among the biggest losers from climate change by the end of the 21st century. But Australians in 1997 were among the biggest winners in the carbon-intensive world economy. We sat with the top beneficiaries alongside the Saudis.

Cheap coal-fired electricity stoked the economy and jobs at home, while the country earned a fortune overseas thanks to the cost-free flow of greenhouse gases into the atmosphere. Australia was the world's largest exporter of coal. Indeed, the majority of its export income came from carbon-intensive goods, including gas, metals, minerals, and of course agricultural products.

Australia was bulldozing more forests and woodlands for farming and mining than any other nation in the developed world. For twenty years, from 1990, Australia began clearing on average 416,000 hectares of forests each year, the rough equivalent of 47 rugby fields every hour of every day, wiping out large numbers of native animals, birds and reptiles.

Australia ranked around twelfth in the world for its greenhouse gas emissions, at roughly 1.4 per cent of the global total, well behind the US and China; but per head of population, we were the highest in the industrialised world. By 1997, Australia had scant hope of meeting the UN convention's aim of winding back its greenhouse gas emissions unless it shifted course. Howard's top advisers were warning that emissions from the energy sector alone were likely to jump 34 per cent by 2010.

Inside the Howard cabinet an intense debate over what to do about Kyoto swung back and forth for eighteen months. The trade and resources ministers put the case that Australia was the number one coal exporter, and any global climate agreement with binding emissions cuts would smash Australia's export trade with Japan, the US and Europe—the 'Annex I' nations in the language of climate diplomacy. A confidential submission warned the cabinet: 'Australia is particularly vulnerable to efforts to address climate change. Emissions reduction efforts by other Annex I countries will have a strong adverse impact on our terms of trade by reducing the price and demand for fossil fuels (particularly coal) and other emissions intensive exports while increasing the price of imports into Australia.'

———

As chief executive of Western Mining Corporation, Hugh Morgan had a lot to lose in Kyoto. His company owned and managed mining, engineering, industrial and chemical plants exporting nickel, copper, uranium, gold and phosphate fertiliser. It also part-owned, with the US giant Alcoa, alumina refineries and the country's big aluminium smelters at Portland and Point Henry in Victoria, fuelled by Victoria's carbon-heavy brown coal.

By Morgan's own calculation, his company was looking at a $100 million bill to meet the cost of its greenhouse gas emission reductions if a deal in Kyoto went through. But Morgan's opposition to the new climate agreement was not just mercenary, it was deeply ideological. He saw environmentalism as a religious movement and the battle over climate change as the biggest threat to liberal private enterprise culture since socialism.

Aside from Rupert Murdoch, Morgan was arguably the most influential businessman in Australian conservative politics. He had the respect of the prime minister and held sway in Australia's Minerals Council and Business Council. Both lobby groups were vehemently opposed to binding emissions cuts.

But Morgan's ideological fervour made some of his allies wary of the maverick powerbroker and his mouthy speechwriter, Ray Evans. His own board at one point told him to tone down his strident rhetoric. Back in the 1980s, Morgan had been a polarising voice in the mining industry's campaign against Aboriginal land rights. In an infamous speech to the Australian Mining Industry Council, he argued that, if the government recognised Aboriginal Australians' spiritual links to the land, it might have to consider sanctioning 'infanticide, cannibalism and . . . cruel initiation rites'. In an incendiary final note, he referred to nineteenth-century claims that Aboriginal cannibals had shown a partiality for 'the particular flavour of the Chinese, who were killed and eaten in large numbers'. After the High Court brought down its historic Mabo decision recognising Indigenous land rights, Morgan had claimed it put at risk 'the whole legal framework of property rights throughout the whole community'.

Morgan and Evans relished political confrontation on multiple fronts. Morgan backed Evans' crusade against Australia's trade unionism and its 'subversive challenge to the authority of the State'. While Evans sat on the Western Mining payroll, he had helped launch an outfit called the HR Nicholls Society in 1986 with the future federal treasurer Peter Costello. At the time Costello was a junior industrial barrister who had already won a major victory against the unions in an ugly dispute known as the Dollar Sweets case. The aim of the HR Nicholls Society was to curb the power of unions, whittle away their legal protections and dismantle the industrial courts that enforced them. The then Labor prime minister, Bob Hawke, attacked the society's members as 'political troglodytes and economic lunatics', but it quickly attracted a hard core of union critics, from business executives and industrial lawyers to political activists with close ties to the Liberal Party.

By the end of the 1980s Morgan was a towering figure in the Victorian Liberal Party. Critically, he sat on the board of its biggest cash cow, a multimillion-dollar investment fund called the Cormack Foundation, whose ultimate role was to dole out generous donations at election time. Morgan's fellow directors on Cormack included Rupert Murdoch's brother-in-law, stockbroker John Calvert-Jones, and the chairman of the ANZ Bank, Charles Goode.

The Cormack Foundation also funded libertarian causes. One of its recipients was the Institute of Public Affairs, the IPA, a right-wing Melbourne think tank set up during World War II by a handful of businessmen, including Rupert Murdoch's father, Sir Keith. By the early 1990s, the Victorian Liberal Party and the IPA had become deeply entwined. Morgan joined the IPA's executive committee for a short time, as did Goode, who went on to become the chair of Woodside Petroleum. The IPA became one of the first serious think tanks in Australia to attack mainstream climate science.

———

Despite Hugh Morgan's growing influence with Howard, Hill won the argument that Australia should accept the scientific advice coming from the UN's IPCC on the threat from climate change. He was backed by his top official, Roger Beale, and their position held as long as Hill was environment minister. 'I saw no sign of Howard saying, "Oh, that's bullshit", or "I don't believe it",' said Beale. 'We, very early on, established a position of entry that we accepted the science.'

This was a strategic decision by Howard during the Kyoto negotiations, according to one of his former advisers, not a wholehearted embrace of the IPCC's climate science. Arguing against the science with the Europeans and Clinton was seen as a dead end both internationally and at home, where protests over climate change were gathering force.

'We were faced with big choices from the beginning and a first big choice was, do we argue about the science of climate change or do we accept that and argue about the impact on the Australian economy?' Howard's senior foreign affairs and trade adviser, Michael Thawley, explained. The answer was obvious. Australia had to plead that its economic case was exceptional; it simply could not afford to cut its greenhouse gas emissions back to 1990 levels. It sounded a bit rich for one of the wealthiest countries in the world, but not to Howard and the government.

Ever since the Rio Earth Summit, Australia's climate policy under both Liberal and Labor governments had followed one rigid rule, crudely called 'No Regrets'. It meant Australia would only reduce its greenhouse gas emissions if it didn't cost us.

There should be no net negative impact on the economy. Howard wanted a Kyoto strategy that kept 'No Regrets' in place.

Six months out from the Kyoto talks, this looked like mission impossible. New advice to cabinet showed that Australia's emissions could jump as much as 40–50 per cent by 2010. If that was the case, binding emissions cuts on Australia would mean real economic pain. By June 1997, a confidential cabinet minute raised a hardline option: Australia could just refuse to sign up for Kyoto. 'There would be some short-term political pain from international criticism,' the minute advised ministers, 'but the effects would not be major and could be quarantined from other international issues.'

As the cabinet debate became more fraught, Howard leant on his adviser, Michael Thawley, a brilliant career diplomat who had also worked in intelligence at the Office of National Assessments and in the Department of the Prime Minister and Cabinet. Thawley had come up through Foreign Affairs and Trade, where safeguarding Australia's exports was a priority. From the outset, Thawley believed Kyoto was intellectually flawed and wouldn't work. He and Beale clashed repeatedly as the deadline for the talks loomed. 'Thawley by nature was a hawk, always,' said Beale. 'Not just on climate but on everything you could think of. But very clever, very competent.'

Beale argued that if Australia wanted special treatment at Kyoto, it needed to go there with clean hands—a credible plan to show it wasn't just asking for a free ride. He and Hill worked up a proposal to bring down Australia's emissions from their steep rising path, the 'business as usual' path, while sticking with the 'No Regrets' rule. The plan pushed energy efficiency, curbing Australia's high rates of deforestation, and replanting native vegetation. In a groundbreaking reform, it also proposed the first national Renewable Energy Target. Australia would get an extra 2 per cent of its energy from renewable sources by 2010. This was on top of the almost 10 per cent already coming from hydro power. It was not much, but a start. 'The policies we had were good policies, but they were very, very modest,' said Beale.

The plan also included one radical idea—that Australia should put a price on carbon pollution. This was already being flagged in the US and Europe, and Beale championed it. It meant setting up a carbon emissions trading scheme that would ultimately set a market price for each unit of carbon businesses wanted to emit. The idea was to give business a financial incentive to work out the cheapest ways to cut their emissions. It was complex and would need a lot of new regulations. Thawley and Howard dismissed the idea, and even Hill was lukewarm.

'Howard wasn't interested in it. He wanted something direct so the public could understand, and that's when we had the birth of the Renewable Energy Target,' said Beale. 'We put that package together which was as good as we could do without emissions trading.'

Less than a month before the Kyoto summit, Howard unveiled the plan as 'the largest and most far-reaching package of measures to address climate change ever undertaken by any government in Australia'. Without qualification, Howard backed the mainstream climate science from the IPCC. 'The world's climate scientists have provided us with a clear message,' he told parliament, 'that the balance of evidence suggests humans are having a discernible influence on global climate.'

But then the prime minister drew his line. It would be 'unfair', he said, for Australia to lose its hard-earned competitive advantage as a carbon-intensive economy at Kyoto. We would reduce our emissions from business as usual, but we would not cut them back to 1990 levels. 'We had very tense negotiations in cabinet up until the death knell, on what the target should be that we would be prepared to commit to,' Beale remembered.

The numbers on the table looked like this. To avoid dangerous climate change, the Europeans argued that rich countries should cut emissions 15 per cent below their 1990 levels by 2020. They could do this in two phases. The so-called 'first commitment period' would start in 2008 and end in 2012. The second would go from 2013 to 2020. Most rich countries were expected to cut their emissions back by 5 per cent of 1990 levels by 2012 to have any hope of cutting them 15 per cent by 2020.

Thawley argued, by contrast, that Australia should ask for a target that kept its emissions at 11 per cent above 1990 levels by 2012. Cabinet settled on 8 per cent above. The 2020 target was put off until future talks.

In the final wrangling over the target, Howard's officials debated a highly contentious plan that would affect how Australia calculated its greenhouse gas emissions. Using research from the Australian National University (ANU), Beale proposed they should include in Australia's 1990 baseline the huge emissions from the country's land clearing which—coincidentally—had peaked in 1990. Land clearing was very high that year, but in the six years that followed it had fallen quite sharply, mainly because of drought and recession. This temporary fall in land clearing could give Australia a major advantage. It meant Australia could claim credits for the reduced land-clearing emissions during the six-year fall without having to do a lot—provided the government could stop land clearing peaking again.

This was a big ask by Australia that would not go down well in Kyoto. Australia was unique among developed countries in having high emissions from land clearing. Indeed, there was controversy over how the emissions from it could even be accurately counted. But Beale believed the 1990 baseline agreed on for

Kyoto greatly advantaged Britain, Europe and the former Soviet bloc because their carbon emissions, mainly from industry and power, were high that year. Using the land-clearing credits was a way of evening up the score for Australia. Beale wanted to keep Australia in Kyoto, and this could give Hill sway to argue Australia could meet its target without damaging the economy. It would also wind up pressure on the federal and state governments to control future land clearing, which was still alarmingly high.

'I think I was really the most convinced person that it was legitimate,' said Howard Bamsey, who was now Beale's deputy at the Department of the Environment. 'If the objective of Kyoto is to reduce the amount of emissions, for goodness sake let the agreement provide some incentive for us to do that in the land sector.'

———

In December 1997, ministers and officials from 150 nations flew into Kyoto for the highly anticipated UNFCCC Conference of the Parties (COP3). Hill led Australia's team including Beale. His key negotiator was Meg McDonald, the Ambassador for the Environment; she was from the Department of Foreign Affairs and Trade and a natural ally of Thawley. On the plane over, Beale argued with Hill that Australia could offer a better target than 8 per cent above 1990s levels. Hill wouldn't budge from the cabinet target.

As the Kyoto talks got underway, the science on climate change quickly took a back seat to the diplomatic horsetrading. All the rich nations were massaging their emissions targets and exploring inventive ways to cut emissions without alarming industry. These included global emissions trading schemes, offsetting schemes in developing countries and preserving forests to store carbon. The wrangling over these schemes—known as 'flex mechs', or flexible mechanisms—was as bitterly divisive as the emissions targets themselves.

In the final days, the talks teetered on collapse and US vice-president Al Gore flew in with a dramatic proposal to break the impasse. He and President Clinton had been 'burning up the phone lines', he told the delegates. He promised the US would raise its target and cut its emissions to 7 per cent below 1990 levels if a comprehensive plan was put in place with realistic targets, flexible mechanisms like emissions trading and, critically, 'the meaningful participation of key developing countries'. This meant America wanted some action from China and India. The message went back to Canberra from the Australian delegation in Kyoto: Gore's gone rogue.

'Gore turned up and condemned everyone for a lack of ambition and said America would show the lead and accept a higher target,' said Hill. 'That meant almost every developed country reconsidered the target that they had said they would accept.' Hill was convinced some of the new targets were unrealistic.

Hill talked to Howard and the message came back loud and clear: keep to cabinet's agreed target of 8 per cent above 1990 levels. Thawley was backchannelling to the Americans, warning Australia was prepared to walk away if it didn't get what it wanted. The US delegates were bunkered with the Europeans and Japanese, trying to hammer out a deal. They needed Australia. Canberra's weak target was ticked off.

The night before the final agreement in Kyoto, Australia extracted its last big concession. Beale and his team drew up the controversial clause to include land-clearing emissions from 1990 in Australia's baseline, giving it a big bundle of windfall credits towards its already weak 2012 target. The next day the so-called 'Australia Clause' was officially read into the Kyoto Protocol. The European delegates were outraged. 'They began to work out what had happened, but by then it was too late,' said Beale.

The outrage was echoed loudly by the climate activists at Kyoto. Dr Bill Hare, a climate scientist then with Greenpeace International, was appalled. He believed the clause killed off any real incentive for Australia to transform its carbon-intensive economy. 'The judgement at the end was that Australia would walk without it. I don't know if it would have,' said Hare. 'It meant that Australia didn't have to do anything for a generation really.'

Howard was delighted with Australia's win, calling it, 'A great result, a splendid result in Kyoto for the environment, for Australia, for Australian jobs.' Under the Kyoto Protocol, Australia had achieved the ultimate exceptional deal, a target 8 per cent above the 1990 baseline. While the rich nations agreed to cut greenhouse gas emissions on average 5 per cent from 1990 levels by 2012, only Australia and Iceland got to keep their emissions well above the baseline. The EU agreed to a cut of 8 per cent, Japan 6 per cent and the US 7 per cent, thanks to Al Gore.

Even before the delegates boarded their planes, it was obvious Gore's target was shaky. The US Senate had already flagged it would not pass laws to cut US emissions, let alone ratify Kyoto, if China and India weren't in the tent. China wouldn't budge and the Europeans had done little to help America. They had blocked critical concessions the US needed to offset its emissions as the world's biggest emitter. 'Kyoto unravelled before we left Kyoto,' was how Hill saw it. The departing ministers kicked the problem down the road to the next round of UN talks.

Back in Washington and Canberra, key corporations and their lobbyists saw a big opportunity to kill off the Kyoto Protocol before it could get off the ground. While the ministers had made their commitments, their governments needed to ratify the protocol to give it any legal effect. There was time to mount a powerful campaign against it.

Myron Ebell wanted to help that campaign. Until Kyoto, Ebell had had a low profile in the climate-change debate. A savvy, middle-aged political apparatchik, Ebell had worked as a lobbyist and Republican staffer; he had gone toe to toe with environmentalists in fights over endangered species before he joined the Frontiers of Freedom. But shortly before Kyoto, Ebell helped set up the Cooler Heads Coalition, a new lobby group that promoted climate-science sceptics.

'The Cooler Heads Coalition very quickly decided, not on the basis of polling but on the basis of thinking about it, that the scientific argument was key,' Ebell explained. 'If the science really was on one side or the other, then that would determine the outcome of the debate. Because if, for example, the science said that global warming really was a major problem, then the people who wanted to spend vast amounts of money to deal with it had a moral imperative on their side.'

Soon after the Kyoto conference ended, Ebell and his boss, Malcolm Wallop, had a meeting with the number two man at the American Petroleum Institute, Bill O'Keefe. The American Petroleum Institute, or API, was the voice of the US oil and gas industry, and the loudest voice in the API was Exxon. O'Keefe was putting together a new committee for the Global Climate Coalition lobby on Kyoto and wanted Ebell to join it. 'They had come up with a budget of what they thought was needed to defeat the Kyoto Protocol and any associated domestic legislation to implement the Kyoto Protocol,' said Ebell.

Ebell was invited to join a small working party of lobbyists for Exxon, Chevron, the API, key power companies and a handful of conservative think tanks, to work up a bold new strategy. The heart of the plan was to ramp up the attacks by climate-science sceptics on the IPCC and other mainstream climate scientists. As Ebell put it, 'We cannot let them create a fantasy crisis.'

The multimillion-dollar 'Global Climate Science Communications Action Plan' was produced in April 1998. A lengthy memo was circulated by the API's in-house lobbyist, arguing that those opposing Kyoto had made a mistake in the lead-up to the summit by pushing the economic case while not attacking climate science. Environmental groups and the government, it said, had got control of the scientific debate. This needed to change. 'If we can show that science does not support the Kyoto treaty—which most true climate scientists believe to be the case—this puts the United States in a stronger moral position and frees its negotiators from the need to make concessions as a defense against perceived selfish economic concerns,' the memo argued.

The plan was to target the media, politicians and executives with briefings by climate-sceptic scientists that homed in on the many uncertainties that still remained in climate science. Under the heading, 'Victory Will Be Achieved When', the memo said the test for victory was when ordinary citizens, the media and industry leaders understood the uncertainties in climate science and people's 'recognition of uncertainties becomes part of the "conventional wisdom"'. The final test was when supporters of the Kyoto Protocol were seen to 'appear to be out of touch with reality'.

There were, of course, many uncertainties in the climate science and the modelling behind it, but the elaborate Action Plan downplayed the consensus among the majority of climate scientists about the growing risks of dangerous climate change.

The API's Action Plan fell over when some corporations in the Global Climate Coalition baulked at funding the campaign. An environmental group got hold of the memo and it was leaked to the *New York Times*. But the strategy behind the plan found fertile ground in American and Australian think tanks.

A year after the plan was drawn up, Ebell landed a new job running energy and climate policy at the conservative Competitive

Enterprise Institute in Washington, a think tank then funded in part by ExxonMobil. From there he chaired the Cooler Heads Coalition, where he prosecuted the strategy to kill off the Kyoto Protocol and other climate policies by attacking climate science.

Ebell's political instincts were acute. If mainstream climate change science was undermined, the imperative for politicians to act was diminished and the economic case for delaying action was boosted. In Australia, Hugh Morgan's man, Ray Evans, had the same instincts and began to collaborate with the Cooler Heads Coalition.

Like their US counterparts, Evans, Morgan and their allies seized the opportunity to run a ferocious campaign in Australia to kill the Kyoto Protocol before it could be ratified. Hill had signed Kyoto in 1998 but Howard held off ratifying while the US stalled. That meant the protocol and its emissions targets had no legal status.

In March 2000, Evans launched Australia's own version of Cooler Heads as a vehicle to attack Kyoto and the climate science that supported it. 'Ray decided that the science needed to be taken on directly and that's why he founded the Lavoisier [Group],' said Ebell. The Lavoisier Group, named after the French father of modern chemistry, showcased the same American climate-science sceptics and ran the same agenda as its US counterpart. In one of his more apocalyptic tirades, Evans warned that if Kyoto came into effect, 'the economic dislocation which must follow its implementation will be unprecedented in modern times. It will be equivalent to the famines of the early nineteenth century in its disruptive power . . .'

Morgan threw his support behind the Lavoisier Group and agreed to give the opening speech at its first conference. He used the platform to hit out at Howard's environment minister, Robert Hill. Lauding Lavoisier's aim to examine the 'scientific uncertainty' around climate change, he told the gathering of climate sceptics: 'In this regard I applaud the objectives of the Lavoisier Group in airing such important issues of public interest. This is particularly so if the senior minister to the Government on environmental issues seeks to ridicule its endeavours.' The Lavoisier Group, Morgan added enthusiastically, 'can rely on my support'.

By now Hill and Beale were facing an uphill battle to get Kyoto ratified. As long as the White House stalled, Australia and the other rich countries stalled. The stalling continued until the presidential election in November 2000, when Al Gore lost to his Republican rival, George W. Bush, in one of the closest races in history.

From Washington to Canberra, climate sceptics cheered Gore's defeat. The politics of climate change had swung decisively in their favour.

CHAPTER TWO

OLD ALLIES

THE PRIME MINISTER was chuffed. The motorcades pulling up at White Oaks, the colonial-style residence of the Australian ambassador, meant the Washington A-list were coming to greet him. The barbecue was smoking as the guests arrived: Vice-President Dick Cheney with wife Lynne; Secretary of Defense Donald Rumsfeld with wife Joyce; Secretary of State Colin Powell; several lesser cabinet ministers and two US Supreme Court judges—all there to mingle with the mob from Down Under. If Ambassador Michael Thawley was out to impress his boss, he'd hit the mark.

'He put on a good party,' John Howard quipped, admiring his new envoy's access to the capital's 'top echelon'. Howard knew this was just the curtain-raiser. The big event, his first meeting with President George W. Bush, was less than 24 hours away, 10 September 2001.

———

Soon after Bush was sworn in as president in January 2001, US business ramped up its campaign to kill off the Kyoto Protocol. ExxonMobil led the charge by challenging the climate science underpinning the UN negotiations. In effect this meant arguing against almost every credible government science agency and scientific academy in the world. ExxonMobil was willing to do that and so were its climate-sceptic allies.

On 6 February 2001, ExxonMobil's senior environment adviser and climate lobbyist, Arthur 'Randy' Randol, faxed a memo to the White House office that was overseeing the new administration's environment strategy, the Council on Environmental Quality. Randol was on full noise. At the top of ExxonMobil's concerns was the IPCC, the UN's scientific and policy advice body on climate change. The chair of the IPCC was a British chemist and former White House science adviser, Dr Robert Watson, who had been, 'handpicked by Al Gore', as Randol put it. 'Can Watson be replaced now at the request of the U.S.?' Randol wanted to know. ExxonMobil was also lobbying for two other 'Clinton/Gore carry-overs with aggressive agendas' to be removed from positions of influence on the IPCC.

Randol recommended the White House promote two high-profile climate-sceptic scientists who were already working with the IPCC; he wanted them in positions of greater influence. In particular, he wanted a pivotal role for Dr Richard Lindzen, a professor of meteorology at the Massachusetts Institute of Technology (MIT).

Professor Lindzen was a highly qualified scientist and also one of America's most cited naysayers on global warming. He didn't dispute that carbon dioxide (CO_2) concentrations were increasing in the atmosphere, or that the earth might be warming slightly. But, like many of his fellow climate-science sceptics, Lindzen strenuously argued that this would not have a significant impact on extreme weather events, sea-level rise and the like. He was the go-to scientist for conservative think tanks around the world. Myron Ebell from the Cooler Heads Coalition said Lindzen had 'the biggest influence' in the global warming debate 'on our side'.

Lindzen had been invited to Canberra in November 2000 by Ray Evans' Lavoisier Group, where he popped up as the star witness at the Australian parliamentary inquiry examining whether the Howard government should ratify the Kyoto Protocol. The MIT professor was adamant that Kyoto should be dumped and, more importantly, that it wrongly made a connection between cutting emissions and the future impact of climate change.

'I would say that it behoves us to kill Kyoto, not only because of its intrinsic defects but to break the perceived connection between these processes and the relevance of science,' Lindzen told the Australian politicians.

He went on to declare that 'a doubling of CO_2 would increase the globe's temperature by about one degree centigrade—not a big deal'. Other climate scientists had attacked Lindzen over his climate analysis, with a few noting the MIT professor had also questioned the link between second-hand smoke and lung cancer.

ExxonMobil's lobbyist Randy Randol was keen to promote Lindzen as soon as possible; he saw the overhaul of the UN's scientific advisory panel as a high priority. The upcoming IPCC report, the third by the UN panel, was expected to be finalised by September 2001, and a strong IPCC report would have a significant impact on future UN climate talks. High on the agenda for these talks was whether the Kyoto Protocol would survive and whether the rich nations would confirm their mandatory targets to cut greenhouse gas emissions.

ExxonMobil already knew the draft IPCC report looked grim for the fossil fuel industry. The draft report confirmed that atmospheric concentration of carbon dioxide had increased by 31 per cent since the beginning of the Industrial Revolution. The concentration had now risen to a level not seen during the past 420,000 years.

The bad news for the coal, oil and gas industry was the draft finding that about three-quarters of the human-induced carbon dioxide emissions pumped into the atmosphere during the past twenty years came from burning fossil fuels. While the IPCC authors stressed that numerous uncertainties still remained, the draft report estimated that average global temperatures were projected to increase by anywhere between 1.4 and 5.8 degrees Celsius over the period from 1990 to 2100.

After five years of gathering new scientific research, the draft report was saying there was 'new and stronger evidence that most of the warming observed over the last 50 years is attributable to human activities'. An increasing warming was 'very likely' going to bring higher temperatures and more extreme weather events, and 'likely' to risk more droughts. Even when the final report took into account all the critical input and qualifications from business, government officials and NGOs, it would still be disturbing news.

The problem for ExxonMobil and its lobbyist was that the IPCC's draft report pretty much contradicted a lot of what the company's chairman and chief executive officer, Lee Raymond, was telling the world and his shareholders about global warming.

By the time of Bush's election, Lee Raymond was one of the most powerful chief executives in the world and ran a top-down empire. By 2001, after Exxon completed its merger with Mobil, the company had hit number one on the Fortune 500 and racked up US$17 billion in profit. It explored for oil and gas across six continents—in every corner of the globe, including in Australia, where it operated the Bass Strait oil and gas project.

'Exxon's attitude is that they're the big boys on the block, and they don't have to bend for anybody,' was how New York senator Chuck Schumer would later describe the oil giant's influence.

ExxonMobil's was the loudest voice in the Global Climate Coalition, the industry lobby calling for the US to ditch Kyoto.

In the lead-up to the Kyoto conference in 1997, Lee Raymond had delivered a landmark speech that sent a message to everyone inside and outside the company about where Exxon stood on global warming.

In that speech to the World Petroleum Congress in Beijing, Raymond asked his audience three questions. Is the earth really warming? Does burning fossil fuels cause global warming? Do we have a reasonable scientific basis for predicting future temperature? His answer was, 'The case for so-called global warming is far from air tight.' Citing 'sensitive satellite measurements', Raymond claimed there had been in fact no warming trend since the late 1970s and insisted, 'The Earth is cooler today than it was 20 years ago.'

He declared the scientific evidence was inconclusive on whether human activities were having a significant effect on the global climate and that cutting greenhouse gases lacked a foundation in climate science. The ExxonMobil chief demanded 'an open debate on the science' of global warming before taking any action to cut greenhouse gas emissions. It was a call-out to challenge mainstream climate science and the IPCC.

By 2001, ExxonMobil was directly and indirectly funding multiple think tanks and websites pumping out attacks on the IPCC and climate science. One of the most influential think tanks was the Competitive Enterprise Institute in Washington, where Myron Ebell, Ray Evans' Washington connection, was now well established. 'We did get substantial funding from ExxonMobil for several years, as long as Lee Raymond was chairman,' said Ebell. 'We typically most years got the largest donations from ExxonMobil of any of the groups on our side.'

Exxon had long wielded big influence in Washington, but with Bush's election the company had a friend right at the top of the administration. Dick Cheney had known Lee Raymond professionally and socially for two decades. Cheney had left his job as chairman and chief executive of Halliburton, one of the world's biggest oil services company, to run as George Bush's vice-president.

The White House did not bow to all of ExxonMobil's demands on the IPCC, but Bush and Cheney quickly launched two reviews that were invaluable for the fossil fuel companies: an energy review pushing greater deregulation for the industry, and a review of the science of climate change.

In March 2001, just two months after coming to power, Bush announced the US would pull out of Kyoto. The president also openly questioned mainstream climate science. His decisive moves came after his new head of the Environmental Protection Agency (EPA), Christine Todd Whitman, had tried to act on one of Bush's own election promises—to use federal laws to cut carbon dioxide emissions from power plants. When Whitman initiated this, the energy industry fought back and right-wing Republicans, led by Senator Chuck Hagel, protested loudly and directly to the president.

In a swift response, Bush announced he was reversing his policy on power plants, and supporting the coal industry and consumers. He also said he would not support the 'unfair' Kyoto Protocol. Significantly, Bush explained that he was acting 'given the incomplete state of scientific knowledge of the causes of, and solutions to, global climate change and the lack of commercially available technologies for removing and storing carbon dioxide'.

Bush not only publicly humiliated his EPA chief, but he gave a big boost to the climate sceptics, who argued against the need to cut emissions because of the uncertainties in the climate science.

A few months later, in response to Bush and Cheney's promised review into climate science research, ExxonMobil's scientists and executives drew up a sweeping proposal. It called for the US to launch its own assessment process separate from the IPCC, to examine climate science uncertainties.

In a detailed paper, the oil giant argued it was vital the US was not relying 'solely on the IPCC', which it said was clearly biased. 'A major frustration to many is the all-too-apparent bias of the IPCC to downplay the significance of scientific uncertainty and gaps, and the role that future research might or might not play in resolving them.' At first glance, the plan's laudable objective was to reduce uncertainties in climate science predictions but it amounted to another attack on the IPCC and the UN climate negotiations.

By July 2001, Bush's pivotal decision to pull out of Kyoto had left the future of the climate negotiations and the Kyoto Protocol hanging in the balance. When world environment ministers and officials met in Bonn that northern summer, many believed the talks would collapse. But in a surprising turnaround, the EU and the UK, with the support of Japan and Australia, rescued Kyoto.

By the end of the Bonn talks, the UK prime minister, Tony Blair, declared the outcome 'the most significant step' on global warming since the Kyoto Protocol was signed. His environment minister, Michael Meacher, called it 'a brilliant day for the environment'. Australia's environment minister, Robert Hill, gave an undertaking that Australia would make the emissions cuts needed to meet its target.

The Bush delegation sat on the sidelines. Its chief climate change negotiator, the State Department's Paula Dobriansky, had little to offer. The *New York Times* scathingly dismissed America's team as 'clueless'. Dobriansky told the delegates 'the Bush Administration takes the issue of climate change very seriously', but she was booed on the floor.

When John Howard arrived in Washington in September 2001 for his first meeting with Bush, US climate policy was in freefall. The corporations that made up the Global Climate

Coalition had won the battle to pull America out of the Kyoto Protocol, but the business lobby was breaking up. Bush had no coherent plan to replace Kyoto in the UN climate negotiations, and was in danger of being isolated. He needed an ally. At that moment, John Howard stepped forward.

———

The leaders' get-together on 10 September 2001 was nicely timed to celebrate the 50th anniversary of the ANZUS Treaty, Australia's military alliance with America, but high on Howard's agenda was the all-important matter of trade. Howard had arrived in Washington with a heavy-hitting business delegation including the influential Hugh Morgan. Soon after Thawley became ambassador in 2000, Morgan and a handful of business insiders had been working with him on a campaign for an ambitious Australia–US Free Trade Agreement. It was now top of the government's wish list in Washington.

Once inside the Oval Office, Howard got his chance to push the trade deal. Flanked by his senior advisers—including Thawley and Max Moore-Wilton, the overbearing head of the Department of the Prime Minister and Cabinet—Howard and the president ran through the hot topics: missile defence; the new Russian leader, Vladimir Putin; China; the Middle East; and the raw deal Aussie farmers got from America. By the end, Bush was ready to flag his tentative support for the new trade agreement.

Speaking to reporters before heading in to lunch with Bush, Howard was effusive. 'The President and I have a great similarity of views on many issues,' he said enthusiastically, 'and it's a great experience to be able to exchange them with somebody who holds the views he does.'

There was one other hot topic on the agenda, at first overlooked by the waiting media—the Kyoto Protocol. A carefully written statement put out by the White House noted the president and the prime minister recognised climate change was a global issue, 'requiring a global approach', and they were committed to developing 'an effective and science-based response'.

The bland language was curious. Ever since Bush had pulled out of Kyoto, French president Jacques Chirac, German chancellor Gerhard Schröder and UK prime minister Tony Blair had urged the president to reconsider his stand. During all that time Howard had equivocated and Australia still had not ratified the Kyoto Protocol.

When Howard emerged from his lunch with the president, he signalled he would come down on Bush's side over Kyoto. This would mean joining America in upending the UN negotiations on climate change. 'We're both of the view that Kyoto in its present form is not the solution, because it doesn't include the United States and it doesn't include the developing countries,' Howard told the waiting media. 'It won't work unless the Americans and the developing countries are part of it.'

Sometime after the lunch, Roger Beale's phone rang in Canberra. 'I got a call from a limo from Max Moore-Wilton, who said something like—he thought that the prime minister might have committed to not ratify the protocol if the Americans didn't.'

Beale was dumbfounded. Cabinet had not made any decision on whether to ratify Kyoto. His minister, Robert Hill, hadn't been consulted. Indeed, Beale and Hill were working hard to keep alive the option Australia would ratify Kyoto, but they believed Thawley had been cautioning Howard against it. In the weeks immediately before and after the US presidential election, Hill had worked closely with Australia's allies in the UN climate talks to try to win concessions he hoped would keep the Americans in the tent. A few climate negotiators in the US State Department were also trying to keep the negotiations going. Thawley believed the concessions undermined Australia and argued strongly against them.

As Hill saw it, Thawley was effectively saying to Howard in Washington: 'You watch that Hill. He'll commit you to something and the American government's going to walk away from this. You're going to be left out on a limb.'

Thawley's position was simple: Bush attacked the Kyoto Protocol in his presidential campaign; the Republicans in Congress were going to kill it after the election; Howard shouldn't ratify a flawed agreement that the US won't. He also believed senior Democrats in Congress wouldn't support it. Thawley infuriated Beale. 'He was running a Republican line. I remember being very cross with him,' said Beale. 'Thawley was going to the Republicans and back to Howard and leaving the Department of Foreign Affairs out of the loop.'

Now it looked like Thawley's advice had won the day. If Howard had made a commitment to Bush in their face-to-face meeting, Australia would never ratify Kyoto. But Beale hesitated. Max Moore-Wilton's hurried phone call was not conclusive. He and Hill could still take it to cabinet. They would wait a few days until Howard returned to Canberra.

The next morning in Washington, shortly before 9 am on September 11, Michael Thawley sat in a room in the historic Willard Hotel, just down the corridor from the prime minister. He was working on the big project, the Free Trade Agreement. Thawley was deep in conversation with Hugh Morgan and David O'Reilly, the head of Chevron, the US oil and gas giant. Chevron was planning a huge investment in the largest single resources project in Australia's history—the development of two new offshore gas fields linked to a massive liquefied natural gas plant on the north-west coast.

Thawley had asked the Chevron boss and Morgan to co-chair a forum that morning at the nearby US Chamber of Commerce, launching the Australia–United States Free Trade campaign. Howard was scheduled to be there at 9.30 am and almost every transpacific heavyweight was coming, including Rupert Murdoch, with whom Howard had dined the night before. Thawley's new deputy chief of mission, Meg McDonald, was already shepherding VIPs into the venue.

Howard was still in his hotel room prepping for a quick media conference when his press secretary, Tony O'Leary, mentioned that one of the towers at the World Trade Center in New York had been hit by a plane. Howard was shocked, but he and O'Leary assumed it was a terrible accident. A few minutes later, O'Leary told Howard a second plane had hit the Twin Towers. They flicked on the TV and watched, horrified, as the carnage unfolded on screen.

Howard went downstairs for his news conference, but he was overwhelmed by the images from the Twin Towers. 'I don't know any more than anybody else,' he told reporters, 'but it appears to be a most horrific, awful event that will obviously entail a very big loss of life.'

O'Leary cut off the questions after learning a third plane had hit the Pentagon. Someone pulled back the curtains in the media room and they saw the smoke rising from the Pentagon fire in the distance. America was under attack. The prime minister and his party were soon hustled out the back entrance of the Willard Hotel and taken in convoy, sirens blazing, to the nearby Australian embassy, where Thawley ordered people into the basement. Hugh Morgan joined the embassy group at some point. Howard's wife Janette and son Tim also arrived.

In the dusty maintenance room of the embassy bunker, Howard was set up with a phone and collected his thoughts.

He called Canberra and then dictated a letter to Bush, pledging Australia's, 'resolute solidarity with the American people at this most tragic time'.

When Howard got back to Australia, he invoked the ANZUS Treaty for the first time in its history and Australia became one of Bush's staunchest allies in the War on Terror. In the dramatic days and months that followed, Howard's loyalty to George Bush was viewed through the lens of his searing experience in Washington on 9/11.

Almost completely forgotten was Howard's decision the day before 9/11 to publicly throw his support behind the president to stymie the Kyoto Protocol. On that day Howard had positioned himself to join Bush as the only other leader in the developed world to reject the global climate change agreement. There was little understanding in Australia of the importance of Howard's choice or what was at stake for his ally in the White House, George Bush.

———

Back in Canberra, the decision on whether to ratify Kyoto was swept to one side as the urgent crisis over terrorism and an ongoing battle over asylum seekers overwhelmed the government. Just weeks after he returned from Washington, Howard called a federal election and declared 'leadership' would be the defining issue of the campaign. His approval rating was soaring. The Twin Towers attacks, and the government's hardline policy to turn back boat people seeking asylum, left Labor and its leader, Kim Beazley, flailing.

The threat from climate change got little traction in the election campaign. Australians brushed off the news that they were the highest per capita emitters of greenhouse gases in the industrialised world. They barely noticed a report by the government's own science agency, the CSIRO, warning that if global emissions kept rising without restraint, temperatures in the country could increase anywhere between 1 and 6 degrees Celsius by 2070.

Beazley did promise to ratify Kyoto, but he was attacked by the government's chief climate sceptic, Nick Minchin, who held the powerful Industry, Science and Resources portfolio. Despite this, as environment minister, Robert Hill still campaigned on a promise to meet Australia's Kyoto target. He and Roger Beale had not given up on getting Kyoto ratified. 'We weren't absolutely sure what was happening so we proceeded to carry on as before,' said Beale, 'and as time went by I assumed I had either over-interpreted Max [Moore-Wilton] or he had got it wrong.'

———

When Howard led the Coalition to its third victory in November 2001 he strengthened his grip over the government. The victory boosted the influence of climate sceptics in the Coalition. In the cabinet reshuffle that followed, Howard moved Robert Hill out of the Environment portfolio and promoted him to Defence. The government's most vocal defender of mainstream climate science was gone. In Hill's place, Howard appointed David Kemp, a former state director of the Victorian Liberal Party and a member of its Right faction; Kemp was also a climate sceptic, although he contested that label. 'I don't see myself in a category,' he said. 'My interest has always been in understanding the science and, as minister, getting policy right.'

Kemp was an intellectual, a former politics professor and, to his staff, 'a really decent guy'. Like many in the climate-sceptic camp, he did not dispute that carbon dioxide emissions were increasing or that there was some warming. But he homed in on the many uncertainties in climate science to question some of the key findings of the IPCC.

The arguments run by the new minister reflected the work of America's leading climate sceptics, Professor Richard Lindzen and Dr Pat Michaels. Kemp highlighted the flaws in the models relied on by mainstream climate scientists to predict future warming. 'Forecasts have often departed from outcomes and observations,' Kemp said. 'In particular the sensitivity of the climate system to a doubling of CO_2 remains contested.'

Kemp was right: the models and forecasts on climate change were regularly contested and qualified. Climate modelling is hugely complex. These arguments went on within the IPCC as well as among its critics. But the central issue for climate scientists was whether the risk of dangerous climate change was increasing or decreasing as these uncertainties continued to be explored. The IPCC and the CSIRO argued that, if emissions were not reduced, the overall risk was very likely to go up, not down.

The new minister brought more than climate scepticism to the job. He brought a belief that climate science in Australia was deeply contested. 'There is a large body of opinion in Australia that has never accepted the assertion that current climate change is different to past climate change and that it is significantly due to anthropogenic [human] factors,' Kemp said.

Kemp knew this was true in the Liberal Party and in the government, where divisions over climate science were growing. He explained later, 'Within the Liberal Party, a significant number question the claims that there is a scientific consensus sufficient to determine the nature of the policy problem.'

Kemp wanted his senior officials to talk to climate sceptics. Specifically, he wanted the head of his department, Roger Beale, and the head of the Bureau of Meteorology, John Zillman, to meet with Ray Evans and the Lavoisier Group at Parliament House. While Evans was still the secretary of the group, the president was the former Labor finance minister and well-known contrarian Peter Walsh.

The meeting did not go well. 'They were a bunch of loonies,' Beale concluded by the end of the session. Beale was especially angry over Evans' treatment of Zillman, a cautious government scientist who sat on the IPCC. 'Ray Evans was hooting when Zillman said anything about the science; it was appallingly discourteous.'

But in Kemp's world, talking to Lavoisier and the climate sceptics was important. He knew that Morgan and Evans were a big influence on climate change policy in the Liberal Party and the government. 'The respect for each within the party, especially for Hugh Morgan, means his views were widely listened to,' said Kemp later.

Morgan's influence in the Victorian Liberal Party was by now entrenched. He still sat on the board of the Cormack Foundation, the party's cash cow, and held a third of Cormack's shares in trust for the party's Victorian branch. In the lead-up to the federal election, Cormack had donated over $1.5 million to the branch. Morgan's own company, Western Mining, had also donated almost $300,000 to the federal Liberal and National parties as well as making big donations to the Western Australian Liberals.

Kemp also understood the influence of right-wing think tanks on the climate debate, especially the Institute of Public Affairs (IPA), which was ramping up its attacks on climate science. Kemp's father Charles had been a co-founder and one of the first directors of the IPA; his brother Rod, who was now Howard's arts minister, had been its director in the 1980s. And of course, Morgan and the Cormack Foundation were funders of the IPA.

Despite Kemp's scepticism, Beale respected him. He thought Kemp was a genuine sceptic rather than a denier of climate science. Kemp also respected the government's climate scientists and accepted that Australia needed to reduce its greenhouse gas emissions, just as long as it stuck to the 'No Regrets' policy—in other words, in the government's view, so long as it didn't have a negative impact on Australia's growth. Beale believed he could work with Kemp to argue the case in cabinet that Australia should ratify the Kyoto Protocol.

More than most, Beale knew the arguments against the Kyoto Protocol, principally its failure to put obligations on China and India. It was, he said, 'the great but probably unavoidable error' of the UN agreement. But the Kyoto Protocol was supposed to be just the first agreement, a vehicle for developed nations like Australia to take the lead on climate change while pushing the two emerging giants, China and India, to act in the next round of treaty negotiations before their emissions overwhelmed every other nation on earth.

As Beale's deputy, Howard Bamsey, would put it: you either have a seat at the negotiating table or you are on the menu. The Australian government had got a very generous 2012 target from Kyoto, the weakest emissions reduction target of any industrial-ised country except Iceland. Australia could meet that target and maintain its influence in the global negotiations with allies like the UK and Japan. That way, Australia had a chance of pushing China, India and other big developing nations in the Asia-Pacific to take on real obligations to restrain their emissions under the treaty that was supposed to follow Kyoto.

Still, Beale knew that getting cabinet support to ratify Kyoto would be difficult. Apart from Howard, the most strident opponent was Nick Minchin, who had been promoted to finance minister after the election. Like Hill, he was a South Australian senator, but Minchin was a powerful player on the Right of the party. 'I think at that stage the US was the biggest emitter and there was a general sense that, if the US wasn't part of it, what the hell was the point of Australia being involved?' said Minchin.

Minchin had argued long and hard against Kyoto when he held the Industry, Science and Resources portfolio. He had done 'a deep dive' into climate science, as he put it. He emerged as a climate change sceptic. He met Hugh Morgan and Ray Evans on the job and clicked with Evans. 'We had a lot in common,' he said. Despite then being responsible for the CSIRO, Minchin found the climate-science sceptics promoted by Evans and the Lavoisier Group more credible than the advice he got from the nation's top science agency.

Minchin saw no reason for Australia to shift from its dependence on cheap fossil fuels, which supported jobs and manufacturing. He was also hostile to any action that could hurt Australia's coal and gas industry. 'The resources sector was part of my constituency too, so I was very conscious of the anti-fossil fuel agenda that was developing, which, to my mind, was potentially dangerous,' he said.

By early 2002, when Beale was still pushing ahead with the case to ratify Kyoto, Kemp got a message from Michael Thawley in Washington. The ambassador was suddenly keen for Kemp to meet with Bush's climate team, who were announcing a new approach to climate change policy. It was aimed at reducing 'emissions intensity' in the US economy. Happily for the White House, 'emissions intensity', the amount of energy used for each unit of output, was falling in the US as its manufacturing moved offshore. Unfortunately, America's greenhouse gas emissions were still going up, so it was still the biggest emitter in the world at around 25 per cent of the global total. Bush was promising to fund new technology and tax incentives to help bring emissions down.

Beale remembered arriving at the Washington embassy with Kemp for a meeting with Thawley and a senior executive of Exxon-Mobil, who was eager to explain the US opposition to Kyoto. 'We met with a deputy head of Exxon, who was claiming credit for turning the Administration around,' said Beale. The Exxon man made it plain the company had lobbied to kill off Kyoto, along with a carbon emissions trading scheme, known in the US as a 'cap and trade' scheme.

'He said to the three of us, "We picked up the phone and called Bush and said, those caps are dead, that treaty's dead,"' Beale recalled. He wondered about the purpose of the meeting. 'I think [it was] to demonstrate how fundamental this was in the US and what the opposition was,' he said. 'He probably thought it would change our minds.'

Kemp would not be drawn on the meeting with the Exxon-Mobil executive; but he said he got the message from Bush's EPA chief, Christine Todd Whitman, and other officials that the US would not shift on its opposition to Kyoto. Instead they offered Australia a one-on-one 'Climate Action Partnership', to cooperate on new technologies to address climate change.

The timing of Kemp's invitation to Washington was revealing. It came at a crucial turning point in the international climate negotiations. Japan, one of America's most important trading partners, had just signalled it would be ratifying Kyoto, despite Bush's opposition.

If Japan had followed Washington and abandoned the Kyoto Protocol, the global climate negotiations would be in limbo. Under the consensus rules of the UN climate convention, the Kyoto Protocol needed to cover at least 55 per cent of the developed world's emissions to come into effect. With Japan now on board, the protocol's future was likely to be secured even without America. With the EU, the UK, Japan and other rich nations making mandatory cuts to their greenhouse gas emissions for the first time in history, the White House had become sidelined on climate change.

In May, the new Japanese prime minister, Junichiro Koizumi,

made his first visit to Australia and lobbied Howard directly to ratify Kyoto. A month later, on 4 June, the Japanese government formally confirmed its own ratification of the protocol, and called on all developed countries to do the same.

The following day David Kemp was on the Indonesian island of Bali. It was World Environment Day and Kemp was being honoured by the World Wide Fund for Nature with the prestigious 'Gift to the Earth' award for Australia's role in helping to preserve natural wetlands. In a generous response, Kemp told the WWF members the award highlighted the fact that the environment 'is everyone's business throughout the world'.

Soon after Kemp left the ceremony, John Howard rose to his feet in Parliament House in Canberra to deliver his own World Environment Day message. In answer to a question from Labor, Howard announced that Australia had decided not to ratify the Kyoto Protocol. 'It is not in Australia's interests to ratify the Kyoto Protocol,' he told the parliament. 'For us to ratify the protocol would cost us jobs and damage our industry. That is why the Australian government will continue to oppose ratification.'

Roger Beale and Howard Bamsey were stunned. 'Neither of us knew a thing,' said Bamsey. 'There was no cabinet process whatever. I'm sure Kemp didn't know. I am pretty sure the first thing he knew about it was when someone from the ABC put a microphone up and said, "What's your reaction to the prime minister's announcement?"'

Beale was with Kemp in Bali. As well as receiving the award, the minister was attending a UN meeting canvassing Kyoto's future. 'Kemp was devastated,' Beale remembered. 'All of a sudden Howard stands up in the parliament and announces—with no prior advice to David, no discussion in cabinet, no submission, no bureaucratic advice to me—that we are not ratifying the protocol.'

Kemp thought the prime minister could have given him some kind of heads-up, but the decision was Howard's alone. 'It was a "captain's call",' Kemp said, but he understood Howard's reasons: the US wasn't in it and developing countries were not obliged to cut their emissions. Kemp also knew the alliance with Bush played a big role.

Kemp's predecessor, Robert Hill, believed Howard's opposition to ratifying Kyoto was largely driven by his loyalty to Bush. 'Where Howard switched, I think, was never really about the science, or the cost or anything,' said Hill. 'He became very uncomfortable about leaving Bush hanging out there. He believed the Americans didn't get a fair go, and that we shouldn't leave Bush to take all the pain alone.'

But underpinning Howard's far-reaching decision was his own climate scepticism about the science of global warming. Years later he would confirm that he was what he called 'agnostic' on the science of climate change: 'Part of the problem with this debate is that to some of the zealots involved their cause has become a substitute religion,' he told an audience of climate sceptics in London. 'And although not given to agnosticism in other matters, I have for a long time been an agnostic on the issue of climate change or global warming . . .'

Howard's captain's call on Kyoto marginalised Australia for the next five years; it joined the US as an outlier nation on climate change in the UN negotiations.

'There was deep shock that Australia moved in that direction,' said Dr Bill Hare, who was then with Greenpeace in Europe. 'Such a fundamental repudiation of the system was a real shock to highly visible allies like the UK.'

Not everyone was shocked. Hugh Morgan was one of the first to applaud Howard. In an opinion piece for *The Australian* newspaper soon after Howard's announcement, Morgan seized the opportunity to double down on his attacks on climate science and the whole UN effort to respond to the threat of climate change. He ended his broadside with a ringing endorsement of the decision, 'The Prime Minister has made the right call on Kyoto.'

Hugh Morgan and his alter ego, Ray Evans, were no longer sounding like maverick voices on global warming. To an increasing number of Liberals in the Howard government, their utterances were sounding more like the conventional wisdom.

CHAPTER THREE

DIVISION IN THE RANKS

SOON AFTER HOWARD'S captain's call not to ratify Kyoto, David Kemp and Roger Beale arrived in London and made their way to 10 St James's Square, a distinguished address synonymous with robust debate. Chatham House is famous for hosting some of the world's great thinkers on contentious issues of the day: Mahatma Gandhi on the future of India, John Maynard Keynes on the gold standard and Lord Carrington on NATO and European security.

Australia's environment minister was now standing in their shoes in the English summer of 2002, trying to explain to the Brits why the Howard government had broken with them and backed the US president's decision to derail Kyoto. Kemp also wanted to explain what Australia was now going to do about climate change.

In the tortured speech, written by Beale and Kemp, two things were clear. Australia would never ratify Kyoto under the Howard government. Yet, despite this, Australia would stick to its 2012 Kyoto target: it would reduce its greenhouse gas emissions. 'Our strategy is to move ahead on emissions reductions in a practical way,' Kemp told his London audience.

Kemp knew how difficult this would be. A number of his government colleagues were, like him, sceptical of the scientific advice about climate change, but some were in complete denial. A lot were hostile to the idea of cutting greenhouse gas emissions because it would mean more government regulation on business, more costs for business and more government spending.

'Climate change was a policy issue that quickly became about another inflation of the role of government,' Kemp explained later. For leading party figures on the Right, the urgent voices calling for action on climate change in Europe and at home were seen as anti-capitalist. 'The issue was readily seen by many as the biggest threat to the liberal private-enterprise culture since social-ism,' Kemp said candidly.

Roger Beale didn't buy into the ideological debate. He was a career public servant. He accepted the scientific advice of the IPCC that climate change was a real and pressing problem. His role, as he saw it, was to give advice on how to cut the country's greenhouse gas emissions without damaging the economy and how to restore Australia's reputation as an honest broker in the global climate negotiations.

After Howard blindsided them with his captain's call on Kyoto, Beale made the argument to Kemp that they needed to regroup with 'some fast footwork' and salvage a climate policy from the wreckage of that decision. 'I said to David that we had to use this embarrassment to at least get a commitment from the prime minister that we were still going to meet our Kyoto target,' said Beale. 'And get him to announce an inquiry into an emissions trading scheme.'

Howard was persuaded to stick to the Kyoto target partly because he'd promised to do so at the last election. He was also on record as praising the weak target wrung out of Kyoto as a win–win for Australia.

With the target accepted, Beale argued it was logical to use an emissions trading scheme to help meet it. Many countries were considering similar schemes, because the idea behind them was to use market mechanisms to prompt business to find the cheapest way to cut greenhouse gas emissions. When Beale returned to Australia, he began work on emissions trading for the second time in his career; but he soon realised he was opening up a hornet's nest.

Emissions trading schemes are often called 'cap and trade'. Put simply, they work like this: the government sets a target or 'cap' on the amount of greenhouse gas emissions it allows a company covered by the scheme to emit and then issues them only enough permits to meet this target. Companies included in the scheme must either buy permits or get free ones issued by the government to cover their emissions each year.

The schemes use a combination of carrot and stick. Businesses can buy permits, or instead invest their money in cutting their emissions by upping their energy efficiency or switching to cleaner fuels. The 'trade' part of the scheme allows companies to buy and sell permits so, if a company cuts its emissions quickly, it can profit by selling its permits. Overall emissions come down because the government gradually lowers the 'cap' along with the number of permits in the market. This forces the price of permits up and with it the cost of emitting greenhouse gases. Ultimately, companies go cleaner or their profits get squeezed.

By and large, in 2002 Australia's big greenhouse gas polluters did not like emissions trading schemes. In particular, Hugh Morgan hated them. He was still sitting on the board of the Business Council of Australia (BCA) and was 'one of the most influential businessmen' with Howard at the time, according to Kemp. His opposition was not surprising, given his climate scepticism. The scheme would put a price on greenhouse gas emissions for the first time, including those from Western Mining, and would potentially hit profits and mean more regulation.

There was also opposition in the government. Managing an emissions trading scheme is hellishly difficult. If the government got the scheme wrong and the permit price went too high, companies in industries with big greenhouse footprints, like aluminium, steel and power, could shut down or move their operations overseas, putting thousands of people out of work or threatening the electricity supply. If, on the other hand, the price was too low and permits were too cheap, companies could trade them to profiteer and greenhouse gas emissions would keep going up.

So it was remarkable really that Beale managed to get senior bureaucrats and ministers to consider an emissions trading scheme in a submission to Howard's cabinet in 2003. 'This is the Beale plan,' recalled Howard Bamsey, who that year was appointed to run the Australian Greenhouse Office. 'Roger [Beale] is chair of the secretaries committee on climate change. Critical is Ken Henry. He's already head of Treasury. He is the strongest voice in support of it.' Henry's boss, Treasurer Peter Costello, signed the cabinet submission along with the industry minister, Ian Macfarlane, and Beale's minister, David Kemp.

Top executives from the big greenhouse gas emitters lobbied furiously against the Beale plan. The climate-sceptics headquarters, the Lavoisier Group, had their president, Peter Walsh, write to Howard in May 2003 warning him off the emissions trading scheme or imposing a carbon tax, which was also on the table. By July, Morgan's allies, including Malcolm Broomhead, head of Australia's largest mining explosives company, Orica, had ramped up the pressure.

Some of the big miners hired the Liberal Party's former federal director, Andrew Robb, to push their cause. Robb was a serious player in the Liberal Party: as party director, he had been the architect of Howard's 1996 election victory, and before that he ran the conservative National Farmers' Federation, the main lobby group for farmers and agribusiness. He later earned a fortune working for Kerry Packer, Australia's billionaire media mogul. By 2002 Robb was a full-time lobbyist and political consultant, and a vocal climate sceptic. Like Minchin and Howard, he believed climate change had become 'a cause celebre for the Left', after the fall of communism. He lobbied to help kill Beale's plan.

Weighing in from Washington, Howard's old adviser Ambassador Michael Thawley also argued against the scheme. The White House and pro-Bush Republicans in Congress were deeply opposed to 'cap and trade' schemes, and so were the leading fossil fuel companies like ExxonMobil.

But despite the opposition from Morgan and his allies, a deep split had developed in the BCA. Behind the scenes a battle was playing out over both Kyoto and the idea of an emissions trading scheme.

One of Morgan's main adversaries was Greg Bourne, then Australasian head of BP. A middle-aged oil company executive, Bourne had worked all over the world, from London to the Middle East and Central America. He had also been an adviser on energy and transport to Britain's Conservative prime minister, Margaret Thatcher, one of the first world leaders to strongly back climate change science. Bourne believed the scientific advice of the IPCC and Sir David King, Britain's chief scientist: climate change was real and greenhouse gas emissions needed to be cut.

Bourne was exasperated by Morgan and his friends. To him they represented the worst of the insular members of the Melbourne Club, the all-male headquarters of the Melbourne establishment, located at the 'Paris end' of Collins Street, where Morgan held court with fellow members of the Victorian Liberal Party.

'Morgan was my nemesis,' Bourne recalled ruefully. 'Hugh Morgan and others of his ilk didn't believe that climate change was real—they believed it was almost a European socialist plot to dud the Australian economy. And they actually spoke about it as such.'

As Bourne remembered it, there were roughly ten big companies wanting to overturn the BCA's long-held anti-Kyoto stance. Their executives were open to the idea of an emissions trading scheme and putting a price on carbon emissions. These 'rebels' included Westpac Bank chief executive David Morgan (no relation to Hugh), who was a former senior Treasury official. He'd even written to Greenpeace supporting a review of the BCA's opposition to Kyoto.

The BCA rebels were convinced that Australia's financial industry would eventually need a price on carbon emissions. Investors had to deal with the uncertainties and risk caused by climate change. Without a carbon price, costing investment in new electricity plants would be a fraught exercise. Without a carbon price, it would also be difficult to drive investment in new technology needed to cut emissions, not only in Australia, but in China, India and the rest of the developing world. But the BCA rebels were still in a minority; they were decisively rebuffed by Hugh Morgan and his allies, who included the chief executives of Rio Tinto, ExxonMobil Australia, Orica and Alcoa.

In early August 2003, two weeks before the cabinet decision on the Beale plan, Howard met with executives from the mining, aluminium, electricity and chemical companies. They had large workforces around the country and their opposition was trenchant. Howard listened sympathetically: these were men he understood. It wasn't simply a case of some big carbon emitters being big political donors to the Liberal Party; Western Mining was, but others gave only modest donations, and some none at all.

Rather, these were Australia's blue-chip companies and their ties to the Liberal Party were deep and meaningful.

The Liberal Party's own cash cow, the Cormack Foundation, where Morgan sat on the board, invested in fossil fuel companies like BHP and Rio Tinto. Howard's nephew Lyall worked for Rio Tinto as their government-relations adviser. Kemp's former chief of staff, John Roskam, had also worked for Rio Tinto before going to run the Institute of Public Affairs, which in turn got donations from the Cormack Foundation. Many Liberal Party officials and staffers moved through the revolving door of politics, big mining and metals companies, lobbying firms and the public service. Some lobbyists were so ubiquitous they earned nicknames: Beale called one pair he clashed with 'The Dodgy Brothers'.

A former Liberal insider, Guy Pearse, spent years analysing the network of ties between the Howard government and the greenhouse industries. In a groundbreaking book, *High and Dry*, written after he left his job as a speechwriter with Howard's first environment minister, Robert Hill, Pearse called the network the 'greenhouse mafia' and identified in detail how their influence permeated the government.

It wasn't surprising then that Howard's own thinking on climate and energy reflected the views of many of the senior executives in companies like Rio Tinto and Western Mining. They all shared what Pearse called 'a quarry vision' of Australia. 'The greenhouse policy advice John Howard has taken seriously depends on the idea that the minerals, metals and energy sectors are the basis of Australia's economic future,' Pearse explained. 'It has been an article of faith across his government since 1996.'

Howard Bamsey believed the prime minister had made up his mind on Beale's plan for an emissions trading scheme before the cabinet meeting to discuss it, and was influenced in part by Hugh

Morgan's opposition. 'What happened was, at a certain point, when the cabinet submission is ready to go, it's all been signed off, there is a meeting between Howard and heavies from industry,' Bamsey recalled. 'The public view is industry is lobbying to kill emissions trading. But one person I spoke to from industry, said, "We were summoned to that meeting by the prime minister and told: this is my inclination." So you could see the hand of Hugh [Morgan] in that.'

By the time Roger Beale's emissions trading scheme was debated in cabinet in late August, new modelling had emerged that now included transport and agriculture, even though Beale had not included them in his proposal. That dramatically raised the estimated cost to the economy and consumers. Ministers swung their support behind Howard and the Beale plan died.

Environment groups were furious at the cabinet decision. 'A minority of big, dirty polluters have won the day and put Australia's national interest at risk' was how the CEO of the Australian Conservation Foundation, Don Henry, put it.

In a coup de grâce, Hugh Morgan's supporters installed him as the new president of the BCA soon after cabinet killed the emissions trading scheme. 'I think it was probably just a lay-down misère,' said Bourne. 'Hugh was considered the man for the times by those who voted him in. It was not a democratic process by the hundred members at all. It was very much "We know who we want" and in came Hugh.'

The Murdoch media's leading climate-sceptic columnist, Andrew Bolt, rejoiced in Morgan's appointment. 'I've seen in his eyes a glint of intent,' Bolt told his readers. Bolt was furious at some members of the BCA for raising the white flag, as he saw it, in the fight with the environment movement. 'This year the BCA even considered backing the greatest green menace—the Kyoto Accord to cut greenhouse gases,' he thundered.

Bolt didn't have to worry about the BCA anymore. Morgan had by now retired from Western Mining, after a major company restructure, but his access to the Howard government was still unmatched by most chief executives in the country. As long as he was president of the BCA, he would block any move to put a price on carbon pollution.

———

At first glance, Beale's defeat looked like a victory for Howard and the big business lobby. On closer inspection, it exposed a giant weakness in Australia's climate policy that would dog Howard until the end of his prime ministership. The government had no serious alternative to reduce Australia's big greenhouse footprint, and many in business knew it.

In a confidential email on 10 September 2003, the head of Australia's peak gas and oil lobby, Barry Jones, bluntly told his members, 'Greenhouse policy is in a state of chaos. It appears there was no fall back option to the emissions permit trading package Cabinet rejected a fortnight ago.'

At the end of 2003 Roger Beale handed in his resignation as head of the Environment Department. In a touching farewell message, John Howard praised the man he'd done so much to thwart and blamed Beale's early retirement on the lingering effects of polio, a disease that had struck Beale down in childhood. This was true—Beale was in increasing pain, and the long hours and frequent travel were getting to him—but he was also completely dispirited. He felt the government was no longer listening to him.

With Beale gone, Howard pushed ahead with a new approach to climate change more in lock step with Washington. The White House had an ambitious plan to lead a new global partnership that would put a priority on the use of 'breakthrough technologies' to reduce greenhouse gas emissions. It had none of the mandatory emissions targets demanded by Kyoto and no price on carbon using an emissions trading scheme. It would eventually morph into the grandly titled Asia-Pacific Partnership on Clean Development and Climate.

This technology partnership would bring together big greenhouse gas-emitting nations and companies from the Asia-Pacific, including the world's fossil fuel giants, like ExxonMobil and Rio Tinto. Australia would be one of the founding partners.

'The fairness and effectiveness of this proposal will be superior to the Kyoto protocol,' Howard claimed when it was announced in 2005. 'It demonstrates the very strong commitment of Australia to reducing greenhouse gas emissions, according to an understanding that it's fair in Australia and not something that will destroy Australian jobs and unfairly penalise Australian industries.'

To its critics, like US Republican senator John McCain, the Asia-Pacific Partnership was 'nothing more than a nice little public-relations ploy', but Howard backed it all the way.

The government was already working closely with the US Department of Energy and Australia's big greenhouse gas emitters to draw up a local version of this strategy in a new climate and energy plan. One breakthrough technology that both Canberra and Washington were very keen to pursue was the promise of 'clean coal'.

Clean coal is known by the technical name Carbon Capture and Storage, or CCS. The basic theory behind it is to capture carbon dioxide emissions from coal-fired power plants and bury them in the ground to a depth of around 800 metres, theoretically reducing much of the world's greenhouse gas emissions from coal-fired power. The idea was being promoted by some of the world's leading coal companies, especially Rio Tinto and BHP.

Howard was excited by clean coal. 'There is no doubt in my mind that John Howard did in a sense dream of a technological solution,' said Peter Shergold, then his top public servant. Howard often sat through meetings of the Prime Minister's Science, Engineering and Innovation Council (PMSEIC) enthralled by the idea. 'I could see from his enthusiasm at PMSEIC meetings that he was a bit of a science junkie,' said Shergold.

The clean coal strategy was heavily pushed in a report called, 'Beyond Kyoto—Innovation and Adaptation', commissioned by PMSEIC's head, Dr Robin Batterham, the government's chief scientist. Batterham was a talented chemical engineer, who once worked for the CSIRO. But he was also the chief technologist for Rio Tinto, a company with huge greenhouse gas emissions in Australia from its aluminium operations along with some from its coalmines. As Batterham saw it, there was no conflict of interest between his job at Rio Tinto and his job with the government advising on greenhouse strategies.

'I'm very careful to separate the jobs there, and the two interests,' Batterham told the ABC. 'So I give my advice covering the whole spectrum of activity, I cannot and do not and, of course, it would be very foolish to put Rio Tinto's interest into the equation.' The leader of the Greens, Senator Bob Brown, was up in arms about Batterham's dual roles; but neither Howard nor his ministers saw a problem.

Despite Batterham's enthusiasm for clean coal technology, some in the energy business doubted it could work on the scale Australia needed, or at a cost that made sense. The basics of CCS had been employed for years in the gas and oil fields. But it had not been used to store huge amounts of carbon dioxide emissions from coal-fired electricity plants. BP's Greg Bourne was dubious from the start. 'So people like myself, who actually drill holes in the ground, who work in refineries and chemical plants, who knew the reservoir engineering, the petroleum engineering aspects, the pipelining aspects, the wet chemistry aspects, knew that this was going to be bloody expensive and really difficult to do.'

But Howard did find more support for the idea from other business leaders. In February 2004, the government invited thirteen executives from the biggest greenhouse gas emitters to help with the new plan for a high-tech solution. The Low Emissions Technology Advisory Group, as it was called, was stacked with the usual players relied on by the Liberal government—top executives from Rio Tinto, Alcoa, BHP, Orica, ExxonMobil, US power company Edison Mission Energy, the gas company Origin and, of course, Hugh Morgan, now with the BCA.

A leaked confidential record of a private meeting—held between some of these executives and John Howard, his industry minister Ian Macfarlane and senior bureaucrats—gives a real insight into their thinking at this time. They talked about a clean coal project, energy efficiency projects and other high-tech solutions for cutting greenhouse gases, but they made no serious case for wind and solar in Australia's cleaner energy future. On the contrary, the key purpose of the meeting seemed to be to stop a push to raise Australia's Renewable Energy Target.

Rio Tinto's senior executive Sam Walsh noted in the meeting that Howard's ideas for 'Super Dooper' technology breakthroughs were 'by no means concrete', but the prime minister wanted the businessmen to come up with some alternative ideas to deflect calls to increase the Renewable Energy Target. The target was still sitting at a low 2 per cent of Australia's energy supply and the renewable energy lobby wanted it raised. Indeed, a new government-commissioned report had urged the lifting of the Renewable Energy Target.

The meeting was at times nakedly political. Howard was facing an election that year and Labor's new federal leader, Mark Latham, was expected to promise an increase in the Renewable

Energy Target. Howard needed alternative ideas before the election. He was not pushing any particular model, he just wanted 'one that would pass the Pub Test'. The only thing he didn't want to do was increase the Renewable Energy Target because of its costs to industry.

Howard unveiled his new climate and energy plan in a white paper in June 2004, a month after that meeting with the senior business executives. Called 'Securing Australia's Energy Future', the white paper was a clear signal to the world that Australia would not be seriously reducing its large greenhouse footprint any time soon. Little detailed thought had been given to a clean energy future. Instead, it was a ringing endorsement of Australia's big-emitting industries.

The speech stared down Howard's climate change critics. He brushed aside the scientific advice from IPCC, refusing to say that climate change was 'real', only that 'the potential for climate change is real'. The subtext was clear: the government did not rate global warming as an urgent threat. 'We are taking action to address greenhouse gas emissions and climate change, but we are determined to do it the smart way—a way that does not threaten our energy advantage and national prosperity,' Howard told the National Press Club.

It was a relaunch of the 'No Regrets' policy, with the promise of silver-bullet technologies to solve the climate problem some-where down the track. Its centrepiece was a $500 million Low Emissions Technology Fund, to attract private dollars to match government grants to develop ideas like clean coal.

Howard's white paper ruled out any increase in the Re-newable Energy Target. In a bid to placate the public and the renewable energy industry the prime minister promised

$75 million for research into 'Solar Cities' and $134 million for the commercial development of renewable energy technologies. The fine print showed much of this money was spread over a wide range of programs over seven years and was hedged with multiple conditions.

Hugh Morgan and the BCA applauded the new plan as a 'win–win' for all sides. But some of Australia's top climate scientists were stunned. One privately called it an arrogant denial of reality. The Australian Conservation Foundation's president, Ian Lowe, described it quite simply as a 'disaster': 'Instead of providing what the Business Council called "a secure energy future",' he said, 'the Howard Government is lurching backwards into the future with its eyes on the past. History will see this Energy White Paper as a tragic missed opportunity.'

———

Howard's strategy glossed over the reality that Australia's greenhouse gas emissions from electricity generation were again on the rise. In 2004 they stood at 37 per cent above 1990 levels. Australians yet again were vying for the title of being the highest emitters of greenhouse gases per head in the developed world.

Within weeks of Howard unveiling his climate and energy plan, Environment Minister David Kemp quietly announced he was quitting. The PM's Office and the treasurer, Peter Costello, immediately contacted Andrew Robb, urging the former Liberal Party director turned lobbyist to run for Kemp's safe seat of Goldstein in Victoria in the upcoming election.

Three months later, Howard won a smashing fourth victory. Labor and its erratic leader, Mark Latham, were decimated. The government not only increased its vote, it won control of the Senate. Howard now had more power than any Australian prime minister in years.

Howard was at the top of his game and his government was riding the greatest resources boom in the country's history. Australia was shipping its coal, gas, iron ore and metals out to the world, and the money was rolling in. It looked like there wasn't a cloud on the horizon.

CHAPTER FOUR

SCORCHED EARTH

UNDER A HOT sun in a cloudless sky Trevor Smith stood in the ruins of his wheat crop looking at the dry stubble surrounding him. 'I think what you've got there is about four or five days of sheep feed,' he said laconically.

It was the first time anyone could remember the crop failing on the Smiths' Pine Park farm since 1948. The prime minister, in a business shirt and tie, sheltered under his Akubra hat and stared at the ground as the farmer ran through his losses from Australia's worst drought in a century.

John Howard was on a listening tour in the spring of 2006 and had flown into Forbes, in the Central West district of New South Wales. He wanted to hear directly from the men and women struggling to cope with the Big Dry. He could have flown pretty much anywhere in south-eastern Australia and heard the same bleak stories. Much of the nation's farmland was in drought and most of New South Wales was drought-affected. Some districts hadn't seen rain for years. Towns and cities in three states were imposing tough water restrictions. In New South Wales, the water minister was debating plans to build Sydney's first desalination plant after the dams supplying the city dropped to alarmingly low levels.

In the big empty grain shed on Trevor Smith's property, Howard reassured the Forbes farmers, the town's businesses and the mayor that the nation would stand behind them. 'We cannot afford to lose our farm sector,' he said sombrely. 'I do not believe that the country would survive the loss of that, and the nation would change its character permanently.'

The Millennium Drought, as it came to be known, was long and severe even for Australians accustomed to their dry, highly variable climate. Multiple weather drivers had all collided to cause the drought, but years later researchers from the CSIRO would also detect the fingerprints of climate change in its depth and scale.

The drought cut a swathe through the Murray–Darling Basin, the nation's most important agricultural region, knocking rural Australia to bits. It also cut the political ground from under John Howard. While at the time no one could say with any certainty the drought was influenced by climate change, that didn't deter many of Howard's opponents. 'What we're seeing with this drought is a frightening glimpse of the future with global warming,' South Australia's Labor premier, Mike Rann, declared as he returned from another emergency meeting to address the water crisis.

The politics of climate change had turned against Howard as the drought tightened its grip on the country. In 2006 almost 68 per cent of Australians thought global warming was 'a serious and pressing problem' and the government needed to do something immediately, even if it involved significant costs.

Howard hadn't picked the shift in the mood. He still doubted the scientific warnings of the IPCC; he called himself 'an agnostic' on climate change, not a 'sceptic', but he used the language of climate sceptics. Howard accepted humans had contributed to the growth of greenhouse gas emissions and that the climate was changing; but, he said, 'instinctively, I doubt many of the more alarming predictions'. Those appeared to include the IPCC advice that dangerous climate change was a threat to the planet and deep cuts to greenhouse gas emissions were needed in the next few decades. 'What makes me suspicious are the constant declarations from the climate-change enthusiasts that the science is all in, the debate is over and no further objection to received wisdom will ever be considered,' he would later complain.

But by the spring of 2006, Howard was finding himself at odds over climate change with top bankers, senior bureaucrats, state premiers and the majority of voters. His determination to stick to the 'No Regrets' policy—no action on climate change that would be a net negative for the economy—was looking stubborn. His spin lines fell flat in the face of political attacks by Labor and the Greens. Worse still, as a few insiders knew, behind the spin lay four years of ineffectual policy that had been shaped by the loud voices of the big greenhouse gas emitters, Howard's climate scepticism and his loyalty to US President George W. Bush.

———

By early 2006, Americans were also rediscovering global warming, much to the surprise of Al Gore, the former US vice-president who turned climate activist after losing the presidential race to Bush. In the wake of superstorms, heatwaves and a record melt of Arctic sea ice, the US media began reporting in earnest the warnings of the IPCC. *Time* magazine ran a seven-page spread on the threat facing the planet from climate change.

In May that year, Gore's grim documentary hit the cinema screens. *An Inconvenient Truth* was a movie-length version of the stark, unsettling slide show that he used on his global lecture circuit; it outlined the dramatic rise in greenhouse gas emissions from burning fossil fuels and the devastating consequences that would follow if this was left unchecked. In Congress, Democrats and Republicans were jumping into the climate change debate with an eye to the 2008 presidential election. Even the former climate sceptic Senator Chuck Hagel, who had led the fight to reject the Kyoto Protocol, had declared climate change 'a top tier issue'.

ExxonMobil's high-profile attacks on climate science, so effective in the past, were being shunned by some of its former business allies. The oil giant's aggressive chairman and chief executive, Lee Raymond, retired at the end of 2005; inside the company his boots-and-all climate-sceptic strategy was beginning to be questioned. The company was under pressure from politicians, shareholder groups, religious activists, the Royal Society (the scientific academy of the UK and the Commonwealth) and environmental activists over its funding of climate-sceptic lobbies.

Greenpeace created a website listing in mind-boggling detail all the climate-sceptic groups, think tanks and scientists that ever got a dollar from Exxon—40 organisations at one count. Exxon began reviewing the flood of money it was pouring into the loud climate-sceptic lobbies, like the Competitive Enterprise Institute in Washington where Myron Ebell was ensconced.

'Lee Raymond was succeeded by Rex Tillerson, who used certain events to very quickly cut us off. And then to cut off all the other groups,' Ebell recalled. 'Exxon was under pressure, as soon as Lee Raymond retired, to stop funding us and to stop funding all the groups.' The public pressure was real, but Exxon and the American Petroleum Institute—the US fossil fuel lobby—would still continue to forcefully fight efforts to put a price on carbon emissions.

In Australia, the split in the business community over climate change finally burst into the open in April 2006. Westpac's chief executive, David Morgan, one of the Business Council of

Australia's (BCA's) old climate rebels, launched the Australian Business Roundtable on Climate Change. It included the head of the Australian Conservation Foundation, Don Henry, who was about to become one of Al Gore's collaborators in Australia. 'We believe that climate change is a major business risk and we need to act now' was the stand-out message in a report they handed out to the media.

The Roundtable included chief executives from two major insurance companies, Swiss Re and IAG, gas giant Origin Energy and the president of BP Australasia, Gerry Hueston. His predecessor, Greg Bourne, had left the oil business to run the Australian branch of WWF, the World Wide Fund for Nature.

The Roundtable executives called for a 'long, loud and legal' framework to put a price on carbon pollution, whether it was a tax or an emissions trading scheme. It was vital, they argued, to push investment in clean energy technology. Australia's ageing, carbon-heavy electricity system needed to be revamped. Labor state governments piled the pressure on Canberra. In a provocative move, they released their own report, canvassing a national emissions trading scheme. Roger Beale, the retired head of the federal Environment Department, was one of the advisers on the scheme.

In June 2006, Howard had tried to seize back the initiative by announcing an inquiry into nuclear power as a way to tackle carbon emissions. This energy was 'the cleanest and greenest energy source of all', he said. He appointed nuclear physicist and former Telstra boss, Ziggy Switkowski, to lead the inquiry. Howard told parliament later he'd talked to the federal treasurer of the Liberal Party, Ron Walker, a few days before announcing the inquiry. Walker and Hugh Morgan had registered a new company called Australian Nuclear Energy, and were looking at how to profit from the industry.

For the ex-BP boss, Greg Bourne, the nuclear option was just one more 'silver-bullet' idea from Howard. Predictably, in the run-up to an election, federal Labor, state premiers, the Greens and the environment movement were all prepared to campaign against any plans for a nuclear plant on Australian soil.

For the first time in years, the prime minister's grip on power was looking shaky. The conservative political hero was now under attack on multiple fronts. A bitter public fight over Work Choices, the government's radical new industrial relations laws, sparked outrage from trade unions and ordinary workers alike, who found their wages and working conditions under threat.

Howard's political capital was running as dry as the country. When he went on his listening tour to drought-ravaged New South Wales, the parched fields and cloudless skies served only to highlight his stubbornness on climate change—a leader out of touch with today's problems.

That spring the rains didn't come; but Al Gore did, and the arrival of the 45th US vice-president put climate change at centre stage of Australian politics. Gore was promoting *An Inconvenient Truth*, and while Howard dismissed the film as a 'slick production', even he had to admit it had 'an enormous impact on the millions of moviegoers around the world who viewed it'.

To Howard, Gore's film was another 'alarmist' missive, the kind he scorned. But Gore knew a bit about politics. He tied Howard to Bush, branding the two leaders climate outlaws, 'the Bonnie and Clyde' of global climate change. He homed in on the drought to reject Howard's claims that 'alarmists' were exaggerating the threat facing the planet.

'He's increasingly alone in that view among people who've really looked at the science,' Gore told Kerry O'Brien on the ABC's *7.30 Report*. 'Look at what's at risk here in Australia. You have an advanced civilisation ingeniously built in the driest inhabited continent on the planet. You have such low water availability on average already. That particular quality is most affected by global warming and, as predicted, you now have growing water shortages in Brisbane, here in Sydney, in Canberra, in the west, in Perth.'

Gore had hit the zeitgeist on global warming. Almost overnight, the debate was switching from wrangling about the cost to the economy to saving the planet, and many in the Labor Party enthusiastically jumped on the Gore bandwagon. 'It is an inconvenient truth that the Howard government is increasingly isolated on climate change,' Labor's shadow environment minister, Anthony Albanese, crowed in parliament.

Gore's intervention in Australian politics was calculated. Just as the climate sceptics had developed deep links between Australia and the US, so too had the climate activists. For three years Gore had worked with local NGOs and business figures to reignite the climate debate. During his speaking tour, Gore would sometimes end his talk with a line from the American poet Wallace Stevens: 'After the final no, there comes a yes, and on that yes the future of the world depends.'

Howard didn't say 'yes', but by November 2006 he said a tentative 'maybe' to putting a price on carbon pollution. His chief of staff, Arthur Sinodinos, later said it wasn't so much the drought that got to the prime minister, it was the banks: 'The tipping point we reached towards the end of '06 was partly what was happening in the community including, you know, all the Al Gore stuff,' Sinodinos recalled. 'But the other thing that was happening is the Business Council of Australia finally came to a more unified position on this. We were getting more and more banks, financial institutions and others talking to us about how they were trying to assess climate risk. And, it certainly struck me and I guess some others that, well, if people in the private sector have to put their money where their mouth is, and are being serious about this, then we should try and get on top of this.'

One of the final straws was the UK government's release of Sir Nicholas Stern's review on the economics of climate change. In a sweeping finding that made headlines around the world, Stern called climate change 'the greatest market failure the world has ever seen'. He warned that unless countries were prepared to sacrifice some of their growth to deal with climate change, the costs to the global economy could ultimately be 'on a scale similar to those associated with the great wars and the economic depression of the first half of the 20th century'.

Howard would later dismiss Stern's analysis as 'beguilingly simple and seductively cheap', but it rattled his government. Howard still had an iron grip on climate change policy. His environment minister, Ian Campbell, had zero chance of persuading him to change direction and put a price on Australia's carbon emissions.

It was left to the head of the Prime Minister's Department, Peter Shergold, to finally cajole Howard into reopening the fraught subject. Shergold's style was the opposite of his brash predecessor Max Moore-Wilton. The English-born academic with a long history of public service was a low-key, unflappable pragmatist. He knew he would never get Howard or the climate sceptics in cabinet to act by lecturing them on the science of climate change, so he tried another approach.

'My argument was always that I don't want to argue the science with you because I had Andrew Robb and Nick Minchin, who were absolute sceptics, who didn't appear to believe it at all,' Shergold recalled. 'I can't argue the science with you. I don't need to,' Shergold told Howard and his ministers. 'All we should do is to say that science shows us that there is significant risk. And what a government should do, as it does in many areas, is to manage that risk for the future. And that the best way to manage the risk for the future, the less costly way of meeting it, was to use market mechanisms to reduce greenhouse gas emissions.'

Shergold was joined in his effort by Treasury secretary Ken Henry, a veteran of the long battle to get Australia to introduce an emissions trading scheme to bring down its greenhouse gas pollution. By this time, said Shergold, Howard had returned from his listening tour of the drought-ravaged properties in New South Wales, where even some farmers were telling him, 'John, look at what's happening. It's real. This is here.'

The wall of opposition in the BCA had also cracked. Hugh Morgan had gone as president. His replacement, Michael Chaney, was a conservative Western Australian who had sat on the board of mining giant BHP Billiton and also chaired the National Australia Bank, a big investor in the resources industry. Chaney had a foot in both camps, and he knew Howard would not embrace anything radical.

After an end-of-year dinner hosted by the BCA, Howard finally flagged a new inquiry into an emissions trading scheme. However, he hedged the inquiry with a strict caveat: there would be no change to the 'No Regrets' policy, nothing that could damage Australia's fossil fuel industry. 'There's one thing I am frozen in time about and that is a determination to protect the industries of this country that give us a natural competitive advantage,' Howard told the reporters trying to digest his volte-face.

Shergold gathered the country's top public servants for the review. But Howard hand-picked every member of the business advisory group that would work with them. Once again, the advisers were mostly executives from the big greenhouse polluters: Peter Coates, senior executive of global mining giant Xstrata; Tony Concannon, managing director of UK International Power (which owned a huge brown-coal power plant in Victoria); Chris Lynch, from mining giant BHP Billiton; John

Marlay, chief executive officer with Alumina Limited; and the chair of Qantas, Margaret Jackson. There was one voice from the financial markets: John Stewart, the chief executive of Chaney's National Australia Bank.

'This is the Prime Minister's coal industry sop,' Greens leader Bob Brown said scathingly, 'it's going to give him the answer they want.' That was pretty much the reaction of the environment movement across the board.

But Shergold wasn't fazed. He knew that without the big greenhouse gas emitters on board, Howard would never agree to any version of an emissions trading scheme. He also knew business wanted a deal on a scheme before one was imposed on them by a new Labor government, and that was looking increasingly likely.

Less than a week before Howard's backflip, the federal Labor Party dumped its unpopular leader, Kim Beazley, and chose a dynamic Queenslander to take them to the next election. Kevin Rudd, a former diplomat and state bureaucrat, was notorious among his colleagues for his frenetic drive and unbridled ambition. He wanted to use climate change to expose Howard as old and out of touch.

Rudd boldly appointed Peter Garrett as his new Shadow Minister for Climate Change, Environment and Heritage and the Arts. The charismatic rock star was the former lead singer of Midnight Oil, one of the biggest bands on the Australian music scene. For years Garrett had been a passionate crusader for Indigenous rights and the environment. Before he was recruited by the Labor Party in 2004, he was president of the Australian Conservation Foundation. Garrett's cut-through message on climate change was much the same as Al Gore's: 'The jury is in and the science is clear. The planet is heating. The time for action is now. Our children and grandchildren deserve no less.'

When Howard returned from his summer holidays in January 2007, his political future looked dire. He was 67 and the years in office had taken their toll. Some of his own ministers, including Minchin, wanted him to hand over to his deputy, Treasurer Peter Costello. He faced a tough election by the end of the year. He also faced another bruising fight over Kyoto.

The UN climate convention, the UNFCCC, was to hold its most important meeting since Kyoto in December that year, in Bali. The world's environment ministers were going to begin negotiations on the new climate agreement that would follow the Kyoto Protocol when it expired in 2020. At stake for Australia was whether it would have a meaningful seat at the table if it still refused to ratify Kyoto.

Howard would not budge on that, but he did try to shore up his climate credentials. In a move that shocked many in the party, he appointed a new environment minister—Malcolm Turnbull. The brash, extremely bright former merchant banker was a surprise decision. Turnbull had only arrived in parliament at the last election, after a career that spanned journalism, the law and finance. But more importantly, Turnbull was a lightning rod for Liberal Party conservatives. Over a decade earlier, he had led the charge for Australia to become a republic. More to the point, he understood and supported the science of climate change.

The climate sceptics were furious, especially their cabinet flag-bearer, finance minister Nick Minchin. It was a fateful decision that would ignite a long, bitter feud between the party climate sceptics and Turnbull. 'Malcolm was all the way with all the green agenda,' Minchin said later. 'And then so Turnbull immediately started running this very green agenda, and we ended up having our own commitment to an emissions trading scheme.'

Over at the Department of the Environment, Howard Bamsey was delighted. Turnbull had already proved his worth to the government by handling the water crisis in the Murray–Darling Basin as Howard's parliamentary secretary. As Bamsey saw it, Turnbull was the first minister since Robert Hill who really understood the science of climate change. 'He was far more into the issues substantively than any new minister I had seen,' said Bamsey. 'He got a briefing on the issues on the first or second day from all the science experts in the portfolio. There was pretty much nothing they could tell him that he didn't already know.'

But Turnbull did not reflect the thinking at the top of the government, and that soon became clear. In February 2007, the UN's scientific advisory panel, the IPCC, released a preview of its latest report on climate change. The 'Summary for Policymakers' contained the bleakest findings since the panel was established. Based on an analysis of thousands of scientific papers, its best estimate was that temperatures could rise by between 1.8 and 4.0 degrees Celsius from pre-industrial levels by the end of the century if nothing was done to slow the rise in global emissions. It was possible the rise could be more than 6 degrees. If temperature rises went above 3.5 degrees Celsius, model projections suggested 'significant extinctions'—of between 40 and 70 per cent of species looked at in the studies. Heatwaves, extreme temperatures and heavy flooding events were 'very likely' to become more frequent, and droughts were also likely to increase. It also found, with over 90 per cent certainty, that increases in global temperatures since the mid-twentieth century were being driven by human-induced global warming caused in large part by burning fossil fuels and deforestation.

A few days after the IPCC summary became public, the ABC's Tony Jones interviewed Howard and asked when he had ceased being a climate sceptic. A wary Howard replied: 'I can't put an exact time on it. It wasn't a Damascus Road conversion.

I've always accepted that greenhouse emissions, carbon emissions were potentially damaging. I think the scale of it has become more apparent as a result of the research.' In reality, Howard hadn't changed his views much at all. He still believed a lot of the advice on climate change was 'alarmist'.

The next day, Labor's opposition leader, Kevin Rudd, walked into parliament and asked Howard whether he accepted the connection between greenhouse gas emissions and climate change. Howard hedged. 'Let me say to the Leader of the Opposition, the jury is still out on the degree of connection,' he replied.

Rudd went straight on the attack. 'The message that came from the Government in today's Question Time, at the beginning of 2007, leaves most of us gobsmacked,' he scoffed. 'We have the citadel of scepticism when it comes to climate change.'

Less than three hours later, Howard was forced to go back into the parliament to clarify his answer. 'Just for the record,' he said, 'I do believe there is a connection between climate change and emissions. I do not really think the jury's out on that. I do think that jury is out on the connection between climate change and drought . . .'

—————

Four months before the election, with his government still sinking in the polls, Howard was forced to announce his own plan to put a price on carbon pollution. The Shergold inquiry had brought down its report, opening the way for Howard to go ahead with an emissions trading scheme. Even though the big greenhouse gas emitters were lukewarm, they knew Rudd would be likely to introduce his own version when Labor came to power, so it was a case of going with the devil they knew.

The Shergold report had cushioned the scheme with recommendations for generous amounts of free permits to protect the big greenhouse gas emitters and the jobs they supported if they were exposed to foreign competitors. Politicians would decide the all-important 'cap' or target for cutting greenhouse gas emissions after the election. In theory, a future government could be lobbied to bring in a weak target that would have little impact on the big polluters or on cutting emissions.

After a decade of fighting the idea of an emissions trading scheme, Howard was still a very reluctant convert. But Turnbull swung behind it enthusiastically. Unfortunately, the former banker got the economics of emissions trading but not the politics. Shergold was worried that Turnbull's enthusiasm was antagonising the opponents of the trading scheme in cabinet.

For some of the government's climate sceptics the report was a huge betrayal. Finance minister Nick Minchin later dismissed it as 'a crock of shit' and 'a picnic for the bankers and financiers' in a bitter, reflective interview with the State Library of South Australia. 'So it didn't really salvage the Government's position at all, and because we'd right up to that point, been attacked as "deniers" and then suddenly do this latter-day conversion, so it was never believed, anyway. So it was a really awful time, particularly for people like me who just thought it was sort of waving the white flag in the face of all this,' Minchin said later. 'I'd rather die honourably than die dishonourably.'

Minchin was right on one thing: it didn't salvage the government. Rudd checkmated Howard. Rudd and the Labor states set up their own inquiry on climate change and emissions trading, headed by the eminent economist Ross Garnaut. Labor also rolled out a string of promises for action on climate change, including a jump in the renewable energy target to 20 per cent of electricity use, support for clean coal technology and the ratification of Kyoto—all moves Howard tried to match—except Kyoto. The embattled PM refused to budge on that.

Howard was still in lock step with President Bush on the Kyoto Protocol. The two were aggressively pushing their alternative strategy for the international technology partnership to help cut greenhouse gas emissions rather than Kyoto's demand for mandatory emissions cuts by developed nations.

In September 2007, Howard and Bush made one last effort to derail Kyoto. Australia was hosting the Asia-Pacific Economic Cooperation (APEC) summit in Sydney. It was likely to be Howard's last outing on the international stage and it was a big deal. APEC brought together the powerhouse economies of the US, China and Japan, along with other regional countries. Bush and China's President Hu Jintao would be there on stage with Howard.

In a provocative move, Howard promoted the APEC meeting as 'one of the most important international gatherings of leaders to discuss climate change since the 1992 Rio conference'. He wanted to produce a Sydney APEC Leaders' Declaration on Climate Change that would bypass the Kyoto Protocol and pre-empt the UN's global climate conference in December.

This move was quashed by the Chinese president. Hu bluntly reminded Howard that the UNFCCC and its Kyoto Protocol were 'the most authoritative, universal and comprehensive international framework' for tackling climate change. The Sydney Declaration ended up as a diplomatic fizzer, forgotten within days of its release.

Howard was a lame duck abroad and at home. The polls showed the government trailing eighteen points behind Labor. In a desperate act on the eve of the APEC summit, Howard asked his foreign minister, Alexander Downer, to test the mood in cabinet for a leadership change. By now Howard believed he would not only lose the election but also lose his own seat of Bennelong, where a former ABC journalist, Maxine McKew, was running against him for Labor. Downer reported back the dismal news that key ministers, including Turnbull, wanted him to go. In the end Howard decided to stay and few thought it was worth the fight to remove him. By the time the APEC leaders left Sydney, Howard's power was ebbing away.

In the dying days of the government, Turnbull made one last attempt to get cabinet to ratify Kyoto; but Howard resisted to the end. Anxious to shore up his own wealthy Sydney eastern suburbs seat, Turnbull had already distanced himself from the prime minister and the Liberal sceptics on climate change. In a memorable clash with Garrett in parliament, Turnbull attacked Labor over Kyoto, but he passionately described climate change as 'an enormous challenge and probably the biggest one our country faces, the world faces, at the moment'.

On election night, 24 November 2007, the Labor Party led by Kevin Rudd swept to power. Turnbull held his seat but Howard ignominiously lost his. The environment movement had campaigned across the country for a change of government and the myth was born that this was the world's first 'climate change election'. It was not. Work Choices, healthcare and education were bigger issues for most Australian voters.

Despite this, the 2007 election was historic. Australians voted into office a prime minister who warned them the planet was at risk from dangerous climate change; he promised them his government would cut the nation's greenhouse gas emissions. On his first day in office Kevin Rudd ratified the Kyoto Protocol.

CHAPTER FIVE

THE STATE OF THE REEF I

ACROSS THE ROAD from Molly Malone's Irish pub at the top of Flinders Street in Townsville, sits one of the great tourist attractions of tropical North Queensland. Every winter in school holidays parents and kids flock to Reef HQ to see the leopard sharks feeding, giant groupers defending their patch or Marlin searching for Nemo in the world's largest living coral reef aquarium.

Reef HQ is the National Education Centre for Australia's Great Barrier Reef Marine Park Authority and it's home to hundreds of species of colourful fish and stunning corals. For those without the money or means to go out there, it offers a glimpse of the Reef's spectacular beauty. The aquarium is also the public face of the Marine Park Authority, the federal agency whose job it is to manage and protect one of the greatest wonders of the natural world.

In a brave move a few months before the 2007 federal election, the Marine Park Authority published a landmark report on the Reef and climate change. It opened with an arresting line: 'Climate change is now recognised as the greatest long-term threat to the Great Barrier Reef.'

The report was signed off by John Howard's environment minister, Malcolm Turnbull, and it catalogued a series of disturbing events. Mass bleachings of corals in 1998 and 2002 had affected more than half the Great Barrier Reef. Around 5 per cent of its coral reefs suffered lasting damage. A more localised bleaching in a small southern section of the Reef in 2006 had killed 40 per cent of corals around the Keppel Islands, a famous holiday destination.

Researchers also recorded the mass die-offs of seabird chicks when the sea water rose to unusually high temperatures and the parent birds were unable to find their usual prey fish. The grim catalogue ended with a warning that rising temperatures at vital turtle rookeries on the Reef islands could change the sex ratio of the hatchlings, increasing the number of females and putting future turtle populations at risk.

A sharp-eyed reader could see the first footnote in the Marine Park Authority's report referenced the IPCC's 2007 'Summary for Policymakers', released in February that year. The Marine Park's scientists were very familiar with the research backing up the IPCC's findings on the Great Barrier Reef. A lot of it was written by Dr Ove Hoegh-Guldberg, an Australian marine biologist, IPCC author and pioneering researcher into coral bleaching.

Ove Hoegh-Guldberg fell in love with the Great Barrier Reef as a kid. He made his first trip there with his Danish grandfather, who was a scientist on a mission to collect butterflies for a natural history museum in Denmark. They went to the Whitsunday group of islands in the heart of the Great Barrier Reef. He was smitten. 'That was when I think I saw my first butterflyfish,' he recalled. 'It's sort of almost like a tattoo on my brain. I can still see it whizzing around the coral and—absolutely—the most beautiful thing.'

Anyone lucky enough to snorkel or dive on the Great Barrier Reef knows why it is one of the seven natural wonders of the world. Stretching along the Queensland coast for 2300 kilometres, from the tip of Cape York in the north almost to Bundaberg in the south, the Reef is the size of Italy, so big it's visible from space—the largest living structure on the planet. It supports an astonishing variety of life: dugongs and dolphins, sharks and turtles, stingrays and sea snakes. The thousands of plants and animals found in its waters include everything from microscopic plankton to 100-tonne whales.

Australians fought one of the biggest environmental campaigns in their history to protect the Reef from oil drilling in the 1970s. In 1975 the federal government made much of it a marine park to help protect it. Six years later it was selected by UNESCO as a World Heritage Site.

The Reef is home to 10 per cent of all the world's coral reefs, some 3000 separate reefs are spread across the marine park. One of the authority's missions is to protect these extraordinary corals. Many of the fish and plants that live on the Reef depend on the corals. If the corals die off in large numbers, the Great Barrier Reef is in serious trouble.

As a young student in 1987, Ove Hoegh-Guldberg was studying coral reefs for his PhD at UCLA, California, when he first heard about patches of 'coral bleaching' appearing around Lizard Island in the northern section of the Great Barrier Reef. At the time, a few outbreaks of bleaching had also appeared on reefs in the Caribbean, but no one was sure why. 'More and more people were seeing these mysterious events where corals suddenly went white and no one really had an explanation,' Hoegh-Guldberg recalled. 'Was it temperature? Was it light? Was it a disease? No one really knew.'

The young marine biologist got a scholarship to go to Lizard Island, where the Australian Museum runs one of the most alluring research stations in the world. The island lies about

100 kilometres off the coast of Far North Queensland in pristine tropical waters and is fringed by spectacular coral reefs. The only other substantial presence on the island is a luxury tourist resort.

The research station was on Lizard Island for good reason. The coral reefs out from it were some of the least likely to be impacted by humans. The island is so remote, it avoids the agricultural pollution that washes down Queensland's rivers into the Reef waters. There's also little impact from tourists. So at Lizard Island it was possible to eliminate obvious causes for the bleaching, like water quality, and explore other reasons why the local corals were turning white.

Most healthy corals get their colour because they live in partnership with single-celled algae, zooxanthellae. The corals provide the algae with protection and nutrients, and the algae in turn produce food for the coral and give it its colour. When corals get stressed, they expel the algae and lose their colour. That's why they look 'bleached'. Importantly, bleached corals don't necessarily die; they can survive days or weeks of bleaching and recover, depending on the species of coral and whether they are robust.

If bleaching goes on too long, the coral will eventually die. If that happens, a reef can take five to ten years or more to regenerate. Successful regeneration depends on there being no new catastrophic disturbance, like another mass bleaching or a cyclone. If a reef is not robust to begin with, because of poor water quality or disease, it might never recover.

After studying the coral bleaching at Lizard Island, Hoegh-Guldberg was convinced of one thing: the bleaching was tied to temperature. He exposed corals to slight increases in light, salinity and temperature and only one of these had a dramatic impact. 'The only change that consistently caused bleaching in the laboratory were relatively small changes in sea temperature,' he explained.

Almost a decade after Hoegh-Guldberg wrote his PhD on coral bleaching, the world experienced its hottest year on record for the twentieth century. That year, 1998, saw mass coral bleachings on tropical reefs around the world, and they grabbed the attention of marine biologists everywhere. Hoegh-Guldberg, by then working at the University of Sydney, wanted to look at the connection between coral bleaching and climate change. He wanted to know the impact of a rise in ocean water temperatures on coral reefs under climate change models used by the CSIRO and the UN.

He and a young climate modeller from Europe factored in effects like El Niño weather patterns and projected temperature rises under different climate change scenarios. The findings were devastating. If greenhouse gas emissions kept rising and climate change continued as predicted, back-to-back bleaching events were likely to happen every year on the Great Barrier Reef, possibly within Hoegh-Guldberg's own lifetime. Until then, he'd had no idea this was even possible.

'I expected the answer would be hundreds of years away and the big shock was that by the mid-century reefs would have annual bleaching events which we knew at that point could be lethal to corals,' Hoegh-Guldberg recalled. 'I remember exploring the risks of bleaching and mortality using a big Excel file and the answer emerging, which wasn't the answer anyone wanted. It wasn't going to be hundreds of years into the future. Rather, annual bleaching, back-to-back bleaching, would occur around mid-century. I thought I must have made a mistake. I remember on the last run on a Sunday afternoon pulling my chair back from my desk and saying, "Oh my goodness".'

Hoegh-Guldberg put his career on the line and submitted his research paper to a journal published by the CSIRO and the Australian Academy of Science. When the findings came out, the blowback was immediate. Some of his own colleagues simply didn't believe him.

'In 1999, when Ove wrote his report on the Great Barrier Reef and climate change bleaching, he was, honestly, heroically brave,' said Dr David Wachenfeld, now the chief scientist at the Great Barrier Reef Marine Park Authority. 'There were other scientists, in other parts of the world, saying the same thing, but in 1999 in the Australian context he was really pushing that agenda and there were people saying it's not that certain, it's not that big a deal, maybe reefs will adapt, maybe climate change won't happen that quickly, maybe you are being alarmist, we're not sure about the data,' Wachenfeld recalled. 'Ove was, certainly in my experience, the first Australian scientist to say we have a very serious problem on our hands.'

The reaction to the paper shocked Hoegh-Guldberg. The reviews by some of his colleagues were scathing. Many thought he had gone too far and there wasn't enough evidence to support him. 'I felt very out on a limb,' he said later.

His problem was not just his disturbing findings. At the time Hoegh-Guldberg was doing voluntary work for Greenpeace, including writing up a version of his bleaching report for the environment group based on his peer-reviewed paper. He had also agreed to be one of their unpaid scientific advisers at the fraught UN climate talks at The Hague in 2000.

When he introduced himself there to Robert Hill, then Australia's environment minister, Hill greeted him with the words, 'Oh you're the author of that alarmist paper.' Hoegh-Guldberg snapped back, 'No, I am the author of an alarming paper, which is somewhat different!'

While he and Hill later laughed about the exchange, the tag 'alarmist' would dog Hoegh-Guldberg for over a decade. Climate sceptics in a raft of right-wing think tanks, as well as columnists in the Murdoch-owned media, scorned his findings and repeatedly attacked him. But unlike many of his colleagues, Hoegh-Guldberg was used to standing his ground. A robust, confident debater with a tough hide, he was more than willing to argue his case. 'I like a good brouhaha,' was how he put it.

In 2002, then the second-warmest year on record, another mass bleaching hit coral reefs, especially the southern section of the Great Barrier Reef. For scientists at the Marine Park Authority it was a turning point. 'That made it undeniable that 1998 was not a freak event and that we really needed to start getting ready for this stuff,' said Wachenfeld, who was then a young scientist. He saw colleagues beginning to swing behind Hoegh-Guldberg.

The two mass bleaching events also rang alarm bells in Canberra. Most of the Reef corals did gradually recover from the bleachings, much to the relief of Hoegh-Guldberg and other marine scientists, but the vulnerability of the Great Barrier Reef was on the federal government's radar. At the time the Reef's tourist industry alone was worth around $4 billion annually to the economy. The advice from the Marine Park Authority was to build up the Reef's resilience.

An ambitious plan was already in the works for the Great Barrier Reef Marine Park Authority to increase the protection of the Reef's corals and fish stocks; this was given the green light by the Howard government in 2003. It turned whole sections of the Marine Park into 'no-take' zones, banning fishing in around a third of it. Despite cries of protest from the commercial fishing industry, Dr Virginia Chadwick, then the chairwoman of the authority and a former NSW Liberal minister, and John Tanzer, the authority's executive director and a veteran of government fisheries, worked closely with the federal Environment Department's head, Roger Beale, to get the rezoning plan up.

Unveiling the plan in parliament in December 2003, the then environment minister, David Kemp, had high hopes for its success. 'It means the Reef is going to be much more resilient in the future than it is at the present time to the impacts of global warming and pollution,' he promised. 'It means that the communities along the Reef who depend on it are going to have their major economic asset preserved for the future.'

Ove Hoegh-Guldberg and other marine scientists cheered when the Great Barrier Reef rezoning plan sailed through parliament with support from both sides of politics. But he was not convinced it would stop coral bleaching inflicting serious damage on the Reef in the long run. 'It's a wonderful story. Australia's response to local threats to the Great Barrier Reef—like overfishing and coastal water quality—has been solid. Unfortunately, it does not solve the underlying climate change issue,' he said.

He was convinced the Great Barrier Reef could only be protected from coral bleaching if greenhouse gas emissions were cut and sea water temperatures stopped rising.

———

Two years after the big Reef rezoning plan, global greenhouse gas emissions were still rising. A third bleaching hit the Reef in 2006. It covered a small area, but in parts it was lethal. Hoegh-Guldberg had moved to Queensland University by then, to set up a marine studies centre. Part of his job was working at the coral reef research station on Heron Island. This beautiful cay in the southern section of the Great Barrier Reef is about 72 kilometres off the coast from Gladstone Port which boasts one of the world's largest coal loaders.

'I remember someone telling me that each of the [coal] carriers lined up outside Gladstone that you see when you fly to Heron Island was worth millions in royalties to the state government,' he said.

In 2006 Australia was still the biggest coal exporter in the world and Queensland's main coal ports sat on the coastline inshore from the Great Barrier Reef: Hay Point, Abbott Point and Gladstone. The coal came from huge open-cut and underground mines in Central Queensland's Bowen Basin. Coking coal for the world's steel mills, and thermal coal for power stations at home and abroad, was cut out of the basin by the global mining giants BHP, Rio Tinto, Anglo American, Peabody Coal and Xstrata. The coal was shipped out through the Great Barrier Reef World Heritage Area. It was a rich business that created handsome profits, good jobs and big coal royalties for the Queensland government.

The idea that there might be a 'coal versus coral' debate over the coal trade was largely ignored by politicians in both major parties. This was easy to do because of the way the UN climate change convention worked. Australia's coal exports were sold around the world and it was up to the customers buying and burning the coal—Japan, the EU, South Korea, Taiwan, China and India—to do something about the emissions they produced. And if they didn't buy the coal from Australia, they would buy it from somewhere else. The problem was that, apart from developed countries like Japan and the EU members, the countries buying Australian coal were not bound to cut their emissions under the Kyoto Protocol. So, as coal exports went up, emissions went up.

Hoegh-Guldberg understood it was 'a mortal sin' to be seen to be making political statements. 'Mentioning that we needed to reduce coal and fossil fuels as the source of the problem might be seen as violating the tenet of being independent and avoiding policy recommendations,' he explained. But in late 2006 he became one of the first marine scientists to wade into the 'coal versus coral' debate that would eventually become one of the most fraught conflicts in Australia's climate change politics.

Even in 2006 it was a sensitive subject for the country's marine scientists. How could you argue the case that the Great Barrier Reef needed to be protected from coral bleaching caused by climate change without saying Australian coal exports are part of the problem? 'We were pussyfooting around each other,' Hoegh-Guldberg recalled. 'People also realised that attacking coal directly would open up the issue of jobs, sidelining climate change in the minds of many Australians. Climate change really was the elephant in the room, which invariably ended up stuck in the corner.'

Hoegh-Guldberg was by now considered a world expert on coral bleaching and climate change. Indeed, he had just made a significant contribution to the IPCC's latest report, due for release in 2007. He decided it was time to cross the boundary and agreed to be an expert witness in one of Australia's first legal cases objecting to a coalmine because of its impact on climate change.

'When you are asked to put in a deposition on climate change and its impacts on coral reefs, you are really being asked for the scientific evidence of a chain of linked events. That is: burning more fossil fuel increases CO_2; increasing CO_2 increases temperature—both land and sea. And increased sea temperatures cause massive changes to coral reefs and many other ecosystems. Essentially, our role as, I guess, established scientific experts, was to give "the real oil on this", pardon the pun,' he said.

The case was brought by the Queensland Conservation Council (QCC) over the planned expansion of the giant Newlands mine in the Bowen Basin. The mine was run by the Anglo-Swiss company Xstrata and the QCC argued the increased greenhouse gas emissions from the mine would add to global warming unless they were offset. Critically, most of the emissions the QCC wanted offset were from the burning of the coal by the mine's customers.

Hoegh-Guldberg agreed to prepare a report for the QCC

outlining pretty much what he had been saying for a while: 'Climate change has grown from insignificance twenty years ago to the major threat facing the Great Barrier Reef.'

The Newlands mine case was a big leap for Australian environmentalists. Xstrata could show the mine expansion couldn't be profitable if it had to offset all the emissions from its overseas customers. The QCC tried to amend their case to cover just 10 per cent of the emissions. In the end, the case didn't make history. Queensland's Labor premier, Anna Bligh, rushed new laws through the state parliament in October 2007 to ensure Xstrata could expand its mine. But before that happened, a remarkable twist occurred in the legal case.

The Land and Resources Tribunal, the state government body ruling on the expansion, began its two-day hearing in early 2007. As part of its case, the QCC relied on Sir Nicholas Stern's climate change review for the UK government, which was making headlines around the world. During the hearings, neither Xstrata nor the Queensland government disputed the link between greenhouse gas emissions from burning coal and climate change. But much to the surprise of the environmentalists, the president of the state tribunal, Greg Koppenol, did dispute the link and specifically attacked the Stern review. To back up that attack, he cited a paper co-authored by an Australian climate-sceptic scientist, Professor Bob Carter.

One of the Lavoisier Group's favourite climate-sceptic scientists, Carter was rocketing to fame not only in the Queensland courts but on the international stage.

Professor Carter piqued the interest of the global online climate-sceptic movement back in 2003 with his punchy critiques of 'warmaholics', as he liked to call mainstream climate scientists.

A palaeontologist and marine geologist, Carter was a well-credentialed scientist with a PhD from Cambridge. He had worked at the University of Otago in New Zealand before he came to Australia and landed the job as the head of the School of Earth Sciences at James Cook University (JCU) in Townsville, a post he held for almost two decades.

When he stepped down from that position, Carter became deeply immersed in the climate-sceptic movement. JCU made him adjunct research professor at its Marine Geophysical Laboratory, a title that gave him gravitas on the sceptic blogs.

Carter had already crossed swords with Ove Hoegh-Guldberg by 2004. That year, *New Scientist* magazine reported on a tough analysis Hoegh-Guldberg had written for WWF and the Queensland Tourism Industry Council, based in part on his coral bleaching research. The report predicted the possible demise of much of the Great Barrier Reef's coral cover if greenhouse gas emissions kept rising apace. 'Even under the best case scenarios, coral cover is likely to decrease to less than 5% on most reefs by the middle of the century. Reefs will be devoid of coral and dominated by seaweed and blue-green algae,' it said. He did qualify in his report that corals could recover in the long term.

Bob Carter was quoted in the *New Scientist* questioning the accuracy of the IPCC's data and the predictions on global warming in the WWF report. His JCU colleague, Dr Peter Ridd, went further, disputing Hoegh-Guldberg's predictions that the Reef was under serious threat. 'Global warming is obviously a concern, but it makes it sound as though we're certain to lose the Reef and I don't think we can say that at all,' said Ridd. He pointed to the recovery of coral reefs from the 2002 bleachings.

These recoveries, of course, did not happen in the face of back-to-back bleachings. But Carter and Ridd were just getting started. Over the next decade they would become two of the most vehement critics of the work of coral reef scientists, including Ove Hoegh-Guldberg.

In 2004, Australia's main climate-sceptic think tank, the Lavoisier Group, still under the control of Hugh Morgan's alter ego, Ray Evans, was quick to recognise Bob Carter's talents. He became one of their favoured speakers. It wasn't long before both Carter and Ridd were embraced by right-wing Melbourne think tank the Institute of Public Affairs. The IPA made Carter one of their 'emeritus fellows', while Ridd joined the board of the IPA-backed Australian Environment Foundation, a new conservative outfit that attacked mainstream climate science and renewable energy.

Carter's global influence as a climate-science sceptic took off in 2006, when he was asked by a leading UK sceptic to help write a scientific paper debunking Sir Nicholas Stern's famous review on the economic impact of climate change. One of Carter's co-authors on that paper was the doyen of the US climate-sceptic movement, Professor Richard Lindzen from MIT. Carter was now in the big league. It was this paper that so impressed the Queensland Land and Resources Tribunal President, Greg Koppenol, in the Xstrata coalmine case, where Ove Hoegh-Guldberg gave evidence.

Soon Carter was being feted by right-wing think tanks in London and the US. The retired professor from a regional Australian university became a rock-star performer for the global sceptic movement. What endeared him to many of his new American friends was his passion. Carter was willing to go over the top in his flamboyant attacks on the IPCC and those who agreed with their findings on climate change.

'Imagine a well-provendered and equipped military fortress in time of war, for that is what the alarmist, pro-IPCC, climate lobby group represents,' he wrote in an essay defending climate sceptics. He would compare sceptic scientists like himself to

Charles Darwin, Marie Curie and Albert Einstein, 'mavericks one and all'.

His reputation in US sceptic circles was sealed in late 2006 when he was invited to appear before the powerful US Senate Committee on Environment and Public Works. The outgoing chair of the committee, Republican senator James Inhofe, was holding his last hearing. The senator, a champion of the oil and gas industry, had famously described climate change as 'the greatest hoax ever perpetrated on the American people' and compared the IPCC with a 'Soviet-style trial'.

Bob Carter's invitation came from Inhofe's communications boss, Marc Morano, a skilled political operative dubbed 'the drum major of the denial parade' by one CNN producer for bombarding the media with reports from climate-sceptic scientists. Morano wanted Carter to give evidence at a special hearing of Inhofe's committee on media bias and climate change. Carter was delighted to oblige. The hearings were clearly aimed at discrediting the media coverage of the upcoming IPCC report in 2007.

In his characteristically exuberant style, Carter titled his testimony, 'McCarthyism, intimidation, press bias, censorship, policy-advice corruption and propaganda'. He opened with a boots-and-all attack on the IPCC and a warning that 'alarmism is reaching unprecedented heights'. He belted the global media's coverage of climate change and the use of lawsuits to fight climate change. The Australian professor also ripped into the California attorney-general and US environment groups for bringing a case against the world's car makers over their vehicle emissions. He defended climate-sceptic scientists appearing as witnesses for the car makers and also put in a good word for ExxonMobil's funding of sceptic think tanks.

Carter argued that global warming had become a big political issue because of 'insistent lobbying from special interest environmental, scientific, political and industry groups'. He was adamant the argument that rising greenhouse gas emissions were linked to global climate change was just wrong. 'Today's dominant paradigm is that human emissions of greenhouse gases, especially carbon dioxide, will produce dangerous warming of the globe,' he complained. 'When tested against empirical evidence, this hypothesis fails.'

But the arguments of Carter and the climate-sceptic scientists were losing their power. By 2007, the ground had shifted in both Australian and US politics. The Democrats had won the mid-term elections the previous November, putting them in control of both houses of Congress. Senator Inhofe lost his position to a Democrat. The Bush–Cheney White House was discredited and unpopular, and the Republicans were widely expected to lose the White House at the next election.

In Australia, the big drought had put Howard and the sceptics on the back foot. The blizzard of publications from the Lavoisier Group and the IPA was getting less media traction, and the loud voices of Bob Carter and Ray Evans no longer carried the same weight. Neither did their political allies in the Liberal Party.

When the IPCC released its full report in September 2007, it created a new wave of political momentum around the world to deal with the threat from climate change. In its many findings on the impacts of global warming, one stood out for Australia's marine scientists—coral bleaching. The report quoted Ove Hoegh-Guldberg's work in detail and noted that yearly or two-yearly coral reef bleachings were projected to occur by the middle of this century if temperatures kept rising.

'Many studies incontrovertibly link coral bleaching to warmer sea surface temperature and mass bleaching and coral mortality often results beyond key temperature thresholds,' the report found. 'Annual or bi-annual exceedance of bleaching thresholds is projected at the majority of reefs worldwide by 2030 to 2050.'

The shift in the public mood on climate change was dramatically underscored when the Nobel Committee in Norway awarded the 2007 Peace Prize jointly to the IPCC and Al Gore. The prize was, the committee said, 'for their efforts to build up and disseminate greater knowledge about man-made climate change and to lay the foundations for the measures that are needed to counteract such change'.

The timing of the Nobel ceremony was impeccable. In December 2007, at Oslo's Town Hall, the IPCC's chairman, Dr Rajendra Pachauri, used his Nobel Peace Prize lecture to put the heat on the world's politicians to act on climate change. As he spoke, ministers and diplomats from around the globe were flying to the Indonesian island of Bali for the most important UN climate talks since Kyoto, the UN's Conference of the Parties 13, (COP13). Dr Pachauri had one question for the Bali delegates: 'Will those responsible for decisions in the field of climate change at the global level listen to the voice of science and knowledge, which is now loud and clear?' the IPCC chairman asked. 'If they do so at Bali and beyond then all my colleagues in the IPCC and those thousands toiling for the cause of science would feel doubly honoured at the privilege I am receiving today on their behalf.'

For the politicians arriving in Bali it could never be just about the science. There were louder voices back home demanding to be heard. But for a brief moment in December 2007, the climate scientists who toiled for the IPCC had their place in the sun.

CHAPTER SIX

THE DEFINING CHALLENGE

BALI WAS SWELTERING, but inside the cavernous International Convention Centre the air con was blasting away. Some 10,000 ministers, officials, business lobbyists, media, environmentalists and climate sceptics from all over the world had descended on Nusa Dua's luxury resort complex for the historic UN climate talks, and many were now sitting in the cool of the centre's auditorium.

The first high-level meeting was underway and the UN secretary-general, Ban Ki-moon, was exhorting the delegates to work together. At the top table with the dignitaries, Australia's new prime minister, Kevin Rudd, looked out at the sea of faces, fiddled with his speech and waited to take the microphone.

A short time before, Rudd had handed the secretary-general Australia's signed ratification of the Kyoto Protocol. It had taken ten long years and countless political brawls for Australia to seal this deal. Now, when his cue came, Rudd was ready to tell the world why he'd done it. 'I did so, and my government has done so, because we believe that climate change represents one of the greatest moral, economic and environmental challenges of our age,' he told the crowd of ministers and officials. 'Climate change,' Rudd pronounced, 'is the defining challenge of our generation.'

Rudd was stunned at his reception. Delegates were clapping and cheering. 'You could have knocked me over with a feather,' he recalled. Rudd, a politician not famous for his oratory, had delivered a speech that would go down in history.

On that heady day in Bali in December 2007 Rudd had no idea just how brutally defining the climate change challenge would be for him personally. He had been prime minister for little more than a week and had written his speech on the flight from Canberra. 'I was not conscious that there was such an international focus on the significance of our changing position,' Rudd later said, somewhat naïvely. 'So when I stood up and they all leapt to their feet I thought, "Sorry?" Seriously I had no idea. I was stunned by the level of international reaction.'

The cheering delegates were the first warning sign that Rudd's decision to act on climate change would inflame passions at home and abroad. The Bali cheer squad was delighted Australia was finally joining the Kyoto Protocol. But they were also delighted the US had lost its most loyal ally in the global climate wars. 'Bonnie went straight, so Clyde is isolated,' as Al Gore quipped.

Rudd was welcomed like a prodigal son onto the top table at the UN climate talks, but he soon found out it was a difficult place to be. On the agenda in Bali were ambitious plans to draw up a new UN climate agreement to follow the Kyoto Protocol. It was hoped this agreement would finally put constraints on all big greenhouse gas polluters, including the US and China, and it was a top priority for Australia.

'Developing countries had to be part of the action,' said Howard Bamsey, who arrived in Bali with the Australian delegation as a seasoned operator. 'There had been zero obligations until now. We were at this really important point in the negotiations.'

But by the time Rudd jetted in, the Bali talks were already in trouble. China and India were baulking again, running their old arguments. Most of the emissions already in the atmosphere came from rich countries that had enjoyed the fruits of industrialisation. They wanted developed countries like Australia to agree to deep cuts in their emissions by 2020 under the Kyoto Protocol before poorer countries acted. On the other side, many rich countries wanted commitments from China and India before they would agree to a new round of cuts. It was a dangerous game of chicken.

By 2007 China had overtaken the US as the world's biggest emitter of greenhouse gases, spewing out 23 per cent of the total. Indonesia, the host of the Bali talks, was then the third-biggest emitter, partly because of its mass deforestation. The year before, a choking smoke haze had blanketed Indonesia's neighbours after large areas of Borneo and Sumatra were cleared and burnt, creating the worst forest fires in the nation in a decade. There was no doubt that in time the emissions from developing countries like China, India and Indonesia would overwhelm everyone else.

The EU delegates and the UN's top climate official, Yvo de Boer, were adamant that the rich, developed countries still needed to take the lead on climate change, because they had the means to drive the new technology shift to clean energy. That meant rich countries needed to make deep emissions cuts by 2020. Without these cuts, they would not get China, India and the developing world on board in a new treaty. 'All countries, all governments, realise that industrialised countries will have to reduce their emissions somewhere between 25 and 40 per cent by 2020,' de Boer told the media.

A target this high was political dynamite for Kevin Rudd. Under Howard, Australia had won a soft, unambitious 2012 target in the first Kyoto round—a rise in its greenhouse gas emissions to

108 per cent of its 1990 level. An ambitious 2020 target would be a complete game changer and much tougher to meet.

Rudd came to Bali determined to avoid making any promises about a 2020 target. Only just elected, he insisted he needed expert advice first and that was coming from Professor Ross Garnaut, whose climate change report was due the next year. Rudd tried to sidestep the issue, saying Australia would cut its emissions 60 per cent by 2050, a date too far away to matter. But he was soon being pushed by the Europeans and environment groups to support the ambitious targets.

These 2020 targets were based on scientific advice from the latest IPCC report—the one Rudd had used to taunt John Howard. Climate scientists warned that global greenhouse gas emissions, from all countries, had to peak and stabilise by around 2030 and then halve by 2050. This was the accepted consensus at Bali on how to keep global temperatures from rising more than 2–2.4 degrees Celsius by the end of the century. Some scientists thought even this temperature rise was too high to avoid dangerous climate change.

Tempers quickly frayed over the targets. The US delegation caused uproar by refusing to accept any reference to the IPCC advice or the 2020 targets in the Bali plan to draw up the new climate treaty. De Boer exhorted delegates: 'We must make the leap forward or be condemned to the *Planet of the Apes*.' But the Americans won support from Russia, Japan and Canada, which were all wavering on new Kyoto commitments for 2020. Anger boiled over at one point and the US delegates were booed on the conference floor.

In the last painful hours of negotiations, the Australians helped broker a compromise producing the Bali Road Map deal. China and India left the door open a crack on the new treaty. The US got its way and relegated the IPCC scientific advice to a footnote in the final agreement that would include an 'Action Plan' to chart the course for a new climate agreement. But with the support of Rudd's new climate change minister, Senator Penny Wong, the IPCC's advice and the ambitious 2020 targets were put in a supporting document for countries that had ratified the Kyoto Protocol. This of course didn't include the US, but it did include Australia. The 2020 targets weren't binding, but they were now a guide.

Australia's environment lobby in Bali rejoiced. Canberra had split with Washington and backed the science of climate change. The head of the Australian Conservation Foundation, Don Henry, waved the document with the IPCC advice in front of the media pack. 'It means we know this road map is going towards a place that would avoid dangerous climate change,' he declared.

The IPCC's advice with the 2020 targets were heralded as a guiding light to help world leaders and ministers negotiate the new climate treaty to be agreed in Copenhagen within two years. But rather than guiding Rudd, they ended up haunting him from the time he left Bali until he was torn down as prime minister.

———

Rudd knew setting the 2020 target to cut Australia's greenhouse gas emissions would ignite a political battle. A weak target would alienate the environment groups that had campaigned for Labor in the recent federal elections and would open the way for the Greens leader, Senator Bob Brown, to wedge Labor on climate change. But an ambitious target would mean economic pain for the carbon club, the big greenhouse polluting industries that had been cocooned by the Howard government. It would force a serious rethink of the 'quarry vision' that had defined both sides of Australian politics for generations.

Rudd was not prepared to do that. Senior members of the new Labor government knew from the last decade of bitter debates that even a modest 2020 target to cut Australia's greenhouse gases would see big business and Labor's own trade union allies up in arms over job losses and threatened industry shutdowns.

Rudd had promised to introduce an emissions trading scheme; in theory it was to achieve cuts to greenhouse gas emissions with less economic pain. The scheme would finally put a price on carbon pollution and make it a tradeable commodity. But if Australia's 2020 target was high, the carbon price would likely be high. If industry had to buy high-priced permits to emit greenhouse gases, that would impact, in the short term at least, Australia's exports, jobs and cheap electricity prices. So it was no surprise that the 2020 target was one of the most sensitive decisions facing the government.

Rudd was acutely aware he needed the opposition's support for the emissions trading scheme and the 2020 target because Labor didn't have the numbers in the Senate and he did not want to negotiate with the Greens. Tellingly, Rudd had stripped the charismatic Peter Garrett of the climate change job as soon as he became prime minister. During the election campaign, Garrett, a political novice, had made some honest remarks about the need for serious emissions targets to cut greenhouse gases. They had provoked a savage backlash from the Liberal Party.

Howard had warned Labor's climate policy would lead to a 'Garrett recession'. The former rock star might have been brilliant at rallying the youth vote for Labor, but Rudd saw him as a lightning rod for attacks from the opposition and the business lobby. 'News Corp usually had an attack dog journo on my heels,' was how Garrett remembered the campaign. 'From that point on Rudd was far from supportive.' Rudd made Garrett Minister for the Environment and the Arts, but handed South Australian Senator Penny Wong the job of climate change minister before they left for the Bali conference.

Wong was a stark contrast to the passionate Garrett. She was not a public crusader on climate change; she was a tough but cautious political operator. Born in Malaysia to a Malaysian-Chinese father and an Australian mother, she came to Adelaide as a child when her parents separated. She dived into Labor politics at university and later worked as an industrial lawyer for some of Labor's most influential trade unions. After she was elected to the Senate she came out as gay, a move that underscored her reputation for personal integrity.

Above all, Rudd trusted her. 'I had formed a deep appreciation of Penny Wong's abilities,' he said. 'Given how fraught this would become at home, with the Senate negotiations, and how fraught it would be internationally with negotiations, I didn't want to be in a position of dealing with a weak link on that.'

———

In many ways Kevin Rudd and Penny Wong were too alike. Both were better technocrats than politicians. They set up a new Department of Climate Change attached to Prime Minister and Cabinet and picked Treasury deputy secretary Martin Parkinson to run it. It was a sound choice. Parkinson had worked closely alongside Howard's top bureaucrat, Peter Shergold, on the Coalition government's report on an emissions trading scheme.

But there was a problem. Rudd's independent climate adviser, Professor Ross Garnaut, was a harsh critic of the Shergold report. He believed the big corporate greenhouse gas emitters appointed by Howard to advise Shergold had been given too much say. 'As you would expect from something where big emitters had a major influence, it looked after, very carefully and elaborately, the interests of those emitters,' Garnaut said later. At the time he was determined the big corporate emitters would not control the advice he gave to Rudd. It was a sharp break with the Howard era.

'I'm not used to giving in to corporate pressure independently of the analysis of the issues,' Garnaut recalled bluntly. 'This is an important point. The way the political culture had developed by then, by 2007, big business was used to getting its own way. We had lost the tradition that we had in the reform era of standing for the public interest. So when I simply started looking at things analytically and just saying what came out of that analysis in a straightforward way, affected industries that were used to getting their own way reacted as if I was upsetting the settled order.'

Martin Parkinson was already beavering away on the new emissions trading scheme for Rudd. He saw Garnaut's main role as advising Rudd on the 2020 target and building public momentum for the whole climate change policy. But in February 2008, Garnaut released an interim report that laid out his ideas on the emissions trading scheme. In it, he took aim at the most politically sensitive greenhouse lobby in the country: he challenged the idea that the privately owned electricity generators in Victoria needed large amounts of government assistance before they could sign up to an emissions trading scheme.

Garnaut's advice sparked a backlash from the generators led by the head of TRUenergy, Richard McIndoe, who warned of a collapse of the electricity system. TRUenergy owned the ageing Yallourn power plant in Victoria, one of Australia's biggest emitters of greenhouse gas. Its Hong Kong parent company, China Light and Power (CLP), was a $22 billion outfit with power plants across Asia, and its chairman, Sir Michael Kadoorie, was the fourth-wealthiest man in Hong Kong. TRUenergy's Australian boss, McIndoe, would quickly become one of the most vocal opponents of Labor's climate policy.

By April, McIndoe had written to Rudd, Wong and Victoria's Labor premier, John Brumby, warning that Garnaut's advice was a threat to the nation's energy supply, a threat to jobs and 'a major threat to the enterprise value of all Victorian coal generators'. Just five months into office, before the 2020 target was even set, the Rudd government was at war with the power industry.

Confronting the Victorian power generators was unavoidable. The privately owned plants were some of the biggest greenhouse gas emitters in the country. The foreign companies that bought them should have been fully aware of their greenhouse footprint. They were inefficient and burned heavily polluting brown coal. The Victorian generators were one key reason why Australia's electricity system was the most emissions-intensive in the developed world. Only eight other countries had a worse record on this: Bahrain, Botswana, Cambodia, Cuba, India, Kazakhstan, Libya and Malta. It was a shocking indictment for a developed country a decade on from Kyoto.

One of the biggest emitting Victorian generators, Hazelwood, was owned by the British corporation, International Power, which also ran Loy Yang B, another brown-coal power plant. Hazelwood had long been targeted by environment groups calling for its shutdown. But International Power's local chief executive, Tony Concannon, was used to sitting at the top table. He had been handpicked by John Howard to advise the Shergold review on emissions trading and he also became a trenchant critic of Garnaut.

A third heavyweight player among the Victorian generators was Ian Nethercote, a veteran of the power industry. Nethercote was CEO of Loy Yang Power, which ran the Long Yang A generator and the brown coal mine that supplied the plant. It was owned by a consortium of foreign and local interests, including Australia's AGL. Nethercote had also been a trusted source of advice for the Howard government, arguing for years against both emissions trading and the Kyoto Protocol. A key figure in the generators' lobby group, Nethercote also sat on the board of the Institute of Public Affairs (IPA), the right-wing think tank pumping out climate-sceptic science in Australia. He joined Concannon and McIndoe in taking up the fight against Labor's climate policy.

The Victorian generators knew they could hang tough. They were the state's economic lifeblood, supplying over 90 per cent of its electricity and fuelling its manufacturing industry. They also employed a large workforce, which was mostly unionised and vocal.

The private generators wanted large numbers of free permits as serious compensation for the impact of the scheme on the market value of their businesses. Garnaut was forcefully telling Wong and Rudd the generators didn't need free permits. He thought they would put up electricity prices to cover their costs from the carbon price and pass that on to consumers. Free permits could give them windfall profits, as had happened with Europe's power generators a few years earlier.

'Economic analysis showed that nearly all of the cost of permits would be passed on to users of power. I was aware that in Britain and Europe the issue of large numbers of free permits to the domestic power generation sector had led to very large transfers from ordinary households to the corporate sector,' Garnaut explained. 'The British Treasury's reference to their own experience reinforced my concerns.'

The generators were just one of the big greenhouse intensive industries up in arms. The Western Australian energy giant Woodside led the charge for the oil and gas sector. Woodside's North West Shelf project accounted for about a third of Australia's natural gas production, with exports to Japan and China. Its joint venture partners were some of the biggest global players in the market: BHP, Chevron, Shell and BP. Woodside's US-born chief executive, Don Voelte, had worked for Mobil before its merger with Exxon, and his lobbying style was not subtle. He made sure Wong heard his very loud voice.

'I flew to Adelaide the other day and talked to Penny Wong for quite a while,' Voelte told reporters that April. 'I think she's probably got every CEO in the country trotting in to see her.' His message to Wong was blunt. An emissions trading scheme that put a price on carbon could have dire unforeseen consequences for Australia's lucrative gas exports. 'The North West Shelf in its own right is a little under 3 per cent of the gross national product,' Voelte said. 'That's huge. If you take that out of play—I'm not going to say that's going to happen—but if you ask me what I stay up at night worried about, it's not that the intention isn't good, it's the unintended consequences.'

Voelte didn't spell out those unintended consequences but hammered the message of impending doom he gave to Wong. 'The good news is that they are worried about it,' he said. But, he complained, 'She says we are all going to have to take a little pain and punishment.'

This was the government's problem in a nutshell. The big emitters wanted to take as little pain as possible. They wanted government assistance, including a chunk of free permits to pollute under the emissions trading scheme so they didn't have to buy them all. Wong's dilemma was acute: the more free permits she handed out to the big emitters, the less money there would be to compensate households and small businesses, which would be slugged with higher gas and electricity prices under the scheme. There would also be less pressure on the big greenhouse polluters to cut their emissions.

Labor's climate policy was under siege by the carbon club, which would soon win support from the new opposition leader, Brendan Nelson. After Howard's defeat, Malcolm Turnbull, ever ambitious and impatient, thrust himself forward for the Liberal Party leadership. But Turnbull's stand on climate change, his support for emissions trading, his push to ratify Kyoto and his former leadership of the Australian Republican Movement, all antagonised the party's conservatives. So too did his brash arrogance. 'He always liked to be the cleverest person in the room,' was how one colleague summed him up.

Nick Minchin and the party's right wing successfully blocked Turnbull. The right swung behind the more affable Nelson, a doctor and former head of the Australian Medical Association. Nelson beat Turnbull in the partyroom ballot, becoming leader of the opposition by just 45 votes to 42.

———

In the middle of 2008 Ross Garnaut released a draft version of his final report and repeated his warnings against giving the power generators free emissions permits. Once again, the backlash was swift, but this time there was also a full-throttle attack from within Labor's own ranks.

One of Labor's most vocal climate sceptics, NSW treasurer Michael Costa, ripped into Garnaut. At the time Costa was trying to privatise the NSW government-owned electricity system to raise money for the state budget, and he was furious. 'The bottom line is if you implement the report as currently proposed by Professor Garnaut and others, you will have a crisis of confidence in investment, particularly in the energy-intensive area, and you could lead to shortages of electricity supply,' Costa told the ABC. Costa's warning that Garnaut's advice could lead to widespread power shortages was matched by TRUenergy's Richard McIndoe, who warned 'the generators will be effectively bankrupt and therefore not able to operate from December 31' because their value would be so diminished.

Rudd and Wong now had warnings that an emissions trading scheme could threaten the nation's power supply in the two biggest states. Inside Wong's Department of Climate Change, Martin

Parkinson was increasingly worried about Garnaut. He thought what Garnaut was saying was economically valid but politically untenable. 'The generators were particularly concerned because they had been hearing Ross in public saying there's no case for compensation. And Ross is right as an economist,' said Parkinson. 'But I was having to think about the political economy of this which was, you know, you can be a purist economist but you're not going to get this thing up.'

At this critical point, the opposition leader publicly dumped his support for the new emissions trading scheme. Brendan Nelson warned Rudd that unless China, India and the US—the big global polluters—signed up to similar schemes, he would oppose it. 'It will be an act of environmental suicide, an act of economic suicide, if Australia were to be so far in front of the world implementing an ill-considered, not yet properly developed and tested emissions trading scheme, if we haven't got a genuinely global response,' said Nelson. This was the same argument the Liberal Party's leading sceptic, Senator Nick Minchin, had pushed when Shergold gave his report to Howard. It was a sure sign the Liberal and National Party opposition would be a big problem in the Senate.

Despite the chorus of outrage, on 16 July Penny Wong, along with Rudd and Treasurer Wayne Swan, unveiled the government's first version of the emissions trading scheme. The sweeping plan was called the Carbon Pollution Reduction Scheme (CPRS). It was put in a 'green paper', which meant it was still up for debate with big business and the Labor Party's powerful trade union allies.

The trio of ministers laid out the scientific advice on the necessity to cut greenhouse gas emissions. By 2030 climate change, unchecked, could cause up to 20 per cent more drought months over most of Australia, more intense and damaging cyclones, and rising sea levels that would hit the nation's coastal properties. Climate change threatened Australia's food production, water supplies, the Great Barrier Reef, the Kakadu wetlands and the tourism industries they supported. 'If we delay action any longer,' they said in a joint statement, 'these costs will be felt even more by not only our generation, but also our children and grandchildren.'

But it was the devil in the detail that the big emitters pounced on. For the first time in Australia's history, the government would limit, or cap, the amount of carbon an industry could emit. The scheme would cover only big emitting industries, around 1000 Australian companies, or less than 1 per cent of businesses.

Beef cattle and sheep farming, a large source of Australia's greenhouse gas emissions, were excluded—but not indefinitely. Rudd flagged the government would begin a review to work out whether it was feasible to cover agricultural emissions by 2015, seven years down the track. This alone provoked a huge backlash from the National Party.

But it was the big companies that would immediately be caught by the scheme which made the most political noise—the leading members of the carbon club. Top of the list was aluminium, the country's most emissions-intensive industry apart from beef cattle. Aluminium needs vast amounts of power for smelting, and in Australia that meant mainly coal-fired power. The US–Australian joint venture, Alcoa, where Hugh Morgan once sat on the board, was one of the biggest greenhouse gas emitters. It ran its Victorian plants on electricity from brown-coal-fired generators. In other states, Rio Tinto's plants were in the frame as well.

Also high on the list of big carbon polluters were the cement industry, the oil and gas industry, coalmines, the chemical industry, the iron and steel industry and, of course, the power generators. Almost every industry targeted had a unionised workforce, supercharging the political sensitivities for the Labor government.

Many of the workers, especially in the aluminium industry, were members of the biggest blue-collar union in the country, the Australian Workers' Union. The AWU was a powerful factional player in the Labor Party; it could often make or break Right-faction Labor candidates in Victoria and Queensland. It covered Alcoa's Victorian workers at Portland and Port Henry. Its federal secretary, Paul Howes, looked at the green paper and decided it was a disaster for the union.

Industries like aluminium were supposed to buy a pollution permit for each tonne of carbon they emitted. To smooth the scheme's introduction, Wong flagged that affected industries would get government assistance. Industries like aluminium, which competed on the world market—'trade-exposed' industries, as they were called—would likely get free permits to cover around 90 per cent of their emissions in the beginning. The electricity generators were also promised some direct assistance, but Wong warned that every free pollution permit for the generators meant less money to compensate households for higher electricity prices.

The crucial number missing from the green paper was the 2020 target for cutting Australia's emissions. A decision on that fraught figure would not be made before the end of the year. Only then would the government determine how much assistance each industry would get. That uncertainty left the door wide open for the big emitters and the unions to lobby like hell for months.

The AWU's Paul Howes was determined to do just that. His union and Alcoa needed each other. The company was Victoria's biggest exporter, but it depended on cheap brown-coal-fired electricity for its smelters. Thanks to both Labor and Liberal state governments, Victorian taxpayers had subsidised that power for years. Howes was not going to let the federal government's new emissions trading scheme jeopardise the plants if he could help it.

Howes turned to an old colleague who was now in a senior role in the Rudd government—Martin Ferguson, the Minister for Resources and Energy. Ferguson was a former president of the ACTU and old-school Labor; he was a champion of blue-collar jobs, with a loathing for environmental activists. As resources and energy minister, he sat on the cabinet subcommittee dealing with climate policy, along with Wong, Rudd and Swan, but he had often been sidelined by the troika until now. After the release of the green paper, Ferguson's office became a workshop for 'Brown Labor'—the unions and Labor politicians determined that climate change policy should not threaten jobs.

Also beating a path to Ferguson's door were the Victorian power generators. Tensions broke out between Ferguson and the climate change minister. Wong thought Ferguson was operating like a shop steward and undermining the policy. When Ferguson commissioned alternative modelling on the impact of the emissions trading scheme, tempers flared. But Ferguson was not alone in his concerns. The government's industry minister, Kim Carr, a Victorian, also worried about the impact on the car industry and the steelmakers.

Howes made it clear to his Labor colleagues that the concessions to business offered by Wong and Parkinson in the green paper were not nearly enough. He was more than happy to see Ferguson push back against them.

———

Despite the attacks on the climate policy from business, from the media and from Labor insiders, Rudd's popularity was still high and so was public concern over climate change. The opposition, fortunately for Labor, was in disarray.

Brendan Nelson's original partyroom defeat of Malcolm Turnbull had not been a decisive victory and Turnbull never accepted that result. He began relentlessly stalking his leader, attacking him over climate change and emissions trading in particular. By September 2008, just over nine months into the job, Nelson threw up his hands and called a new leadership ballot. Turnbull won by four votes. It signalled deep divisions in the party.

Turnbull's rise was a double-edged sword for Labor. There was no doubt the former merchant banker was a tougher opponent for Rudd, and that worried Labor's hard heads. But with Turnbull, Rudd had a real chance to get Labor's climate change policy through the Senate. It was a unique opportunity to secure a bipartisan deal that could put to rest Australia's decade of bitter division over climate change.

But the omens were not good. Just as Turnbull took the opposition leadership, America's fourth-biggest investment bank, Lehman Brothers, filed for bankruptcy. The global share markets began their long dive, heralding the beginning of the Global Financial Crisis (GFC). Climate change, the greatest challenge of our generation, was suddenly a second-order issue for the Rudd government. 'It changed everything,' said Martin Parkinson, then head of the Department of Climate Change.

With the world financial system on the brink of collapse, the government went into crisis management. Rudd, his top public servants and a kitchen cabinet—which included Deputy Prime Minister Julia Gillard, Treasurer Wayne Swan and Finance Minister Lindsay Tanner—were consumed with managing the meltdown. In October 2008, they were forced to guarantee the country's bank deposits. To forestall a recession, they spent big to keep the economy afloat, shovelling cash to pensioners and low-income families from a huge $10.4 billion stimulus package.

Delivering a climate change policy was suddenly a whole lot more difficult, but Rudd decided to press ahead. With the backdrop of the economic crisis, the demands for generous government assistance from the affected industries were much harder to resist. The threat of job losses was too sensitive. As the AWU boss Paul Howes later admitted, 'The GFC was my greatest ally.'

Europe, Japan and the US were heading into recession, and Australia's manufacturing export industries, like aluminium, were in trouble even without a carbon price. The Business Council of Australia (BCA) issued a dire warning that three manufacturing companies would shut their doors immediately if the emissions trading scheme was introduced.

Garnaut had given his final report to Rudd in late September. It included his advice on the highly sensitive question of the 2020 target to cut Australia's emissions. By now he was pessimistic about the global will to act on climate change, which in his report he called 'a diabolical policy problem'. He offered Rudd three options.

If an effective climate treaty was reached at Copenhagen the following year, Australia should agree to an ambitious 2020 target—a 25 per cent cut to its emissions from 2000 levels. This was within the range of the IPCC advice and the Bali conference, which had called for emissions cuts of between 25 and 40 per cent by 2020 for rich countries. Garnaut believed an effective agreement in Copenhagen was in Australia's interests but was unlikely. He advised Rudd the outcome would probably be a weak climate agreement, one that would lead to a 3 degree Celsius temperature rise by the end of the century. On the IPCC's projections, a 3 degree rise would risk dangerous climate change.

Despite this, he argued that if this was the deal in Copenhagen, Australia should offer only a 10 per cent emissions cut by 2020 and wait for a better agreement down the track. This meant the world would 'overshoot' the 2020 target. Australia shouldn't get out in front of other developed countries, Garnaut advised.

If there was no deal at all in Copenhagen, Australia's 2020 target should be just a 5 per cent cut.

What was almost overlooked in Garnaut's report was the minimal cost to the overall economy. Garnaut described the cost as 'manageable', whether Rudd went for either the lower or higher target. It amounted to a loss in annual economic growth to 2020 of either 0.1 a year for the lower target or 0.2 per cent for the ambitious target.

Some of Australia's leading climate scientists who had worked with the IPCC tried to lobby Rudd to go for the ambitious target of a 25 per cent cut. Sixteen of them wrote an open letter to Rudd urging him to support it and to push for a tough global agreement in Copenhagen. 'Failure of the world to act now will leave Australians with a legacy of economic, environmental, social and health costs that will dwarf the scale of national investment required to address this fundamental problem,' the scientists told Rudd. Among the signatories on the letter was Ove Hoegh-Guldberg, IPCC author and Australia's leading coral reef expert.

The scientists were rebuffed and so were 50 environment groups that wrote to Rudd with the same message. In mid-December, Rudd and Wong unveiled the 2020 target. It was a cut of just 5 per cent of Australia's emissions from 2000 levels. They promised the target would rise to 15 per cent if there was an ambitious new global agreement in Copenhagen, but it would have to include China, the US and all the big polluting nations. The 25 per cent cut was off the table. The Bali goal of emissions cuts for developed countries of between 25 and 40 per cent was gone, and with it the belief that an effective climate agreement in Copenhagen was possible.

When Rudd stood up at Canberra's National Press Club to sell the new climate change policy he was met with heckles and screams from protestors. He doggedly mustered his old rhetoric, labelling climate change 'one of the greatest, enduring challenges that we face' and 'nothing less than a threat to our people, our nation and our planet'. But his pragmatism was clear: 'We are not going to make promises that cannot be delivered.' He touted the new plan as responsible, saying repeatedly, 'We have got the balance right.'

The 2020 target was released in the government's white paper on the CPRS, which also set out new details on the emissions trading scheme. It would begin with a carbon price around $25 a tonne in 2010. Low-income households would be generously compensated for increases in electricity and gas prices. And the scheme now included a large pot of money for the power generators—$3.9 billion worth of free permits to pollute over five years. As well, the trade-exposed industries, like aluminium and steel, would get a boost to their free permits flagged in the green paper. The gas export business, Liquefied Natural Gas, would also be assisted, along with coalmines. The big greenhouse industries could also buy emissions credits from overseas schemes if they wanted to avoid cutting their own emissions in Australia.

The generosity towards the generators and the failure even to consider a 25 per cent target to cut Australia's emissions angered Garnaut, who went public with a savage critique: 'Never in the history of public finance has so much been given without public purpose, by so many, to so few,' he said.

Wong was furious with the criticism. But as Rudd saw it, Garnaut seemed to be 'oblivious' to the impact of the financial crisis. 'Garnaut was operating in a different policy zone to the rest of us just then because of the all-consuming nature of the climate crisis engulfing us all,' was Rudd's take. But the reality was that, with or without the financial crisis, the Labor government did not want an ambitious 2020 target then.

Wong and her top adviser, Martin Parkinson, firmly believed the priority was getting the emissions trading scheme passed by the Senate. The concessions they had to make to the big polluters to do that could be sorted out later. Once the scheme was in place and working they could look at increasing the emissions target if greenhouse gases weren't coming down or the rest of the world agreed to more ambitious global targets.

'I knew that we just needed to get a mechanism in place and that we could ratchet up as required,' Wong recalled. 'I also knew what a 15 per cent reduction would mean for the Australian economy, given how carbon-heavy we were. That was a very massive turnaround. And it would have sent a signal through the economy around investment which would have had long-lived effects.'

But the yawning gap between Rudd's rhetoric and his pragmatism confused and alienated many of the government's own supporters, who found it hard to follow his logic. Rudd had endorsed the scientific warnings of the IPCC but then argued it was okay to have a weak 2020 target, because somehow Australia would be able to make its big emissions cuts down the track.

The Greens and their supporters were quick to attack Rudd and wedge Labor from the Left. 'Mr Rudd has betrayed the science, betrayed the community and betrayed the next generation, who will have to live with climate change impacts,' Greenpeace climate campaigner John Hepburn scoffed. Even the moderate leaders among the environmentalists were stung by the low 2020 target. 'A very disappointing day' was how Don Henry from the Australian Conservation Foundation put it.

That December was a critical turning point for Rudd and Wong. They began losing public support on climate change just when they needed it most. The big emitters, like the generators and the coal industry, had no intention of being bought off by the concessions in the white paper. With their allies in the opposition ranks, they would unleash a campaign against Labor's climate policy that would put the government under almost daily attack in the media. Once it started, it was unrelenting.

Rupert Murdoch's newspapers set the agenda as the main outlet for those attacks but how much of this was influenced by the media baron's own views was unclear. Rudd had no doubt that Murdoch was deeply sceptical of the mainstream science on climate change. 'My conversations with Murdoch over the years made it absolutely clear in my mind that he denied the climate change science and denied therefore the need for any real or substantive policy action to reduce carbon emissions,' said Rudd.

Whether it was Murdoch or his editors who were driving the coverage, it put Penny Wong under siege on a daily basis. 'We had report after report, with *The Australian* running front-pages about how we were shutting down industry, shutting down factories, shutting down communities, shutting down towns, destroying the Australian economy. For two years I just remember so many mornings being called at five-thirty because there was another thing on the front page about how this was destroying us,' Wong recalled. 'There was a campaign being run by those who had an interest in no change.'

She was dead right about that. But at the time neither she nor Rudd understood just how organised the campaign was, or how unstoppable it would be.

CHAPTER SEVEN

THE HEAT GOES ON

SENATOR CORY BERNARDI was perspiring and it was not a good look for the cameras. The air con had broken down and he was launching Ray Evans' tract *Thank God for Carbon*, his latest climate-sceptic attack on global warming. In the bleak Adelaide office block that served as headquarters for the city's dogged conservative forces, it was as hot as hell. Outside, the mercury was hovering around 40 degrees Celsius. It was January 2009 and one of the worst heatwaves in South Australia's history was gripping the City of Churches. 'The ABC was there and I remember it was so hot,' Bernardi recalled, barely suppressing a laugh.

Always dressed in a well-cut suit, Bernardi looks like a cross between an evangelical preacher and a financial adviser, which in many ways sums him up. His Italian heritage blessed him with good looks and a healthy ego; even his wife, Sinéad, famously quipped it was lucky she and the senator were both in love with the same man. His years as a competitive rower gave him an athlete's build, tall and lean. After a back injury ended his rowing days, he became an investment adviser. But his passion was politics.

Bernardi got his place in the Senate thanks to his mentor, Nick Minchin. Bernardi, like Minchin, worked his way up through the party machine. He personified a new Australian breed of Christian politicians, defined by their deeply conservative social values and their bone-dry economics. The Adelaide launch was the perfect platform to rail against the evils of government regulation and high taxes being forced on Australia by Kevin Rudd's new emissions trading scheme.

'Populist of the year, Kevin Rudd, pronounced global warming as the "most important moral issue of our time" but only offers a solution that guarantees all Australians will lose out through higher tax and a sprawling bureaucracy,' Bernardi warned the crowd before landing his punchline: 'A bit too much like socialism for my liking—but perhaps that is the ultimate goal of those who worship at the altar of global warming?'

The senator was performing for the home crowd. The hardline Christian lobby, the National Civic Council, was hosting the event and Bernardi was a local hero. He'd shot to national fame after fingering Gordon Ramsay for saying 'fuck' 80 times in one particularly hot episode of *Kitchen Nightmares*. That won him a Senate inquiry into swearing on prime-time TV. He would also campaign against recognising same-sex relationships, arguing it could legitimise bestiality and polygamy; and he crusaded long and loud against abortion rights.

But one particular cause captured Bernardi's attention early in his Senate career. He became a vocal defender of Australia's climate sceptics. One man who was guiding that interest was Hugh Morgan's factotum, Ray Evans, still Australia's most influential campaigner against climate science. Not only did Evans continue to steer the sceptic think tank, the Lavoisier Group, he was diligent at maintaining the deep connections between the Australian and US climate-sceptic movements.

Back in April 2007, soon after he got into the Senate, Bernardi wrote a remarkable opinion piece for Adelaide's tabloid newspaper, *The Advertiser*. Under the heading 'Cool Heads Needed on Global Warming', Bernardi launched a sweeping attack on climate science, much of it inspired by key climate-sceptic scientists promoted by the Lavoisier Group and America's Cooler Heads Coalition.

In his opinion piece Bernardi made a point of quoting from the Oregon Petition, an infamous climate-sceptic document. The petition was supposedly signed by 17,000 scientists who backed the claim that 'there is no convincing scientific evidence that human release of carbon dioxide, methane, or other greenhouse gases is causing or will, in the foreseeable future, cause catastrophic heating of the Earth's atmosphere and disruption of the Earth's climate'.

The petition had long been discredited by the US National Academy of Science. As well, some of the early signatures had been exposed as obvious frauds added by pranksters, posing as the stars of *M*A*S*H*—H. (Hawkeye) Pierce and Frank Burns—and (Dr) Geri Halliwell, aka Ginger Spice of the Spice Girls. Bernardi's use of the petition was chancy, but so was the timing of his opinion piece. It appeared just a month before his then prime minister, John Howard, was due to release the Shergold report on the emissions trading scheme. That earned him a rebuke from one of Howard's loyal cabinet ministers. 'You're absolutely right, but it's not very helpful,' Bernardi was told.

Bernardi is fuzzy about Ray Evans' role in that first opinion piece, but he admits that both Evans and Morgan played a role in his campaign to champion climate sceptics and fight against emissions trading. 'I can't pretend that I knew Ray Evans' and Hugh Morgan's long-standing history,' he said, 'but these were people whose names I was familiar with, who made contact, and were supportive of it and would supply information as well.'

Bernardi began developing his own links in the US with right-wing Republican activists by 2008: 'I think around that time I went to America and I met with a whole range of different organisations over there, you know, the activist organisations in the tax space, in the freedom space, in the smaller government space, in leadership development,' he explained. 'I kind of became their go-to person in Australia.'

In the US, Bernardi signed up for a training course with the Leadership Institute in Virginia, run by a senior Republican Party operative called Morton Blackwell. The Institute prides itself on training 'freedom fighters' to combat the radical Left. Bernardi would learn about using voter databases and online technology to mobilise grassroots campaigns. These tools would prove vital in the fight to block Labor's climate change policy.

Bernardi knew mounting a successful campaign to stop Rudd's emissions trading scheme would be tough. The public wanted action on climate change; so did his own leader, Malcolm Turnbull, and Bernardi was now the Shadow Parliamentary Secretary for Disabilities in Turnbull's opposition team.

So on 27 January 2009, that hot day in Adelaide when Bernardi launched *Thank God for Carbon*, he was sending out a big signal to the Liberal Party climate sceptics. Despite being in Turnbull's team and despite his leader's views, he would fight against the emissions trading scheme. Bernardi wrapped up the launch by lavishing praise on the report's authors—Ray Evans and the Lavoisier team. 'I am pleased to say "Thank God for carbon" and I also say "Thank God for you".'

That day Ray Evans was seething over the Liberal Party's failure to hammer Rudd over the emissions trading scheme. Weeks before, he had written a stinging letter to Andrew Robb, Turnbull's Shadow Minister on Emissions Trading Design, over the party's position. Robb, once a leading climate-sceptic voice in the Liberal party room, was taunted by Evans: 'The Liberal Party doesn't know which side to take. How embarrassing it is to be a member of that party at this time.'

When Evans took the microphone at the launch he went straight on the attack over Rudd's scheme, echoing the dire warnings of the brown-coal generators in Victoria: 'The first issue,' he told the crowd, 'is the impending bankruptcy of the Latrobe Valley Power Stations.' Evans' key talking points could have been written by the private generators and their lobbyists: Labor's emissions trading scheme would bring them financial ruin, causing a power crisis not only in Victoria but in the other states linked to the national power grid.

The feisty climate sceptic was well informed. He spelt out the generators' high debt levels (around $5 billion), the rollover date for those debts, and their prospects for refinancing. He then bitterly complained about the opposition's 'apparent acquies-cence' with Rudd's climate policy. Why? He asked. 'Because this fantasy about anthropogenic carbon dioxide and controlling the world's climate has seized the minds of the political classes in the West, and the rent-seekers see huge gains to be made from trading tax receipts in the form of carbon credits . . .' Evans peppered his fiery speech with Lavoisier's trademark abuse of Australia's climate scientists at the CSIRO and the universities for supporting the mainstream science on climate change. More and more people were aware, 'it was always a scam,' he said.

Soon after the sceptics' sweaty launch, Adelaide had its hottest night on record. Half an hour north of the city, the RAAF base at Edinburgh recorded an extraordinary 41.7 degrees Celsius at 3 am on 29 January. 'Such an event appears to be without known precedent in southern Australia,' the weather bureau said. Over the next few days and nights, hundreds of people overcome by the heat arrived at the state's hospitals.

An extreme weather event had gripped south-eastern Australia. After a slight temperature drop in early February, the heat returned with a vengeance. On Saturday 7 February, the temperature at Melbourne airport hit 46.8 degrees Celsius for the first time on record. That day the worst bushfires in Australia's history roared through Victoria.

The Black Saturday fires burned for weeks. By the end, the shocking toll was 173 people dead and another 414 injured. The fires rocked the country to its core. Around 450,000 hectares of land went up in flames along with 3500 buildings, including more than 2000 houses. Up to a million wild and domesticated animals died in the disaster.

Australia's leading climate scientists cautioned that the events of that summer in 2009 could not, on their own, be taken as evidence of climate change. But most of the scientists made the same points: the warming trend over Australia in the last 50 years was consistent with human-induced climate change, and the southeast of the country was experiencing a long period of unusually dry weather that could also be related to climate change. Both of these trends would increase the number of days when bushfire risk in Australia would be more extreme and fires more intense. And unless global greenhouse gas emissions were curbed, these trends would get worse.

'The projections are based on climate models that include increases in greenhouse gases and that tell us that we can expect higher temperatures and much drier conditions over southern Australia,' explained CSIRO bushfire researcher Kevin Hennessy. 'Not only are we estimating there will be an increase in the frequency of extreme fire danger days, but the duration of the fire season will be longer and the intensity of some of the biggest fires may increase.'

The CSIRO's work made absolutely no impression on Australia's climate sceptics. In March that year, Ray Evans' Lavoisier Group and the Institute of Public Affairs (IPA) proudly co-sponsored an international climate sceptics conference at the Marriott Marquis Hotel in New York under the banner, 'Global Warming: Was It Ever Really a Crisis?' The IPA was by now getting big grants from the Cormack Foundation, the Liberal Party cash cow where Hugh Morgan sat on the board. Not surprisingly, the Australian speakers at the conference included some of the IPA's and Lavoisier's favoured scientists: Professor Bob Carter, who had spent the previous year attacking 'warmaholics', and Bill Kininmonth, a former head of the Australian Bureau of Meteorology's National Climate Centre. His book, *Climate Change: A natural hazard*, was admired by Morgan.

The New York sceptics' conference showcased high-profile American speakers well known to their Australian friends, including Myron Ebell from the Cooler Heads Coalition and the Competitive Enterprise Institute. Ebell proudly noted he'd been named in Greenpeace's *A Field Guide to Climate Criminals*. Dr Pat Michaels, a veteran from the anti-Kyoto campaign, was also there as a senior fellow from the Cato Institute, a think tank co-founded by right-wing billionaire Charles Koch whose brother David sat on its board.

Hosting the big sceptics shindig was a little-known outfit called the Heartland Institute, which advocated a range of causes, from lower taxes to protecting private healthcare. Heartland had got funding from ExxonMobil for years, before the oil giant cut back its grants to sceptic groups. Among Heartland's current donors was the Mercer Family Foundation, set up by a reclusive right-wing hedge-fund operator, Robert Mercer. It gave its first $1 million grant to Heartland in 2008 when that organisation held its inaugural global climate conference. Robert Mercer, years later, would be revealed as the top donor to Donald Trump's presidential campaign.

Former Greenpeace activist Kert Davies also knew Heartland for its work promoting climate sceptics but only later discovered its connection to Mercer. 'We had never heard of the Mercer family in 2008 and there was no way to do a global search of all the tax forms,' said Kert Davies. 'We only learned of their years and years of funding of Heartland when Trump was elected.'

In March 2009, the Heartland sceptics' conference was buzzing with excitement. The new Democrat president, Barack Obama, was promising to regulate greenhouse gas emissions as pollutants. In Congress, two seasoned Democrats, Henry Waxman and Ed Markey, were also working on an emissions trading scheme much like Kevin Rudd's.

The Democrats were pushing to pass their 'cap and trade bill' through the House within months. Environmental activists were hoping the Waxman–Markey bill would then pass the US Senate in time for the UN's Copenhagen Climate Conference in December. If that happened it could be a game changer in the global climate negotiations. America could finally end its long stand-off with China and secure a new climate treaty that included both countries. The stakes were huge.

For many of the Heartland delegates what was happening in Washington and Canberra was all wrong. They needed more than ever to persuade enough politicians and voters that global warming was not a serious threat to the planet. If they could do that, it would undercut the political pressure to regulate carbon emissions in America and Australia. And the idea of a new climate treaty in Copenhagen would again be seen as a noose around the neck of the carbon economy.

Myron Ebell was well aware that killing off the 'cap and trade bill' in Washington would take more than an argument about climate science. It would also take a political strategy that appealed to voters' hip pockets. Speaking at Heartland's first climate conference a year earlier, Ebell spelt out his simple message: 'If you want to fight this stuff,' he said, 'just remember: *It's a tax. It's a tax.* Just keep repeating that, *it's a tax.*' Ebell was convinced politicians would buckle if that message got through to voters. 'They didn't understand it was going to raise their energy prices,' he said. 'So we started calling it "cap and tax". And by 2009, when Waxman–Markey hit the floor of the House, virtually everybody was calling it "cap and tax", and at that point we knew we were going to win.'

When Cory Bernardi found himself digging in for what he called the fight of his life, to stop the emissions trading scheme in Australia, he sought advice from Myron Ebell, the Heartland Institute and their American political allies. He would hammer the same message they hammered. Putting a price on carbon emissions was just a big new tax imposed on the economy for no good scientific reason. Not only would it hit their hip pockets, it would destroy jobs and shut down whole industries. It was a message the carbon club would embrace in Australia.

On the outskirts of Newcastle, Greg Combet was in his electorate office preparing for a big day. Combet was the MP for a coal district in the NSW Hunter Valley, in one of Labor's safest seats. A few kilometres away, Lake Macquarie was dotted with small boats, but many of his constituents were hard at work in the nearby mines.

In early 2009 the prime minister had given Combet a new job, Parliamentary Secretary for Climate Change. His task was to cut a deal with the powerful coal industry to help Penny Wong get the emissions trading scheme through parliament that year. It was a delicate business for an MP from the heart of coal country, but Combet wasn't fazed. Tough negotiations with corporate chiefs had long been his bread and butter. Combet was a former secretary of the Australian Council of Trade Unions, the ACTU, and had handled some of the country's most bitter industrial disputes.

This day, the coal lobby was coming in to put its case and its members were not happy. Under the proposed emissions trading scheme, the mining companies would need to buy permits to pollute because some of their mines released large amounts of the potent greenhouse gas methane during drilling of the coal seams. The government had offered the industry $750 million over five years to help it adjust, but it wanted a lot more.

The Australian Coal Association (ACA) heavies, including senior executives from the global mining companies, led by a former trade diplomat, Ralph Hillman, arrived at the office. 'Arrogant and patronising' was how Combet remembered them, but he thought the talks went pretty well, right up until the end. 'As we are all leaving the meeting, one of my staff came up and said, "They've released this report while we've been in the meeting and said nothing about it,"' he said. Unbeknown to Combet, the ACA had released its latest withering analysis of Labor's scheme. 'One of those dodgy things, shit-canning the government,' as Combet put it.

'I grabbed Hillman and I said, "What the fuck are you doing?" No shame whatsoever—it was just a tactic to put me on the back foot,' said Combet. 'Penny Wong was on the phone saying, "What the hell?"'

That first meeting set the tone for brutal negotiations that went on for months. Combet found it bizarre. He was the last person who wanted to shut down coalmines. The miners' union, the CFMEU, was one of his strongest backers and, as a young engineering student, he'd worked in coalmines. But Combet was also an economist. He knew some mines were methane-intensive and some were not. He didn't want to hand out money to companies that couldn't put up a good case.

The coal lobby was adamant that some mines could be crippled by the cost of Labor's emissions trading scheme. Ultimately, the companies made it clear that if they didn't get what they thought they needed, they would run a political campaign aimed at MPs in coal seats around the country. Their message was simple, said Combet: 'You're going to kill the industry and smash jobs, and they are in marginal seats.' It was designed to put fear in the heart of the Labor Party.

When Mick Davis, the head of the Anglo-Swiss mining giant Xstrata, swung through Australia he gave the government the same message. 'Contemptuous, he was,' said Combet. 'He came in with his bejewelled cufflinks, so condescending and patronising.'

Combet believed that for most of the global miners, the cost of Australia's scheme was not a big burden. 'A carbon price in Australia is a piddling piece of their liabilities,' he said. Combet became convinced they were more worried that pricing carbon emissions would go global and hit their customers who imported the coal from Australia.

'While we send hundreds of millions of tonnes of coal overseas, it's combusted in Japan, Korea and China. Under international rules, they account for the emissions from burning the coal, not us,' Combet explained. 'We only account for the fugitive emissions—the methane gas that's released during the mining process, which is a minor part of it. They weren't so much worried about the impact in Australia of the scheme that we were putting in place. They were looking at it in a global context.'

The wrangling between the government and the coal industry was an omen of the fraught times that lay ahead for Labor as it tried to put in place a meaningful climate change policy. Rudd was publicly proclaiming climate change as the greatest moral challenge of our time, but his government did not want to jeopardise Australia's position as the world's largest coal exporter, employing miners on good wages. That contradiction left Labor's MPs wedged between the Greens on the Left and the opposition on the right. And the coal companies knew how to exploit their exquisite dilemma.

Coal exports were worth $54.8 billion that year and the companies had big expansion plans on the drawing board. State Labor governments in New South Wales and Queensland, Rudd's home state, were anxious to demonstrate their enthusiastic support for the industry. Labor's Queensland premier, Anna Bligh, later that year would announce she was opening up the vast new region of the Galilee Basin in Central Queensland for coal exports. Two powerful billionaires were interested in coal exploration licences there: Gina Rinehart, Australia's richest woman, and Clive Palmer, one of Queensland's richest men. The last thing Labor wanted was to go to war with the coal industry. But that was now very much on the cards.

———

By May 2009 the Rudd government was under attack from all sides and needed to shore up allies. That month it planned to bring into parliament the Carbon Pollution Reduction Scheme (CPRS) bills, which would create the emissions trading scheme. To regain the momentum, Rudd and Wong unveiled a new multipronged plan to win back support both at home and abroad. In a surprising move, they reached out to the environment groups that had been attacking them for months over their low 2020 target to cut emissions by just 5 per cent.

As part of that plan, Rudd did an extraordinary backflip. He announced a new target to cut Australia's emissions 25 per cent by 2020. This was back in the range called for by the environment groups, the Bali summit and climate scientists. Rudd was now aiming high. He said he wanted a Copenhagen agreement 'consistent with Australia having the prospect of saving the Barrier Reef'.

The backflip came after Wong returned from a meeting in Washington, where it became clear that, unless the rich countries accepted cuts of around 25 per cent or more, China and India would not be interested in a deal in Copenhagen. Rudd later denied his backflip was just a negotiating ploy. 'Did I see it as pie in the sky? No. Because I'd already begun to have conversations with Obama about this in the course of 09, and the great variable at play in all this was how far would Obama go?' he said.

There were, of course, big caveats: the new target would only come into force if there was an ambitious climate deal in Copenhagen that included the US and China. Rudd also made it clear he would stick to his low 2020 target announced back in December, a 5 per cent emissions cut, if there was no deal in Copenhagen. He might also go back to his old maximum 15 per cent cut, depending on the level of ambition shown by other big economies in Copenhagen.

Despite these stipulations, Rudd's backflip worked on two fronts. It kept Australia in the game at the UN climate talks and it won back support at home from the three most influential environment groups on climate change: the Australian Conservation Foundation, the WWF and the Climate Institute. With the generators, the coal industry, the climate sceptics and the Greens all attacking Labor's policy, the government really needed well-funded friends in the climate lobby out in the media defending it. But under Rudd's new plan, the environment groups had to swallow more concessions to some of the big greenhouse gas emitters, in particular more free permits for industries with big workforces exposed to foreign trade markets, like aluminium and steel.

To reassure the coal industry, in the May budget the government unveiled a huge $4.5 billion 'clean energy' package, with over half of it slated for 'clean coal' research projects. Like Howard, Rudd and his resources and energy minister, Martin Ferguson, wanted to send a strong signal that they had faith in coal's long-term future. To shore up his green credentials, Rudd put up almost $2 billion for renewable energy and promised to get the long awaited 20 per cent Renewable Energy Target through parliament that year.

But there were some very important losers in the new plan— the coal-fired power generators. The environment movement had been scathing about the compensation offered to the generators in December. In this round the generators' assistance was cut from $3.9 billion to $3.3 billion.

Their reaction was swift and powerful. Within weeks, their peak lobby, the National Generators Forum, wrote to every federal MP warning there would be a 'systematic failure of the electricity market' unless they got more. They were now demanding an astonishing $10 billion, arguing some generators would be left 'technically insolvent' if the government didn't come good.

Combet had been promoted to Minister for Defence Personnel, Materiel and Science as well as Assistant Minister for Climate Change with Wong. He began tentative talks with the Victorian private generators. He got nowhere. All of them had the same message, but he remembered the warnings of International Power's Tony Concannon most clearly: 'You're going to shut us down, put everyone out of a job. There's going to be blackouts,' Combet recalled. 'He'd just do that and do that aggressively in the media all the time.'

The generators' campaign was causing serious rifts inside the government. Martin Ferguson was sympathetic to their demands and so was his new head of the Resources and Energy Department, John Pierce. Pierce had come from running the NSW Treasury, where he had played a part in the Labor government's unsuccessful attempt to privatise that state's electricity industry. He backed the generators' claims that the emissions trading scheme could seriously impact their balance sheets. Combet was taken aback by his advice.

'John Pierce took a very aggressive stance against the CPRS not taking sufficiently into account, in his view, the financial impact on the generators. Basically, "If you do this you will destroy investment in the energy-generating sector",' Combet remembered Pierce saying. 'Rudd's exposed to that from the head of the Energy Department? For Rudd and the rest of us, it was difficult to assess how alarmist that message really was. That's a pretty full-on message to get from your senior public servant in that area.'

As the generators piled on pressure in the media, Rudd got the head of his own department, Terry Moran, to adjudicate between Wong and Ferguson and their public servants. They agreed to hire merchant bank Morgan Stanley to investigate the generators' balance sheets and their case. It was a big win for the generators, but they were not going to shut up.

———

Kevin Rudd made it clear he did not want to work with the Greens to get his climate policy through the Senate before Copenhagen. His relations with the Greens leader, Senator Bob Brown, were non-existent. Labor needed a deal with Malcolm Turnbull, and Penny Wong had a strategy to do just that. 'We have to make it about him,' she told her staff. 'He's got an intellectual pride. That's not a criticism, he actually wants a coherent position and we need to make it something he feels he has to do for his own consistency. And we have to make it about him until he sits down with us.'

Unfortunately for Wong, by the end of June 2009, the political debate was all about Turnbull, but for the wrong reasons. In a spectacular political blunder, Turnbull had relied on a leak from a mole in the Treasury Department to accuse the prime minister and the treasurer, Wayne Swan, of being involved in a corrupt deal that would allow a government program to benefit one of their local supporters, a Brisbane car dealer. Turnbull called on Rudd and Swan to resign over the scandal. But Turnbull's source was a mentally unwell Treasury officer, Godwin Grech, and the corruption claims were based on a fake email.

Rudd tore Turnbull to shreds in parliament, saying he had disqualified himself 'from ever being fit to serve as leader of this country'. Turnbull's approval rating crashed in the polls and Labor leapt ahead of the opposition, 58 per cent to 42 per cent. Turnbull and many in the opposition feared Rudd would call an early election on climate change and the opposition would be decimated. Turnbull wanted to do a deal on climate change, but getting his own party room to support him was now a whole lot harder.

Unknown to Turnbull, Cory Bernardi was already organising a serious insurrection inside the Liberal Party against the CPRS and its emissions trading scheme. Some months earlier, Turnbull had sacked Bernardi from his shadow ministry over an internal party dispute. That gave the conservative warrior 'a free voice', as he put it, to speak out against the emissions trading scheme, and he was using it at full throttle.

Bernardi's strategy was to whip up the public campaign against the scheme while organising the internal partyroom revolt. By July he was well underway. 'There was a group of us who would meet quite regularly here in parliament,' said Bernardi. 'We were essentially a group of people who believed in the cause that we were doing, who trusted each other implicitly, which was a very rare thing in politics. And we forged very firm friendships.' This core group, said Bernardi, was convinced 'If we go along with Labor's policy, we will destroy the Liberal Party.'

The group included three fellow Liberal senators: Mathias Cormann and Michaelia Cash from Western Australia, and Victorian Mitch Fifield. Cormann and Fifield were both parliamentary secretaries. Some of them would meet at least once a week in Bernardi's office, and often speak daily 'to discuss the strategy of how we can defeat the emissions trading scheme,' said Bernardi.

One of Bernardi's most vocal allies was the maverick National Party senator from Queensland Barnaby Joyce. The Nationals were stalwart supporters of the mining industry, and many of them, like Joyce, were deeply sceptical of climate science. Despite their coalition agreement with Turnbull, Nationals leader Warren Truss broke ranks early, calling the scheme 'a job-destroying rabid dog that should be put down'. Joyce was in a pivotal role to campaign against it. He was the National Party's leader in the Senate, where Labor needed the numbers to pass the bill.

Bernardi and Joyce began barnstorming the country—in meetings in rural towns, on talkback radio and online—branding the scheme a massive tax grab while attacking climate change science. 'We explained that the policy would have no environmental benefits but would see tens of thousands, possibly hundreds of thousands of jobs, and industry move offshore,' said Bernardi.

Rudd had ruled out putting agriculture into the scheme until 2015, and only if a government review found it was feasible. But this allowed Joyce to tell farmers agriculture was 'potentially' in the scheme. It let him push his best one-liner everywhere he went—a roast lamb dinner would end up costing $100. 'If you live on a diet of naturally grown wild berries and lentils, which you scavenge for in your back yard, then you'd also probably be OK,' he quipped.

That August, the Senate was set to vote for the first time on the CPRS bills, which included the emissions trading scheme. In the days before the vote, one of the top men from the US sceptic group the Heartland Institute arrived in Australia. Its senior scientist, Jay Lehr, joined one of the IPA's favourite local climate sceptics, Professor Bob Carter, on a 'whistle stop' tour of Australia to campaign against the trading scheme. Carter was by now one of the most high-profile climate-sceptic scientists in Australia, thanks to the IPA and the Lavoisier Group, and he personally lobbied politicians to stop the bill.

Many in Labor and the Greens dismissed the sceptics as 'mad uncles'. Bernardi and Joyce were derided as recalcitrant rebels. But in Canberra, the Liberal Party's climate sceptic in chief, Senator Nick Minchin, was standing right behind them, encouraging the campaign. Minchin's role would now become central. He was the opposition Senate leader as well as Turnbull's Shadow Minister for Communications. 'I was Senate leader so in the hierarchy I was the most senior member of the Coalition who had doubts about us supporting the CPRS as it was,' said Minchin. 'And I was also the de facto convenor of the conservative faction of the Liberal Party. So I wore two hats. That meant that I was in a critical position, I suppose.'

He and Andrew Robb, the opposition's Shadow Minister on Emissions Trading Design, secured Turnbull's agreement that the party would not consider supporting the emissions trading scheme in the Senate until a host of preconditions had been met by Rudd and Wong, including more money for the Victorian power generators. It was a nerve-racking strategy for Turnbull, who feared that ultimately Rudd would call an early election on climate change after Copenhagen and the opposition would lose it.

'Malcolm was desperate to avoid a fight with Rudd over the CPRS,' said Minchin. 'We all knew that he just wanted to support it.' Minchin was smarting that Turnbull had already won the Liberal Party's approval to back Rudd's 20 per cent Renewable Energy Target, which he thought was 'madness'. But Minchin was determined to block the emissions trading scheme until after Copenhagen, which he was convinced was going to be a failure. That would pave the way for the party to vote against it in the new year.

Robb was also deeply hostile to Labor's scheme. In many ways he was an odd choice as Turnbull's shadow minister on the scheme. Robb was still a climate sceptic, although not a denier, or as he later explained, not an 'alarmist', but definitely a sceptic. Robb thought maybe Turnbull had put him in the job to have someone who could 'bridge' both the conservatives and the moderates in the party room.

Robb also had the advantage of knowing some of the big greenhouse gas emitters from his lobbying days, and he met them regularly to hear their very vocal objections. He was convinced the scheme was 'very bad policy', being pursued for largely political reasons. 'If you set out to design the most bureaucratic, high-taxing, interventionist, government-directed socialistic policy, Labor's ETS [emissions trading scheme] would fit the bill,' he said later.

That July, Robb went to Washington and Beijing and came home believing neither country would adopt emissions trading. By now, however, Robb was also suffering badly from depression and so he decided to step aside as shadow minister. Negotiations between the opposition and the government were going nowhere.

Wong brought the CPRS bills to the Senate that August knowing they faced certain defeat. She was hoping this was just round one and that the serious horsetrading could now begin. Wong still believed the bills could pass before Copenhagen. As she sat in her office in Parliament House monitoring the debate, she suddenly noticed Barnaby Joyce about to speak. She stopped to watch with one of her staff.

Joyce was in full flight, defending the coal industry and denouncing climate science. But there was one point he really hammered: Labor's emissions trading scheme was going to be a great big tax. 'Every time there is a problem, their solution is a new tax,' Joyce declared. 'And this is a supertax; it is a supertax that people have to pay whether they are making a profit or not. This is a tax that people are going to have to pay merely by reason of the fact that they exist.'

At that moment, Wong realised Labor's whole climate policy was capable of being boiled down to one simple idea: a great big tax on everything. 'John who worked for me said, "He's found the line that will kill us." And I thought, you're right. If that gets traction, it's so easy. It's such an easy line.'

CHAPTER EIGHT

TRIUMPH OF THE SCEPTICS

MALCOLM TURNBULL WAS in London and in his element. He had just met with Britain's young Conservative Party leader, David Cameron, and found a soul mate—a modern Tory politician willing to lead on climate change. 'One of the great achievements that David has been able to do, or effect, is to position the Conservative Party as being environmentally credible,' the Australian opposition leader enthused.

Turnbull had flown to the UK in September 2009 looking for advice on navigating the political minefield of climate change. He was not disappointed. The Conservative's rising star was boldly promoting his plans for Britain's new low-carbon economy. Cameron had famously toured the warming Arctic with the World Wide Fund for Nature, hugged a husky on a dog sled, and, more importantly, persuaded his party to pass the UK Labour Government's Climate Change Act. He was now campaigning on the slogan: 'Vote blue, go green'. 'He's done that very well,' said Turnbull, 'and of course I have a common commitment, a similar commitment to the environment.'

But while Turnbull was in London, lapping up advice on compassionate conservatism, his nemesis Cory Bernardi flew in the opposite direction. He went to Washington for talks with right-wing Republicans and climate-sceptic warriors. 'I have no doubt as to which one of us got the better advice,' said Bernardi. He was absolutely right.

Bernardi met up with Myron Ebell from the Cooler Heads Coalition, who introduced him to two hardline Republicans leading the fight against the emissions trading legislation on Capitol Hill. One was James Inhofe, who was still on the Senate Environment and Public Works Committee. While the Republicans had lost control of the committee, he was determined to win it back in the mid-term elections. The second was the top Republican on the House Energy Committee, Jim Sensenbrenner, who had fought tooth and nail against the Waxman–Markey bill on emissions, calling it the job-destroying 'cap and tax' bill.

Bernardi left convinced that the Democrats' plans for an emissions trading scheme were in deep trouble. While the Waxman–Markey bill had passed the House of Representatives in June by a handful of votes, that win had come at a horrible political cost to the Democrats. The chances of a bill getting through the US Senate before Copenhagen were fading rapidly.

Republican attack ads were already targeting vulnerable Democrats who'd voted for the bill. The American Petroleum Institute sponsored rallies across the country. Fox News was in overdrive, with one of its star hosts, Glenn Beck, calling supporters of the bill 'treasonous'.

The campaigning by right-wing pressure groups was unrelenting. One in particular stood out—Americans for Prosperity—an outfit co-founded by David Koch, one of the billionaire Koch brothers. The group reportedly organised some 80 'grassroots' events against the 'cap and trade' bill. It famously launched a Hot Air Tour of key Senate seats with a giant balloon emblazoned with the slogan 'Lost jobs, higher taxes, and less freedom', warning voters what would happen to them if the climate bill passed. Greenpeace was now calling the Koch brothers' empire, 'a kingpin of climate science denial'; their funding of sceptics had overtaken ExxonMobil's.

In Washington, Bernardi hooked up with a young Australian who knew more than a bit about the Koch brothers' grassroots organising. Tim Andrews, a former Liberal Party activist, had trained in a Koch brothers' internship program designed to skill up defenders of free enterprise. Intelligent, charming and seriously zealous, Andrews was a protégé of the conservative crusader and climate denier Ray Evans. Andrews was still a student with the Young Liberals at the University of Sydney when he first met Evans and was captivated. 'I am one of the many people Ray inspired to do a lot of things,' recalled Andrews. In no time, Andrews had become involved with the trade union antagonists, the H.R. Nicholls Society co-founded by Evans. After a particularly exciting meeting of the group in Melbourne, he remembers walking out and calling his mother: 'Mum, I know what I want to do with my life.'

On a break from Sydney University, Andrews went to Washington to become a Koch Associate. In 2004 he was given an internship at the Koch-backed Cato Institute. By the time he met up with Bernardi in Washington, Andrews was a true believer. He had been honing his skills by working with Americans for Tax Reform, an advocacy group deeply opposed to the 'cap and trade' bill.

'One of the things I was able to do over here was to try to learn how you effectively mobilise people,' said Andrews. This gave him exactly the kind of political skills Bernardi valued. Andrews would now become Bernardi's confidant and helper in the campaign to defeat an emissions trading scheme in Australia, including introducing Bernardi to a range of right-wing advocates in the US.

Soon after Bernardi left Washington, *The Australian* newspaper ran an extraordinary survey of Liberal backbenchers on Rudd's scheme. It revealed two-thirds were opposed to negotiating the legislation before the Copenhagen climate conference in December. Some were members of the core group Bernardi had been working with assiduously since July, but there were a lot more besides.

Tim Andrews was with Bernardi before the story broke. He had little doubt Bernardi had played a role in persuading some of the backbenchers to talk. 'Cory was the one who arranged it behind the scenes,' said Andrews. 'That was my understanding.'

The story sent shockwaves through the Liberal Party. Turnbull's deputy leader, Julie Bishop, read the report, stunned. 'If this is half true,' she said to Turnbull, 'we have a problem.'

———

Turnbull was fired up when he returned from London and met his shadow ministers in Adelaide on 30 September. He was ready to stare down the doubters. He was convinced he needed to do a deal with Labor on climate change before Copenhagen.

The politics were brutally simple. The government's emissions trading package was in the Carbon Pollution Reduction Scheme (CPRS) bills that were going up to the Senate again. If the opposition knocked it back, Rudd could have the trigger he needed for a double dissolution election. That meant Rudd could ask the governor-general to dissolve both the Senate and the House of Representatives, and go to an election after Copenhagen. He would most likely win and Turnbull, having lost the election and his credibility on climate change, would then almost certainly lose the Liberal Party leadership. His political career would be over.

At the shadow cabinet meeting, Turnbull laid down his position, which he would bluntly sum up on radio the next day:

'I will not lead a party that is not as committed to effective action on climate change, as I am.'

Turnbull's stand tore open the divisions in his shadow cabinet. They all knew the National Party backbenchers would not vote for the emissions trading scheme in the Senate. Minchin now believed some Liberals would also cross the floor in the Senate to oppose it, at least until after Copenhagen.

The Adelaide shadow cabinet meeting galvanised a very important player, Tony Abbott. One of the most cunning political animals in the parliament, his sceptical views on climate science were well known. The natural inclination of the 'Mad Monk', as Abbott was known by his enemies in the party, was to line up with the conservatives, but so far he had been a weathervane on the emissions trading scheme. He'd backed his hero, John Howard, when he proposed it in 2007. As Shadow Minister for Families, he had publicly backed Malcolm Turnbull on it.

Indeed, that July, Abbott had written an opinion piece in *The Australian* warning that the opposition would lose an early election on climate change. They should pass Rudd's bill, he argued, 'to save the Coalition from a fight it can't win'.

Minchin had been appalled when he read it. 'What the hell is this, Tony?' he asked.

Abbott was pragmatic. 'Mate, we have got to be loyal about it. We have got to support the leader and do the right thing.'

Minchin tried to explain the big picture to Abbott: before they supported anything, they needed to see if the emissions trading bill crashed out in the US Senate and whether the UN climate talks fell over in Copenhagen. 'We're crazy to get ahead of the game,' Minchin told him, 'so just hold back a little bit so we don't give Malcolm carte blanche on this.' Abbott agreed to think about it.

After the shadow cabinet meeting in Adelaide, Abbott sniffed the political wind. That night he was doing the star turn at a local

Liberal Party gathering in the small Victorian town of Beaufort. He was there promoting his new autobiography cum political manifesto, *Battlelines*, recently launched by Sarah Murdoch, Lachlan Murdoch's wife. Many in the party saw the book as Abbott's application for the post-Turnbull leadership of the Liberal Party.

During a barrage of questions from the Beaufort locals opposing the emissions trading scheme, Abbott decided to say what he really thought. 'I gave a speech then there were questions. And a lot of questions were really critical of an emissions trading scheme—the great big new tax on everything,' he recalled. The debate turned to the science of climate change.

'The argument is absolute crap,' Abbott told the crowd. 'However, the politics of this are tough for us. Eighty per cent of people believe climate change is a real and present danger.'

The editor of the local newspaper, the *Pyrenees Advocate*, jumped on the 'crap' quote. While he would later claim his words were 'a bit of hyperbole', that line would define Tony Abbott's position on climate change for the rest of his political career.

Abbott credited Minchin with turning him against the emissions trading scheme, but his switch was largely political. When Abbott drove to Melbourne airport the next day, he and Minchin talked on the phone and worked up their strategy. Minchin warned Abbott that if they stuck to Turnbull's plan to vote on the emissions trading scheme before Copenhagen, it would split the Liberal Party.

'By that stage,' said Abbott, 'I'd come to the view that by far the best way to maximise our chances at the next election was to campaign against the emissions trading scheme—the great new tax on everything.'

A week later, the Liberal Party was in turmoil. The media was running leaks about a serious challenge to Turnbull's leadership, and the story was off and running. The shadow treasurer, Joe Hockey, had been approached. Hockey and his leader were forced into an embarrassing press conference, where Hockey pledged his 'absolute, unqualified support' for Turnbull.

The leaks came just as Turnbull and his Shadow Minister for Energy and Resources, Ian Macfarlane, were about to ask the shadow cabinet to sign off on their strategy to negotiate with Penny Wong over the emissions trading scheme. With Andrew Robb on sick leave for his depression, Ian Macfarlane had taken over his role.

Macfarlane was a forthright ex-farmer from regional Queensland. He knew all the arguments and all the players around the emissions trading scheme after spending six years as the Industry and Resources Minister in Howard's cabinet, when he had held his own sceptical views on climate science. But he was a pragmatist, and right now he was loyal to his leader: he wanted a deal with Wong that excluded agriculture from the scheme, got more industry assistance for coalmining and, above all, got more compensation for the Victorian power generators.

Turnbull and Macfarlane struck a truce with the backbench. Macfarlane could negotiate with the government for the next five weeks, but any deal had to come back to the joint party room for sign off.

Bernardi and the opponents were fine with that. They had no intention of agreeing to any deal with Rudd. 'I saw the opportunity to negotiate as a means of highlighting all the flaws that were in the government policy,' Bernardi said.

Bernardi would use the five weeks to mobilise an extraordinary grassroots campaign among Liberal Party supporters against the bill. 'Over those five weeks, a number of us put together an online petition, where we asked people to submit their details, saying they didn't want an emissions trading scheme,' Bernardi explained. 'I didn't realise how important the tens of thousands of email addresses and contacts that we got out of this nationally would become.'

Actually he did have some idea. And so too did Tim Andrews in Washington. This database would soon come to play a role in killing Labor's climate policy in the Senate and Turnbull's leadership.

———

In early October the lobbying campaign against the emissions trading scheme went into overdrive. The Australian Coal Association made good on its warnings to Combet earlier in the year. Backed by its biggest members—Xstrata, Peabody, Anglo-American, BHP and Rio Tinto—the coal lobby signed off on a multimillion-dollar attack on the emissions trading scheme.

Under the banner 'Let's cut emissions, not jobs', the lobby blitzed ads on regional TV, radio and newspapers in coal seats in Queensland and New South Wales. A website directed miners to email their local MP to protest. The coal lobby hired some of the best in the business for the campaign: Lynton Crosby, the former Liberal Party director whose firm Crosby Textor still worked closely with the party; Tim Duncan, the former head of media at Rio Tinto, who also sat on the board of the Institute of Public Affairs; and the famous creative director Neil Lawrence, who had produced the inspired Kevin 07 slogan for Labor's winning election campaign.

The Victorian power generators were also revving up their lobbying in the media and getting support from Macfarlane. 'If International Power's balance sheet gets destroyed, they are shot,' Macfarlane told the ABC. The generators wanted at least another $4 billion on top of the $3.3 billion on offer, warning that without it the electricity network was in jeopardy.

Meanwhile, the climate sceptics' ground campaign was back in full force. The National Party's Senate leader, Barnaby Joyce, was on his very own hot-air tour around regional Australia, albeit without a balloon. In the Queensland town of Roma, home of the biggest cattle saleyards in Australia, Joyce was whipping up farmers and small business against the emissions trading scheme: 'I believe Australia has drawn a line in the sand over this. And the blue is on.'

Joyce was barnstorming with Bob Carter, the sceptic professor who had become his intellectual bedrock on climate science. At a Roma community hall, Carter assured the locals there was no link between rising carbon emissions and the threat from global warming. 'How many of you in this room are under 50? There's been no global warming in your lifetime. None. Zip. Zero. None,' Carter told them. 'Temperature has gone down and carbon dioxide has gone up. How is it possible to have a prime minister that believes that increasing carbon dioxide emissions are causing dangerous global warming?'

Carter insisted he was a scientist without any political leaning; however, he backed Joyce to the hilt. 'But as Barnaby said,' Carter lectured them, 'you have to beat down the door of every voting senator. Forget the Labor ones—every voting Liberal senator. Forget the National ones, they're going to vote against it anyway. The Liberal senators have to be convinced this bill's got to be defeated a second time.'

By early November, the media debate on climate change and the emissions trading scheme was febrile. In Sydney, Alan Jones, the city's top radio shock jock and outsized influencer in the NSW Liberal Party, pounded his listeners with attacks on climate science. When Abbott was driving home from a school bus drop-off, he heard Turnbull take on Jones in a hectoring exchange. Jones was again denouncing global warming as a hoax. 'Don't you think you sound like the old lady who says the whole world is mad except for thee and me, and I have my doubts about thee?' an exasperated Turnbull asked Jones.

Abbott was dumbfounded. He thought Turnbull had blown his leadership.

———

With some anxiety Penny Wong and Greg Combet brought the CPRS bills back to parliament with a deadline to pass them by the end of November. Wong was trying to get the negotiations with Macfarlane moving and push the bill through the Senate before Copenhagen. But she was unsure whether her own prime minister wanted the same thing. She turned to the deputy leader, Julia Gillard, for advice. When Wong came to see her, Gillard said, tears were welling up in her eyes. 'She revealed that she was simply at her wits' end because even as she was doing the negotiations with the opposition's Ian Macfarlane, she did not know whether her political instructions from Kevin were to get a deal or to crash the prospects of a deal,' Gillard recalled.

Combet was equally frustrated. His office was getting calls from Rudd's staff, telling him to go hard on Turnbull in parliament over climate change. He thought it was just plain dumb. 'Why shit on someone you're trying to do a deal with? Shouldn't we nurture him until we get the CPRS through?'

Some in his own party believed Rudd was weighing up a brutal political strategy. Turnbull was weakened by the Liberal Party infighting and a few hard heads in the Labor Party, including Rudd, were already considering an early election on climate change. If Labor kept stalling on a deal and the Liberal infighting kept going, it could damage Turnbull irreparably.

The Labor Party national secretary, Karl Bitar, was already kicking around election plans, targeting both Turnbull and his likely successor, Shadow Treasurer Joe Hockey. But Gillard thought this was a bad idea—she wanted a deal with Turnbull, and made her views very plain.

Rudd finally sat down with Wong and agreed to give her what he called 'maximum negotiating leverage' to get a deal done with Turnbull. 'We had a core conversation, Penny Wong and I, in my office, which was—do we want this thing passed even though we have to give more ground to Turnbull?' Rudd recalled. 'And it was an existential conversation about would Turnbull survive. And, is it better for us to have an agreement with Turnbull, which is legislated, which is sustainable, as opposed to the uncertainty on the climate change front if Turnbull bit the dust or we found ourselves without an agreement going into a federal election?'

Getting a deal with Turnbull, Rudd knew, would mean giving more concessions to the big corporate emitters and alienating the environment groups. Rudd told Wong it couldn't be a deal at any price but, 'Our conclusion: We would seek to politically accommodate Turnbull on the way through to land this thing.'

On 6 November, Rudd gave a speech to the Lowy Institute think tank, castigating the sceptics in the opposition; he called out Nick Minchin, Cory Bernardi and Barnaby Joyce, but he also pressured Turnbull to cut a deal. 'You've got to know when to fold 'em—and for the sceptics, that time has come,' Rudd said heroically. But the sceptics had no intention of folding. They were just about to up the stakes.

Three nights later, in a move that rocked Turnbull's authority, his shadow ministers, Nick Minchin and Tony Abbott, along with Cory Bernardi and Barnaby Joyce, appeared on the ABC's *Four Corners* program attacking climate change science and questioning the emissions trading scheme.

Asked what proportion of the Liberal Party were sceptical of climate change science, Minchin replied, 'If the question is, do people believe or not believe that human beings are causing, are the main cause of the planet warming, then I'd say a majority don't accept that position.' It was a slap in the face for Turnbull, who had publicly insisted he would not lead a party that was not as committed to action on climate change as he was.

Abbott backed Minchin in pushing the views of the Lavoisier Group's leading climate-sceptic scientists. 'It seems that the world has cooled slightly since the late 1990s,' Abbott told the program. 'One of the things which I think has disconcerted a lot of people is the evangelical fervour of the climate change alarmists because they haven't pursued their case with the kind of careful moderation that you normally associate with the best scientists.'

More disturbing for Turnbull was Joyce's prediction. 'I think we're going to win on the ETS [emissions trading scheme],' he said. 'I think it's going to be blocked in the Senate and we'll end up with a double dissolution.'

———

After weeks of frenzied lobbying on all sides, Macfarlane and Penny Wong struck a deal on the emissions trading package. It cost the government a lot, not just in dollars but in political capital. The price of a bipartisan deal with the opposition was Labor angering its allies in the environment movement. The Australian Conservation Foundation's chief, Don Henry, urged the Senate not to pass it unless it was strengthened. 'The excessive long-term industry handouts will make the shift to a low-carbon future slower and risk locking us into a heavy polluting economy.'

The Rudd government had agreed to double the coal industry assistance to $1.5 billion and give more assistance to the booming liquefied natural gas (LNG) producers. The vulnerable steel industry also got a boost. More controversially, the heavily polluting Victorian brown-coal generators were the biggest winners. With the support of the resources minister, Martin Ferguson, the generators' compensation was upped from $3.3 billion to $7.3 billion.

Labor thought the generators were on board. So did Macfarlane. They were utterly mistaken.

Turnbull's former emissions trading spokesman, Andrew Robb, knew the generators were still fighting Labor's bill, and so was he. The privately owned brown-coal generators believed the compensation was split too thinly with other generators, and that the true value of the losses to their companies would not be covered. Robb agreed with them.

Despite being on sick leave, Robb somehow got hold of a confidential copy of Macfarlane's presentation on the day before it went to shadow cabinet and the opposition party room. 'It was a total sell-out,' Robb concluded. He sat up late that night working furiously to rip it to shreds. He spoke confidentially to Minchin and Abbott, and agreed he would let loose at the joint partyroom meeting the next day.

Turnbull's shadow cabinet, which included both Liberals and Nationals, approved the deal, fourteen to six on the morning of 24 November. Only three Liberals voted against it: Minchin, Abbott and their Tasmanian ally, Senator Eric Abetz, the deputy opposition leader in the senate.

The opposition MPs and senators then headed into a marathon meeting of the joint party room for Macfarlane's presentation. 'I was immersed by that stage in a battle of life and death to protect Malcolm's leadership,' said Macfarlane. 'I'd done the negotiation and we'd got it to a point where we believed that people would accept it.'

Then Andrew Robb turned up at the meeting and put his name on the speakers list. Turnbull saw him, but kept avoiding his eye. Worried that Turnbull would sideline him, Robb wrote an urgent note and passed it up to his leader. It read, 'The side effects of the medication I am on now make me very tired. I'd be really grateful if you could get me to my feet soon.'

With little option, Turnbull gave him the nod. Robb stood up and proceeded to savage the deal and, by implication, Turnbull's and Macfarlane's credibility. Robb attacked the package for the coal industry as too little, and said the brown-coal generators still believed their compensation was 'a dog'.

'I had poked holes all over their proposal,' he recalled triumphantly.

Macfarlane was gobsmacked at the betrayal. 'If I was in that position I would have rung my leader before the meeting. That phone call never took place,' Macfarlane said later. 'Right up until the moment he spoke in the party room, Malcolm and I thought that Andrew Robb was accepting of the position that we'd negotiated.'

Robb left the meeting soon after his tirade and went to his office, where Turnbull called him, incandescent with rage, accusing him of treachery. 'I got the most uncomplimentary character assessment that I've ever had,' Robb would write later. He didn't care. 'It was water off a duck's back to me', was how he saw it. 'This issue was bigger than either of us.'

By the end of that scarring day, Turnbull knew the rebels were now a majority in the joint party room. They were demanding the emissions trading package be blocked until after the Copenhagen climate talks, or more likely indefinitely. But Turnbull decided to crash through. He had a majority of Liberal backbenchers, if not the Nationals, and a majority of shadow cabinet. That afternoon Turnbull met Minchin and Abbott, and refused to budge. His credibility and his leadership were on the line.

In the conservative bunker Bernardi and his colleagues met for talks. One of their inner group, the Christian conservative MP Kevin Andrews, agreed he would challenge Turnbull for the leadership the next day in the party room. Andrews was a straw man. His job was to test the waters and destabilise Turnbull. Bernardi was convinced Abbott would make a run for the leadership. 'Actually yes, I spoke to him that night and said, "Well, you know, if this takes place I'm with you,"' Bernardi said.

The next morning Turnbull was hit with the Kevin Andrews leadership challenge and the resignations of the opposition parliamentary secretaries in the Senate, Western Australia's Mathias Cormann, Victoria's Mitch Fifield and Queensland's Brett Mason. While all three senators were careful to say that 'climate change is a global problem that requires a global solution', all said they would vote against the emissions trading legislation in the Senate unless it was put off until after Copenhagen.

Turnbull tried to head off Andrews' challenge by calling a partyroom vote on the motion to spill the leadership. It was defeated by not that much, 35 to 48. It was a show of strength by the Minchin forces. Turnbull tried to put it behind him, but he was now on notice: dump the emissions trading scheme until after Copenhagen or he would be dumped.

'It was obvious in the party room that Malcolm was on a suicide mission,' said Abbott. 'Kevin [Andrews] did his thing on the Wednesday and then the situation kept deteriorating.'

Bernardi was now turning up the pressure on Turnbull with a bit of help from his new friend in Washington, Tim Andrews. For weeks Tim Andrews had been using his political blog to air the internal Liberal Party ructions with leaks from the party room. 'I had people from the partyroom meetings—because there were no real bloggers out there in Australia—leaking me the numbers on who voted which way,' he recalled.

Working with Bernardi and others, Tim Andrews now helped deluge Liberal Party offices with emails, faxes, letters and phone calls opposing the emissions trading scheme, although he insisted his own role was small: 'Did I try and lobby and advocate for people to contact their members of parliament? Yes. Do I deserve credit? No. But yes, I was involved in both media and back-grounding people, and trying to mobilise things.'

Bernardi would later provide details of his campaign to a meeting of the US Heartland Institute sceptics conference, reveal-ing the turning point came after the first challenge to Turnbull's leadership: 'I contacted the tens of thousands of people on my email database, as did a number of my colleagues. And we said to them the fight is not over yet. There is a real opportunity here for us to change and make a difference to the policy. So those people all contacted their local members of parliament and said, "I'm a Liberal Party voter, I'm a Liberal Party member, I support the Liberal Party, but I cannot support you if you go down this path and you pursue this track."

'There was hundreds and hundreds of emails per hour going into people's offices. There were a number of offices that couldn't cope with the email deluge and the phone calls that they were getting,' said Bernardi. 'It was the first real mobilisation of grassroots support on the conservative side of politics I think in Australia's history.'

Bernardi and his helpers also cleverly targeted the Liberal Party apparatus. 'The party headquarters right around the country were hit with similar sentiments,' Bernardi recalled. 'And frankly, the very viability and future of the Liberal Party was at stake.'

At the same time Minchin and Abbott moved to demolish Turnbull's dwindling authority. They resigned from his shadow cabinet, triggering others to follow. Turnbull still refused to back down. Instead he gave a defiant press conference calling for the Liberal Party to rally around the climate change bills: 'We must retain our credibility of taking action on climate change. We cannot be seen as a party of climate sceptics, of do-nothings on climate change,' he warned. 'Australians expect their political leaders to act responsibly, to take action on climate change, to protect and safeguard the future of our planet, the future of our children. That is the challenge for us now and I'm committed to it. We must be a party committed to action on climate change. Anything else is irresponsible.'

But it was too late. Bernardi was rounding up the signatures needed for a leadership ballot to remove him. By then, the consensus among the rebels was that Joe Hockey was likely to win against Turnbull or Abbott. Although Abbott had not declared his hand, even his allies knew he was unpopular in the party and, more importantly, unpopular with voters.

In a meeting with Hockey, Minchin insisted the vote on the emissions trading scheme had to be delayed until after Copenhagen, and then defeated because Copenhagen would be a failure. To the surprise of the climate sceptics, Hockey told them he wanted every member and senator to have 'a conscience vote' on the emissions trading scheme. 'I cannot walk down the street and have people think I'm a climate change sceptic,' Hockey told Bernardi.

Minchin was beside himself. A conscience vote would almost certainly give Rudd the numbers to pass the scheme in the Senate. 'For you to make your first act of leadership to say, "I do not have a clue and you can all do whatever you like" is nuts,' Minchin said. 'And I remember Howard ringing me and saying, "What is Joe doing? This is crazy!" Either you have a position or you don't.'

For Minchin the entire battle had been about stopping the emissions trading scheme and Hockey was not going to guarantee that. Minchin swung behind Abbott and did everything he could to get him up.

'Joe couldn't make up his mind,' said Abbott. 'And so I went out and said to Sky, "Well, there should be one candidate who is against this ETS and I'm prepared to be that candidate."'

Meanwhile Ian Macfarlane was loyally counting numbers for Turnbull. But the night before the ballot, he went to see Turnbull in his Canberra apartment and gave him the unvarnished truth: 'Malcolm, if you give Minchin a Senate inquiry to review this legislation, you will stay as leader tomorrow. But I can't get the numbers for you to jam this plan through the Senate. You'll get rolled by three votes.'

Turnbull still refused to budge. 'If that's what they want, I'm not going to back down,' Macfarlane recalled him saying. The two men had a glass of wine together and Macfarlane went back to his hotel knowing it was over. While they slept, Turnbull picked up another vote but it was not enough.

In the leadership ballot the next day, Hockey was knocked out in the first round. In the contest with Turnbull, Abbott won by just one vote, 42 to 41. Almost immediately, Kevin Andrews put up a motion for a secret partyroom ballot to reject the emissions trading scheme.

It was a brilliant move by the sceptic forces. It passed 54 votes to 29. It was a resounding win for Minchin and his supporters. Any hope of Australia getting a bipartisan policy on climate change in the foreseeable future died that day. It was a huge win for the carbon club.

Down in Melbourne, the godfather of the Liberal Party's sceptics, Hugh Morgan, rejoiced that Australia had been saved from becoming 'a green despotism'. He would later tell members of the Lavoisier Group that the deluge of emails and calls that rained down on Coalition MPs in the fortnight before the leadership ballot had swung the vote against Turnbull. 'This was the first time in Australian political history that a party leader had been deposed by the rank and file of the party,' said Morgan.

He gave credit to the climate-sceptic movement for destroying Turnbull's leadership over his support for the emissions trading scheme. 'This revolt by the party rank and file did not happen because of a press campaign against Malcolm Turnbull. Over the years preceding these events, but particularly in the preceding twelve months, the voice of the sceptics began to be heard through the land, despite the efforts of the chattering class media and the government to dismiss that voice as deluded or even criminal.'

Morgan praised the efforts of Professor Bob Carter for stumping the country and for the other sceptic scientists the Lavoisier Group promoted. He also singled out the efforts of three prominent News Corp columnists for holding the line against the media consensus on climate science: Andrew Bolt, Miranda Devine and Terry McCrann.

That year, 2009, Morgan was elevated to chairman of the board of the Cormack Foundation, one of the most important funders of the Liberal Party federally, and of its Victorian and Western Australian divisions. It was a position than ensured he remained one of the most influential figures in the party.

A week after he lost the Liberal Party leadership, Malcolm Turnbull offered his own bitter analysis of the party's stand on climate change under its new leader, Tony Abbott: 'As we are being blunt, the fact is that Tony and the people who put him in his job do not want to do anything about climate change. They do not believe in human caused global warming,' he wrote on his blog. 'As Tony observed on one occasion "climate change is crap", or if you consider his mentor, Senator Minchin, the world is not warming, it's cooling and the climate change issue is part of a vast left wing conspiracy to deindustrialise the world.'

He ended with a prediction: 'Any policy that is announced will simply be a con, an environmental fig leaf to cover a deter-mination to do nothing.'

One day after Turnbull lost the leadership, Wong and Rudd's CPRS was defeated in the Senate, 41 to 33. Two Liberal senators voted for it, the rest against. All the Greens voted against it and so did the independents.

From this time on, Turnbull would see climate and energy policy as toxic in the Coalition. Having lost the leadership over the issue, he sank into a deep depression and would later write of being dogged by suicidal thoughts.

With their strategy to pass the climate change policy blown up, Rudd and Labor turned their anger on the Greens. Rudd had never once picked up the phone to negotiate with the Greens' leader, Bob Brown, but he now blamed them for failing to help Labor pass the bills and rebuff Abbott. 'If the Greens saw Conan the Barbarian (aka Tony Abbott) rampaging through the wilderness bellowing out his own battlecry for a jihad against climate change action would that cause the Green Party to become amenable to passing CPRS through the Senate to just get it done, and get carbon mitigation underway?' Rudd said bitterly later. 'Of course not. The Greens didn't change track, given the good Leninists that they are, always playing the political game.'

In truth, Rudd hoped to pull victory from the jaws of defeat. Labor was ahead in the polls after the bloodletting in the Liberal Party. Like many in his cabinet, he was convinced Abbott would be deeply unpopular with Australian voters. If the Copenhagen climate conference was a success, he could return triumphant, call an election, win decisively and secure his climate change policy.

It was a big if.

CHAPTER NINE

COP-OUT

IT WAS WELL into the night and Kevin Rudd was trapped in a small crowded room with some twenty heads of state and their staff in the cavernous Bella Center in Copenhagen. Outside, a light snow was falling, but inside it was overheated and the air was stale. The room was named after the Danish functionalist architect Arne Jacobsen, and, having sat in it for hours, Rudd thought it was Scandinavian minimalism gone mad—a room designed to inflict maximum discomfort on the weary leaders.

That night, 18 December 2009, was supposed to be a celebration, the triumphant end of the UN Copenhagen climate change conference, COP15. Instead, a core group of leaders called the 'Friends of the Chair' were in this room trying to salvage a climate agreement from a diplomatic train wreck. More disconcerting for Rudd, the chair, Danish prime minister Lars Rasmussen, had disappeared without explanation. Fearing others would follow, Rudd, seized by desperation and chutzpah, told them he had been asked to take over the role. 'This was not true,' Rudd admitted later, 'but for all I knew, Lars was simply responding to a call of nature; I was not about to allow this vital negotiation to fall in a heap through a combination of procedural technicality and biological necessity.'

The Chinese delegates had already blocked any mention in the agreement that greenhouse emissions needed to be at least halved by 2050 to avoid dangerous climate change. Rudd and the others were now working on a new draft of a weak Copenhagen Accord that at least aimed to keep global temperatures from rising more than 2 degrees Celsius. But this too was being resisted.

Earlier that day, the UN climate talks had virtually collapsed. More than 100 heads of state, the largest gathering ever seen outside UN headquarters, had come to Copenhagen to make a deal on a new climate agreement, along with 40,000 delegates, lobbyists, environmentalists, business leaders, politicians, scientists, sceptics and journalists. But now the conference venue, the Bella Center, was in partial lockdown: the NGOs had been banished and police with dogs were guarding the main doors. The summit was mired in recriminations and chaos. Any hopes for an ambitious climate treaty were dead.

That morning, US president Barack Obama had flown in to Copenhagen and joined the small meeting with Rudd and the core leaders, sending a rush of adrenalin through the room. But the Chinese premier, Wen Jiabao, refused to sit with them, sending his negotiators instead. The stand-off between the US and China was sinking the chances of any deal, even a weak one, getting up. By late afternoon, Obama had had enough. He left in search of the Chinese leader. 'I want to see Wen,' he insisted.

That night, Obama and Secretary of State Hillary Clinton found Wen holding his own closed-door meeting in the Bella Center with three pivotal leaders from the developing countries: India's prime minister, Manmohan Singh; Brazil's president, Luiz Inacio Lula da Silva; and South Africa's president, Jacob Zuma. Obama decided to crash it.

'Jaws dropped when they saw us,' Clinton recalled. 'Now the real negotiations could begin.'

The US and China were the two biggest greenhouse polluters on the planet, but neither of them was bound to reduce their emissions under the Kyoto Protocol. Copenhagen was supposed to change that and bring them into a new global climate treaty. But both Wen and Obama had come to Copenhagen offering very little. At this stage, neither wanted an ambitious climate treaty binding their countries. Obama could not sign a treaty or make pledges on big emissions cuts without the support of Congress, and it was clear he wasn't going to get it. That left him promising a weak, non-binding target to cut American emissions 17 per cent by 2020 from a new baseline of 2005. It was far short of what the IPCC said was needed, or what China wanted.

China in turn did not want any treaty that would jeopardise its own economic growth. As a developing country, it was not expected to cut its emissions by 2020, but it had promised to curb the fast growth in its emissions by reducing its 'emissions intensity'. This meant reducing the amount of greenhouse pollution it emitted per unit of production.

Both China and the US were in agreement there would be no binding climate treaty at Copenhagen. But China was baulking at demands from the US and the developed countries that, in any new deal, its promises to reduce its emissions intensity needed to be submitted to the UN climate body to be measured and verified. This was a red line. For the developed countries it was the first serious attempt at putting real obligations on China and the big developing countries to verify their actions to curb emissions in a global climate agreement.

In the meeting with Wen and the three leaders, Obama insisted China needed to agree to some version of verification of its commitments. 'That was resisted down to the very, very end,' said Todd Stern, the US special envoy for climate change, who was with Obama and Clinton in Copenhagen. Wen finally agreed to move a little. China would agree that the information reported by developing countries on implementing their climate actions would be subject to a form of review, called 'international consultations and analysis'. 'That was the final piece that got the deal done,' said Stern.

The deal between Wen, Obama and the three other leaders that night was a top-down political fix. It was a breakthrough but for a weak deal. Even Clinton admitted it was 'far from perfect', but she believed it saved the summit from failure. Obama and Clinton then called a handful of European allies, including Danish prime minister Rasmussen, to a meeting and told them there would be no binding treaty in Copenhagen. This was the best they would get.

'They wanted a legal treaty out of Copenhagen and didn't like our compromise,' Clinton recalled. 'However, they reluctantly agreed to support it since there was no viable alternative.'

Rudd was not invited to that meeting. He had no idea what was happening until later. He was still negotiating hard in the Arne Jacobsen Room.

Worse still, Obama announced his deal at a pre-emptive press conference before he boarded Air Force One to head home: 'For the first time in history all major economies have come together to accept their responsibility to take action to confront the threat of climate change,' Obama told reporters as he ramped up the breakthrough he'd got with Wen and the other three leaders. 'And that's where we agreed to list our national actions and commitments,' he said. 'We agreed to set a mitigation target to limit warming to no more than 2 degrees Celsius, and importantly, to take action to meet this objective consistent with science.'

Alden Meyer, from the American Union of Concerned Scientists, said Obama's backroom deal that night left a lot of bruised feelings. 'Obama cut the deal and was heading to Air Force One to take off,' said Meyer. 'That left the EU in particular, and others, feeling that they were on the sidelines and it was a take-it-or-leave-it thing.'

The Danish prime minister was completely humiliated; the Europeans were angry; and Rudd was in the dark, 'because I'm still in the room, negotiating away,' Rudd recalled. And in that room there was no final text for the Copenhagen Accord and Rudd was determined to get one. The Chinese and Indian negotiators 'were being absolute bastards', he said.

Rasmussen, Rudd and the core group of leaders still had to labour on in the small room for hours to finalise an agreement that would be acceptable to the nearly 190 countries that had to approve it after Obama flew out. They finished at 1 am. When Rudd and his climate change minister, Penny Wong, finally emerged to face the media, they both looked ashen. 'This was really, really hard,' said Rudd. He put on a brave face; Wong could not—she was shattered.

'This represents a significant global agreement,' Rudd said heroically. 'A huge amount of work still remains to be done. But the alternative, which we confronted, staring into the abyss at midnight last night, with these negotiations collapsing altogether [was] throwing back all progress that has been reached in recent times in global climate change action.'

In reality the deal was a terrible defeat for Rudd. Behind the scenes, he had been working for months with the Danish prime minister: cajoling leaders, lobbying Obama, trying to get on board the Small Island Developing States, like Tuvalu, which were demanding a climate agreement that kept the temperature rise below 1.5 degrees rather than 2 degrees Celsius. The Australian delegation under Wong had worked doggedly on the technical details of the accord to get a formula for both developed and developing countries to register their commitments to reduce emissions.

The weak result was not what either of them wanted. 'I remember seeing him in the room, the small room,' Wong recalled. 'There was just he and I for Australia. He was outstanding, he was committed, he was determined and he worked really hard. And he was really devastated.'

Exhausted and angry, Rudd held a background briefing for a group of Australian reporters, where he lashed out, savaging the Chinese delegation in language he thought would never see the light of day, but it did. 'I did not say these "ratfuckers",' Rudd insisted later when the briefing was leaked. He did admit accusing the Chinese and Indian delegations of ratfucking the talks. 'That's different from me saying the Chinese people or government were a bunch of RFs,' he said. Rudd had not slept for two days, and insisted it was a throw-away line. 'I said how we'd got ratfucked on this, that and the other.'

The environment groups, which had come to Copenhagen with high hopes, slammed the unambitious accord. They handed out a leaked UN report showing that the emissions cuts volunteered by the leaders at the summit would put the world on a path to 3 degrees Celsius warming, enough to risk dangerous climate change.

The final humiliation came when the Danish prime minister went to the main conference floor at the Bella Center to get all countries to endorse the Copenhagen Accord. He was shouted at on the stage, accused of pushing an 'illegal' agreement, a backdoor deal. A small group of developing countries, which many believed were acting on behalf of China, blocked its passage. A diplomatic face-saver was found with the help of the Brits: the Copenhagen Accord was 'taken note of', but it had no legal status.

Rudd was right about one thing: the Copenhagen Accord

kept the UN climate talks alive. It included a $30 billion fund for poor countries to adapt to climate change, and promises of a $100 billion a year fund in the future. More importantly, it would pave the way for the next attempt at a real agreement in Paris. But on that bitterly cold morning in December 2009, Rudd had to confront the reality that Copenhagen, in the eyes of the world, was a failure. At home it was a huge political blow.

The opposition's gamble that Copenhagen would be derailed had paid off. A double dissolution election on climate change and the emissions trading scheme would now be much harder. The prime minister got that message loud and clear while he was still in Copenhagen. After leaving the world leaders at 1 am, sleep deprived and dispirited, Rudd took a call from Sydney. The Labor Party's national secretary, Karl Bitar, was on the line.

Bitar had been working away on the early election plans with Labor's veteran political strategist, Bruce Hawker. Bitar put Rudd on speakerphone. They were worried. He told Rudd they'd been watching focus groups of voters in the southern Sydney suburb of Hurstville. 'Copenhagen was a disaster,' said Bitar—that was how it was being seen at home. The headline in Rupert Murdoch's *Daily Telegraph* said it all: 'COP OUT'.

Tony Abbott's attack lines were working. The carbon price in the emissions trading scheme was being seen as 'a great big tax', said Bitar. 'The debacle of Copenhagen had cut through to the average voter who'd had very high expectations. As it fell apart the sentiment was, "How is a carbon price in Australia alone going to have any impact?"' Bitar recalled.

Rudd had a recollection that Bitar was with Senator Mark Arbib, a junior minister and party powerbroker who worked closely with the national secretary, and that he told Arbib to 'calm down and fuck off'. But Bitar said later that Arbib was not there. He and Hawker remembered a very different conversation with

Rudd. Bitar said the prime minister listened carefully to what he had to say. He told Rudd, 'If Copenhagen is not going well and you want to ditch the CPRS, here is your opportunity to do it, but you would need to replace it with something serious.' This was Hawker's view as well. Rudd agreed to talk when he got home.

A few days later, Rudd arrived at a meeting of Labor's top political strategists in his Sydney office. It was 23 December, two days before Christmas. He looked exhausted. Around the table were Bitar and Arbib, along with Deputy Prime Minister Julia Gillard, Treasurer Wayne Swan and Senator John Faulkner, the defence minister and the party's most trusted operator. The climate change policy, the CPRS, Copenhagen and an early election were on the agenda.

Bitar had a detailed election brief: data on marginal seats, advertising, polling and research from focus groups. Labor would win an early election, he told them, but not increase its majority. Labor's primary vote was still solid and Rudd still had a big lead over Abbott as the preferred prime minister, but Labor was losing votes to both the Liberals and the Greens.

Labor was still vulnerable to the opposition's attacks over asylum seekers. But Abbott's assault on the emissions trading scheme, a great big tax on everything, was also making a real impact on voters. If Rudd wanted to go to a double dissolution election on climate change and the CPRS he should go now, Bitar advised. 'If you stick with it for six months you will bleed.' That meant calling an election straight after the summer holidays, at the end of January.

The early election strategy was backed by Faulkner and Swan. 'My view was that if we were going to progress it, we should go to a double dissolution,' said Swan. But he was worried about Rudd's condition after Copenhagen. 'Rudd was all over the shop.'

The deputy prime minister held her counsel. Gillard was not convinced they should risk an early election on climate change.

She read Bitar's advice differently. 'Karl's report told us that carbon pricing was not breaking the government's way,' she recalled later. Labor was being wedged between the Liberals, who argued Labor should drop the policy, and the Greens, who were saying it was too weak. She doubted Rudd 'was in the zone to fight an election' after Copenhagen.

Gillard also had a technical worry about whether there were grounds for a double dissolution election. The final bill Labor had put to the parliament was quite different from the original because of all the concessions they'd made to Turnbull. That, she believed, could be a problem.

Faulkner was beside himself. He was adamant Rudd needed to go to an early election and get the climate change policy through. 'I was pleading with him to go to the people on the issue,' said Faulkner. Rudd refused to make a decision—except to keep the early election option open.

In early January, Rudd and Gillard met again. Gillard was about to go on leave and there was a backlog of problems to be dealt with, as well as the early election issue. When she arrived at Kirribilli House, the prime minister's Sydney residence, the sun was sparkling on the harbour and the two of them sat on the verandah for Christmas cake and coffee. But Rudd said there was little warmth between them.

What happened at that Kirribilli meeting would be disputed ever afterwards, but Rudd claimed Gillard made it clear she would not support an early election on climate change. 'She has a definitive conversation with me about never supporting a double dissolution on the basis of the failure to pass the CPRS,' said Rudd. He also claimed Gillard told him that day they needed to abandon Labor's core climate change policy altogether, 'otherwise we'll be walking into an election year with a huge target painted on our foreheads'.

If this was true, it was a stunning demand. Dropping the policy would be a humiliating backflip for Rudd on top of the Copenhagen failure. It would be hoisting a white flag and surrendering to the climate-sceptic opposition leader, Tony Abbott, rather than fighting for his own policy at an election.

Gillard disputed this version of events. She believed Rudd did not want an early election when they met that day. 'I was left with the clearest of impressions that he neither wanted to nor was in the right state to go out and fight an election.' Whatever was said between the two of them that afternoon, it was the beginning of the end of the climate change policy and Rudd's government.

Throughout January, Faulkner remained convinced Labor's best chance to keep the climate policy and defeat Abbott was an early election. He asked Rudd to call a cabinet meeting; he canvassed the cabinet members. Wong was supportive. 'I had a discussion with Penny about this, Penny was on the side of the angels,' said Faulkner. But enthusiasm among most others dropped away. Rudd said nothing to support it.

Faulkner warned anyone who would listen that they were making a terrible mistake. 'I said at the time, "I fear I am part of the greatest strategic fuck-up ever made by a Labor government," words to that effect. I could see the writing on the wall for us not taking the opportunity to have an early election that we could win.'

On 7 January, Rudd went on a much-needed break. When he returned, the deadline to call the early election, Australia Day, 26 January, was looming. Still Rudd made no decision. He turned his mind to health policy and launched a children's book he had co-written with the actor Rhys Muldoon. *Jasper & Abby and the Great Australia Day Kerfuffle* was billed as a tale about the PM's pets averting a disaster on Australia Day at the Lodge. Rudd was photographed holding Jasper the cat in his lap. Gillard and Swan were incredulous.

Swan gave up. The only real opportunity for an election was March, unless they were going to crash the May budget. After Copenhagen he believed Rudd just didn't have it in him to fight an early election to save his signature climate change policy. Rudd had called climate change 'the defining challenge of our generation', but when faced with risking his prime ministership, he choked. 'There is absolutely no question he choked,' said Swan. 'And he went off and wrote a children's book. And he wouldn't talk to people about it.'

Rudd insisted he was holding open his options because he did not have the support of his senior cabinet members, most importantly Gillard. By the end of January, any real prospect of calling an early election was disappearing.

Faulkner, a former environment minister, was by now deeply worried about the politics around the climate change policy. He felt Labor was losing the public momentum. 'The more we moved away from the science, the more we struggled. The debate was not about climate change anymore but almost exclusively about the mechanism,' said Faulkner. 'The CPRS was just the mechanism to do something. Virtually no one in the public had a clue about the mechanism; the real problem was the planet was getting hotter.'

―――――――

The new opposition leader, Tony Abbott, wasn't convinced the planet was getting hotter. Even if it was, he wasn't sure it was caused by burning fossil fuels. The scientific advice Abbott relied on came from a cantankerous geology professor at the University of Adelaide, Ian Plimer.

'Now not everyone agrees with Ian Plimer's position,' Abbott told the ABC's *Four Corners*, 'but he is a highly credible scientist and he has written what seems like a very well-argued book refuting most of the claims of the climate catastrophists.'

Plimer's book, *Heaven and Earth: Global Warming—The missing science*, had been launched in 2009 at the grand old Windsor Hotel in Melbourne. The evening was hosted by John Roskam, the head of the Institute for Public Affairs (IPA), whose board now included key Victorian Liberal Party powerbroker, Michael Kroger. The guest speaker at Plimer's launch was the former chair of BHP, Sir Arvi Parbo, Hugh Morgan's old mentor.

Plimer challenged the very basics of climate science—that humans could cause global temperatures to rise. 'The hypothesis that human activity can create global warming is extraordinary because it is contrary to validated knowledge from solar physics, astronomy, history, archaeology and geology,' Plimer said.

While the Adelaide professor had the ear of the opposition leader, the IPA and the Lavoisier Group, he did not impress many mainstream scientists. Reviewing *Heaven and Earth* in *The Australian*, Professor Michael Ashley, from the Department of Astrophysics at the University of New South Wales, wrote that Plimer's book had done an enormous disservice not only to climate scientists but to science in general. 'It is not "merely" atmospheric scientists that would have to be wrong for Plimer to be right,' said Ashley, 'It would require a rewriting of biology, geology, physics, oceanography, astronomy and statistics.'

Despite this withering criticism, by January 2010 Plimer and his fellow sceptics were getting a big run in the Australian media, thanks in part to a rolling global 'scandal' that was hailed by some sceptics as proof that climate science was backed up by dodgy studies.

'Climategate', as it was called, broke a few weeks before the Copenhagen summit, when someone hacked thousands of emails and documents from the Climatic Research Unit at Britain's University of East Anglia and secretly uploaded them to a Russian server. Within days the emails were picked up by sceptic websites in Australia, Canada, the US, New Zealand and other countries around the world. Within two weeks, the 'Climategate' story had appeared on 28,400,000 internet web pages.

Many of the hacked Climategate emails came from two internationally recognised IPCC authors, Professor Phil Jones and Professor Michael Mann. Jones ran the Climatic Research Unit at East Anglia University while his American colleague, Mann, worked at Pennsylvania State University. Jones's work dealt with the instrumental record on global temperature rise, while Mann had created the famous 'hockey stick' graph used by the IPCC in its 2001 report to illustrate the temperature rise after the Industrial Revolution. Both scientists had been hounded by sceptic scientists for years. As Mann put it, he was on the front lines in the climate wars.

At first glance, some of the leaked Climategate emails between the climate scientists were disturbing. They used words like 'trick' in a discussion about inconsistent data; they raged about a sceptic scientist and talked about sceptic science papers getting rejected. The emails also revealed that the Climatic Research Unit had frustrated some Freedom of Information requests from sceptic scientists.

Britain's *Telegraph*, the *New York Times*, the *Washington Post*, the BBC and Fox News all ran with the story. Fox News declared it would 'undercut the whole scientific claim for man's impact on global warming'. Climategate ran right through December and January, especially in the Murdoch media. For the more extreme climate sceptics, the story was their Waterloo in the climate wars. It ramped up just as senior figures in the Rudd government were beginning to have doubts about sticking with Labor's climate policy.

In Australia, News Corp's Andrew Bolt jumped in early on the Climategate story in his popular column that ran in Melbourne's *Herald Sun*, Sydney's *Daily Telegraph* and Brisbane's *Courier-Mail*.

Bolt had no doubt what Climategate really meant: 'It means the "consensus" of scientists we hear of may not actually exist and the IPCC reports cannot be trusted to be balanced,' Bolt thundered. 'It means that claims we've never been hotter are false or unproven, and much research now needs to be rechecked. And at heart it means global warming theory is too weak to accept, being contradicted by a decade of climate.'

For Bolt, Climategate was also the nail in the coffin for Labor's emissions trading scheme. 'Tell me why Kevin Rudd says the science is still good enough to hit us with a colossal tax on gases to "stop" a warming that's actually stopped already?' Bolt ended by commanding his readers: 'Heed this fraud. End this farce.'

Numerous independent inquiries would later make clear Climategate's leaked emails were often published out of context and sometimes deliberately misinterpreted. The emails did not undermine the IPCC findings or the scientific evidence for global warming. A UK parliamentary inquiry and an independent government review cleared Jones and his team of any scientific misconduct. Michael Mann was also cleared by a board of inquiry at Pennsylvania State University.

But the sceptics' case got another boost that January with the revelation that a study on Himalayan glaciers, referred to in one of the IPCC's 2007 reports, gave the year for the glaciers' possible disappearance as 2035. This was not based on a peer-reviewed paper and considered so unlikely the error was picked up by other scientists. But it was a howler, and it was jumped on by the sceptics.

The Climategate controversy got a big run in the Australian media. But as it kicked on, Professor Will Steffen at the Australian National University's Climate Change Institute called for calm, predicting that Climategate 'will have zero impact on the scientific case for climate change'. Those pushing the Climategate story, said Steffen, were overlooking the obvious. After decades of climate research, the evidence for the warming of the planet was not relying on the work of a few scientists, but on thousands upon thousands of scientific studies from around the world.

'This evidence,' said Steffen, 'includes rising ocean temperatures, reduction in Arctic sea-ice thickness and extent, the melting of permafrost, the satellite measurements of rising atmospheric temperature, the loss of ice mass in Greenland and more recently Antarctica, and thousands of ecological case studies on land and in the ocean showing changing times for ecological events like the flowering of plants and mating of organisms, the migration of fish, plants, birds and many others in response to the warming environment.'

The sceptics were dismissed as mad uncles by many climate scientists and environment groups, but their Climategate campaign that December and January supercharged the failure of Copenhagen. The two events appeared to have a big impact on the Australian public. By 2010 a Lowy Institute poll found the number of Australians who believed climate change was a serious and pressing problem had dropped to 46 per cent from 68 per cent just four years earlier.

Sceptics like Plimer were now being feted by some of the most powerful figures in the country. In February 2010, the IPA organised an Australian tour for Plimer and the British sceptic contrarian Lord Monckton, who famously called climate science 'essentially a baseless scientific scam'. When their tour hit Perth in Western Australia, the local contact was listed as Daphne Dhimitri, the personal assistant to Gina Rinehart, Australia's richest woman.

Rinehart was a climate sceptic and a big fan of Professor Plimer's work. The daughter of the eccentric Australian mining entrepreneur Lang Hancock, Rinehart had parlayed her inheritance into a billion-dollar fortune. And she had no qualms about using her wealth and influence to promote her own interests.

Apart from her rich iron-ore holdings in Western Australia, Rinehart's Hancock Prospecting held large coal exploration licences in the Galilee Basin in Central Queensland. The licences had languished for years because the deposits were too far from any port or rail line to profitably develop a mine. But that year Rinehart hoped this was about to change.

With the support of Queensland's Labor government, Rinehart was banking on at least one coalmine, the Alpha Project, getting the green light. Her plan was to get the necessary government approvals and persuade a rich foreign investor to buy into the plan to build one of the biggest new coalmines in Australia. One thing Rinehart didn't want was an emissions trading scheme putting a price on carbon. In 2010 the richest woman in Australia threw her support behind the climate sceptics. The carbon club had a powerful new member.

———

When federal parliament returned on 2 February 2010 the future of the Labor government's emissions trading scheme hung in the balance. Wong and the assistant climate change minister, Greg Combet, again introduced the CPRS bills. Everyone knew they were headed for defeat. Rudd would now have to make the final decision: step up and call the double dissolution election or defer the policy indefinitely. Without the CPRS, there would be no emissions trading scheme and no price on carbon pollution in Australia.

When Rudd faced Abbott in question time that day, the new opposition leader had one line of attack—the CPRS, the great big tax on everything. He taunted Rudd to go head to head with him on climate change. 'If this, as the Prime Minister says, is the greatest moral issue of our time, it deserves to be debated,' said Abbott.

That morning Abbott had released his own climate change policy. It ditched the emissions trading scheme and a price on carbon. Instead, he promised a new 'Direct Action' plan that would pay the big companies to reduce their emissions. It also had incentives for farmers to store carbon in soil, for household rebates on solar panels and hot water systems, and for a lot of tree planting. It was, in effect, a return to John Howard's 'No Regrets' policy—do nothing that could put a burden on the big greenhouse-polluting industries.

Abbott claimed his plan would achieve the minimum target Rudd promised for Copenhagen, to cut Australia's emissions 5 per cent by 2020, but without putting a price on carbon. Turnbull slammed the plan as 'a recipe for fiscal recklessness', while Rudd scorned it for 'letting polluters off scot free'. Many doubted it could meet the 5 per cent target. But the politics were clever and the contrast stark. 'Our policy will deliver the same emissions reductions as the government, without the government's great big new tax,' Abbott claimed.

Abbott's Direct Action plan owed a great deal to his shadow climate change and environment minister, Greg Hunt, a super-smart law graduate who had been a parliamentary secretary in the Howard government. Hunt was the Liberal Party's credible face on climate change. He believed in climate science and renewable energy targets, and had supported an emissions trading scheme under Turnbull. But Hunt was a very pragmatic and ambitious politician. He had come up with a plan that allowed Abbott to campaign against Rudd's 'great big tax' right through to the next election.

By the end of that first parliamentary week, Labor was spooked. Rudd's chief of staff, Alister Jordan, called the party's national secretary and asked him confidentially to do voter research on the CPRS. They needed to know the reaction if Labor put its signature climate change policy on hold, possibly until after the 2013 election.

Bitar quietly commissioned a round of focus groups in Brisbane. The response was brutal. While voters didn't like the scheme, they hated the idea of Rudd betraying his policy. If it wasn't handled very carefully, walking away from it could seriously damage Rudd's credibility. In an email to Jordan, Bitar laid out his advice: 'Given how hard KR has supported and defended the CPRS it is important to get the timing, justification and language right or delaying it might neutralise the CPRS as an issue but do KR a lot of damage in terms of what he stands for.'

Bitar said Rudd should wait until the opposition blocked the CPRS in the Senate a third time before announcing that he was putting the scheme on hold. Critically, Bitar advised, Rudd needed to be ready with a big new package on renewable energy as an alternative policy. A few weeks later, the prime minister's staff quietly began exploring the options with Gillard, Wong and Swan about what to do when the Senate blocked the CPRS.

Gillard wanted it dumped altogether until the opposition was willing to give it bipartisan support. Bitar briefed Wong on polling showing voter opposition to the scheme. She reacted angrily. 'Instead of just polling on the CPRS, they should poll on what does it mean if the prime minister stands for nothing?' she warned.

Rudd kept stalling until April, when Swan and Gillard demanded a decision. Swan needed to finalise the May budget and the CPRS was a $9 billion item that couldn't be ignored. By now, both ministers wanted it dropped from the budget. 'We can't afford it politically. And we certainly can't afford it financially,' Gillard told Rudd.

Rudd wanted Wong to have a final say before he agreed to put the policy on hold. Rudd, Gillard, Swan and the finance minister, Lindsay Tanner, got Wong on speakerphone while she was in transit, returning from climate talks in Washington. She burst into tears, warning them all of the damage it would inflict on Rudd personally and the government once the CPRS disappeared from the budget: 'the immediate conclusion of the public and the commentariat is that we no longer care about climate change,' she said.

They all agreed the backdown needed to be handled with extreme care. But that didn't happen. A few days later, on 27 April, the story leaked before Rudd could even get his act together. 'I was perfectly poleaxed,' he said later, blaming Gillard for the leak, an allegation never established but pushed by Rudd. 'As a piece of pure Machiavellian politics, Gillard scores ten out of ten,' he said bitterly. The *Sydney Morning Herald*'s Lenore Taylor had sensationally revealed the CPRS had been put on hold for what looked like naked political reasons, 'in a bid to defuse Tony Abbott's "great big new tax" attack in this year's election campaign'. Rudd was touring a hospital in western Sydney when the media pack caught up with him, forcing him to stumble through an unconvincing press conference.

Wong watched the debacle unfold, paralysed: 'Our phones were going nuts and I'm sitting in my office saying there's nothing I can do, I just have to sit here because we didn't have a position. I couldn't do any media. It wasn't a great day.'

It was hugely damaging for Rudd, said Wong. 'You can't tell someone for years how important this is and then act in a way that is utterly contradictory to that because it's not just the thing—it's actually saying something about you. And this was his problem. It was inexplicable,' she said. 'I just had this feeling that we would never recover.'

Her grim reaction was shared by the head of the Climate Change Department, Martin Parkinson, who had fought for more than two years to get the policy over the line. He was completely gutted. 'It, to me, was the final nail in the coffin. That we'd just completely failed on dealing with this,' he said. 'Rudd had made this the moral challenge of our generation and then, when he walked away from it, never gave any explanation to the public about how this is still the greatest moral challenge of our generation but he was walking away from it. And I just kept watching that and thinking, "He's completely lost the public now."'

The reaction from voters was immediate. Rudd's approval rating plummeted from 50 per cent to 39 per cent in Newspoll. To many it looked as if Labor had surrendered its climate change policy under attacks from Tony Abbott, the man who called climate science 'crap'.

It was less than five months since Rudd had flown to Copenhagen, carrying the country's high expectations. Now his credibility on climate change was in tatters. His government had one of the weakest 2020 targets to cut emissions of any developed country and no plausible way to meet even that.

The collapse of Rudd's credibility triggered the rapid unravelling of his government. Just months later, in June 2010, Gillard launched a leadership coup against him, backed by key members of the party's right, including Bill Shorten, the former head of the Australian Workers' Union and a parliamentary secretary in Rudd's government. Gillard's rise to the prime ministership left deep and bitter scars that would inflict serious damage on her and her new government. Rudd and his supporters would ruthlessly destabilise her leadership.

When Gillard called the election for August that year, Labor's climate change policy was still in limbo. In a bizarre move, she promised a 150-person 'citizens' assembly' to work out what the policy should be after the election. When this was met with derision, Labor scrambled to find a credible stand. In one of her last election interviews, Gillard said she was prepared to look again at putting a price on carbon in an emissions trading scheme if she had community support, but she ruled out a carbon tax.

A few days earlier, in the face of a new round of attacks from Abbott, she made another promise in an unscripted interview with Channel Ten's news: 'There will be no carbon tax under the government I lead.'

Gillard would forever regret those words. They would be used with devastating effect to launch the next battle in Australia's climate wars.

CHAPTER TEN

BATTLELINES

WINTER IN CANBERRA had never been this exciting for the Greens. The bitter 2010 election had delivered a hung parliament and the joint was in turmoil. Neither big party had won enough seats to form government. The Greens had snatched the vital inner-city seat of Melbourne from Labor, so if Julia Gillard wanted a shot at power, she needed their support. That's why, a few days after the election, the Greens' deputy leader, Senator Christine Milne, was striding across the polished walkways of Parliament House on her way to the prime ministerial wing. She was going to seal a deal with Gillard that would change history.

The daughter of a Tasmanian dairy farmer had come a long way before ending up in Gillard's office. As a young school-teacher, Milne had thrown herself into the campaign to save the state's wild Franklin River from being drowned by a huge dam. That campaign, spearheaded by the Greens' leader, Bob Brown, became the most iconic environmental battle in Australian history. Milne was one of 1500 protestors arrested and was imprisoned in Hobart's Risdon gaol along with the British naturalist David Bellamy and novelist John Marsden. That was over a quarter of a century ago, but it was the defining experience of her young life and catapulted her into Green politics.

Despite her reputation as a strident activist, Milne had already worked with minority governments in Tasmania. Those experiences had ended badly and she was keen not to repeat them. When Bob Brown and Gillard first met after the 2010 election, they both wanted a deal to help Labor form a minority government. But Milne wanted the details nailed down before she would go along with it.

The Greens had one overriding demand—they wanted a climate change policy that put a price on carbon pollution. Milne had two other conditions. She wanted a multi-party committee to negotiate the climate policy, and she wanted a personal assurance from Gillard the carbon price legislation would be put through the parliament by July 2012. This deadline was critical, because until 2013 the Greens' nine senators were going to have the balance of power in the upper house. Together with Labor they would have the numbers to finally put a price on carbon, so long as it was done in this term of parliament.

What the Greens were proposing guaranteed another head-on battle with the carbon club. But Gillard agreed without batting an eyelid. 'That's what she knew was going to be the price of government and so she didn't argue,' Milne recalled. 'She was desperate to get into government. She had taken over the leadership from Rudd, gone into the election and ended up hanging on by her fingernails with minority government.'

Tony Abbott was just as desperate to become prime minister, but he was outmanoeuvred by Gillard and her negotiating skills. Like Abbott, Gillard had lived and breathed politics since university. But unlike Abbott, she also had a warmth in private that could win over opponents and keep supporters loyal. Before coming to parliament, she had also been a sharp industrial lawyer and a canny political adviser. She knew how to cut a deal.

Gillard used the Greens to seize the political momentum and secure minority government. With the Greens' lower house seat in the bag, Gillard won over three of the four other independents. Two of them, Tony Windsor and Rob Oakeshott, both from regional New South Wales, agreed to work with her on climate change.

From the beginning of their improbable political relationship, Milne was impressed with Gillard. 'I always found her personable, respectful and pleasant, even when we totally disagreed on things,' said Milne. 'She didn't resort to nasty personal abuse or anything like that. The thing I liked about her was when she said that she would do something, she then went ahead and did it.'

True to her word, Prime Minister Gillard chaired the first meeting of the multi-party committee on climate change just weeks after forming government in September 2010. She persuaded a reluctant Greg Combet to go back into battle and take on the high-stress job of climate change minister. He was pessimistic that a climate agreement with the Greens was possible, but he and Milne were now deputy co-chairs of the new committee, two extremely unlikely political allies.

As part of their deal, the Greens were allowed two advisers on the committee. Milne picked Professor Ross Garnaut, Labor's erstwhile economic adviser on climate change, and Professor Will Steffen, the head of the Australian National University's Climate Change Institute. Steffen was a critical choice for Milne.

Under Kevin Rudd, Labor had signed up to the Copenhagen Accord promising to keep global temperatures from rising more than 2 degrees Celsius, and Milne was determined to use this as a battering ram in the negotiations with Labor. 'I needed a scientist there to keep everybody on the straight and narrow on what 2 degrees actually meant,' said Milne.

Predictably, the first big fight on the committee was over

Labor's weak target to cut Australia's emissions—just 5 per cent by 2020. The 5 per cent target was far less than the average target for rich countries proposed at Copenhagen if the global temperature rise was going to stay below 2 degrees. While Labor claimed it could increase the target to a 15 per cent cut, this was never really on the cards. Milne wanted a 25–40 per cent cut in Australia's emissions by 2020, in line with the range of the IPCC advice on targets for rich countries. Combet, Gillard and the treasurer, Wayne Swan, dismissed the upper range out of hand as a recipe for wrecking the economy.

After months of wrangling, both sides agreed to an uneasy compromise. The 2020 target was put to one side. A new agency, the Climate Change Authority, would advise on it later. In the meantime, the price on carbon pollution would be set within a new emissions trading scheme. To avoid the need for a target, the government would put a fixed price on the permits to pollute for the first three years of the scheme. The three-year fixed-price period would give everyone a chance to see how it was working. After the three years was up, the government would move to a market price guided by the new 2020 target, which hopefully would be agreed to by then.

The only problem with this clever compromise was that the three-year fixed price could be portrayed by the opposition as a carbon tax. Gillard was on camera before the election promising voters: 'There will be no carbon tax under the government I lead.' That made the compromise solution very difficult politically, but it was the best Labor and the Greens could come up with.

By now Combet was desperate to start negotiations with the big corporate emitters. He knew from bitter experience the talks would be rough, and he was on a tight deadline. When parliament returned in February 2011, he urged Gillard to announce the deal so he could get started.

Gillard war-gamed a strategy with her staff, knowing the announcement would trigger a ferocious political assault from Abbott and the opposition. It would also upset some on her own side, including some of the big unions. Even so, she agreed to unveil the deal at a joint press conference with the Greens, Windsor and Oakeshott in the prime minister's courtyard.

'I'm determined to price carbon,' Gillard told the media as she stood shoulder to shoulder with Bob Brown. 'History teaches us that the countries and the economies who prosper at times of historic change are those who get in and shape and manage the changes. The time is right and the time is now.'

Brown was beaming at her as the cameras rolled. 'This agreement is the Greens in action,' he said, 'delivering certainty to the Australian economy, community, investors and the environment after productive negotiations with the Government.'

It was a political gift for Abbott. He pounced on the images of Gillard and Brown standing together and slammed the deal as 'a carbon tax' imposed on Australia by the Greens. 'Well it's pretty clear who is leading this government—it's Bob Brown,' he said.

In parliament that afternoon, Abbott tried to censure Gillard 'for breaking her solemn promise to the Australian people'. He threatened to bring down her minority government. And in a taste of the personal attacks to come, he compared her to the murderous Lady Macbeth—'in the still, small hours of the night when she considers what she has said and done, like some latter-day Lady Macbeth would consider the statement "There will be no carbon tax under the government I lead" and say "Out, foul spot!" "Out, foul spot!"' he piled on.

Gillard brainstormed the attacks with her advisers. She was due on the ABC's *7.30 Report* program that night. Both Combet and his chief of staff, Allan Behm, joined the session. Behm was a long-time Defence Department official with years of government experience behind him. Combet and Behm argued there was a critical difference between a carbon price and a carbon tax. You could avoid a carbon price if you reduced your emissions through energy efficiencies or cleaner energy, but you can't avoid a tax. If Labor conceded it was putting on a carbon tax, it would play right into the opposition attacks.

Gillard went into the ABC studio that night and blew it. When reporter Heather Ewart asked her bluntly, 'You do concede it's a carbon tax, do you not?' Gillard barely hesitated, 'Oh, look, I'm happy to use the word tax, Heather. I understand some silly little collateral debate has broken out today. I mean, how ridiculous. This is a market-based mechanism to price carbon.' When she tried to explain her confused answer, it just got worse: 'It has a fixed price period at the start, a price that will be fixed. That is effectively a tax and I'm happy to say the word tax,' she said.

When Behm heard the prime minister, his heart sank. 'The most profound mistake we made was when Julia called the price a tax in her conversation with Heather Ewart,' he said. 'In that one interview, Julia managed to pull the capstone out of our climate change policy architecture.'

Unlike Rudd, Combet and Wong, Gillard was a latecomer to the brutal politics of climate change. While she understood Abbott would try to call any carbon price 'a great big tax', she didn't fully understand that the strategy—to deliberately conflate a carbon price with a 'tax' on ordinary voters—had been workshopped for years in Washington and Canberra by right-wing think tanks, US Republicans and the Coalition. In one interview, the prime minister had surrendered the argument to them.

As Gillard saw it, she was being up-front with voters by explaining that a carbon price on the big corporate emitters would be passed on to consumers. That's why the government planned to compensate low-income earners for rises in their electricity bills. Gillard would later see it as an epic mistake but she also believed whatever she said that night would not have stopped the unholy campaign that was unleashed against her and the government.

Her deputy, Wayne Swan, agreed: 'If you stand back and look at the nature of the campaign, the sheer weight and the bastardry of the people that were going to bring that down, that wouldn't have mattered,' he said.

That was probably true, but Gillard handicapped her own defence. The opposition set out to mobilise a 'people's revolt' against the climate change policy by slamming Gillard as a liar, which quickly morphed into 'Ju-liar'. Blooded by their success against Malcolm Turnbull and Kevin Rudd, Abbott and his inner circle believed a populist 'Axe the Carbon Tax' campaign would kill off this latest attempt to put a price on Australia's greenhouse gas pollution.

'I don't believe it's going to happen because I think there will be a people's revolt,' Abbott predicted. 'They will see this as an assault on their standard of living, which is exactly what it is. Every time you turn on the lights, you will pay under Labor's carbon tax. Every time you go to the petrol pump, you will pay under Labor's carbon tax.'

If Abbott was right, with this campaign he could destroy the climate policy and bring down Gillard's minority government. He was determined to give it everything he had. 'We will fight this every second of every minute of every hour of every day of every week of every month,' he promised.

Abbott launched his people's revolt against the carbon price at exactly the right time. That summer the Millennium Drought had broken in spectacular fashion. In December and January the rain poured down. Widespread flooding hit Queensland, where 33 people died and much of the state was declared a disaster zone.

The rest of the country, too, was deluged. For the next two years, Australia would have its wettest period on record as the country experienced an unusually strong La Niña. The extreme weather was in keeping with IPPC predictions, but for many Australians it washed away the urgency of climate change that had peaked during the years of crippling drought and bushfires.

———

Within a short time of Julia Gillard announcing the deal with Bob Brown, a new website with the domain name stopgillards carbontax.com popped up on the internet. One of the people behind it was sitting in Washington. Tim Andrews, the conservative activist who had trained with the Koch Brothers' internship program, had once again joined forces with Liberal senator Cory Bernardi. They had worked together to help stop Rudd's carbon emissions trading scheme and now they would take up the battle against Gillard.

The two helped create the 'people's revolt' against the climate policy, using the power of social media and the tactics of the Tea Party movement that was sweeping through the US Republican Party. One of the driving ideas behind the campaign was to exploit the anger and disaffection among ordinary voters towards politicians.

'That was a unique thing that was so different about this campaign,' Andrews recalled. 'And it was somewhat of a precursor of what you're seeing even more now by people who felt disenfranchised by the political system and thought that they were not necessarily being listened to by a lot of political elites.' As Andrews saw it, their most powerful weapon was the message that Gillard had lied. '"There will be no carbon tax under the government I lead" was the death knell,' he said. 'It was the issue of being lied to that infuriated people.'

Andrews and Bernardi were central players in the 'people's revolt', along with a core group of conservative activists across Australia. 'There were a lot of people that deserve credit,' said Andrews. 'I think some people tried to give me too much credit because I was living in the United States, and so these evil Koch people are doing things from behind the scenes.'

Andrews' talent was, in fact, working behind the scenes. He brought together a broad front of fractious right-wing groups in Australia to create a powerful show of populist unity against Gillard's policy. The strategy was straight from the handbook of America's most successful conservative political operator, Grover Norquist, founder of Americans for Tax Reform, who notoriously once quipped, 'I'm not in favor of abolishing the government. I just want to shrink it down to the size where we can drown it in the bathtub.' Norquist became an inspiration for a generation of American conservative activists.

'This was very much a Grover Norquist-style campaign,' Andrews explained. 'There's no use in having everyone in silos, not talking to each other, doing their own thing.' Andrews had spent 2008 working with Norquist's outfit in the US, which is funded by some of the country's wealthiest conservative donors. Norquist had a take-no-prisoners style of campaigning and famously praised the role of 'throat slitters' in politics—activists who could get things done.

Throughout 2011, Andrews and Bernardi helped unite the Christian conservative lobby the National Civic Council with the Christian Democratic Party, the Young Liberals, the Liberal Students' Federation, shooters, farmers, the Lavoisier Group, the Democratic Labour Party, the Libertarian Party of Australia and a host of fringe climate-sceptic groups. Together, they formed one broad anti-carbon tax campaign.

From his home in the US, Andrews would coordinate

conference calls every week with twenty or more of the groups. 'It was a somewhat surreal experience organising protests in Sydney at two o'clock in the morning in Washington DC, getting a skywriter to do the carbon tax ad,' he said.

Bernardi helped fund Andrews' political blog, which was called Menzies House; Andrews ran it with a group of like-minded activists including a Bernardi staffer. They set up stopgillards carbontax.com as the hub of the campaign against the emissions trading scheme; it linked to other websites that highlighted what they called the 'Big Lie'.

In Australia, Bernardi's own fundraising and training organisation, the Conservative Leadership Foundation, backed a local activist group, CANdo, to organise ground operations like guerrilla marketing campaigns, sticking carbon tax price rises on supermarket products and printing T-shirts for rallies. Within six months the campaign fostered by Andrews, Bernardi and their conservative allies would culminate in big demonstrations of activists and their supporters in front of Parliament House. While helping foment the people's revolt, Senator Bernardi was officially Abbott's shadow parliamentary secretary; but at the same time he was a boots-and-all activist for the climate-sceptic movement.

The links between the US and Australian climate-sceptic movement had been deepening since Bernardi's first trip to Washington. In October 2010, influential US climate-sceptic group the Heartland Institute held its first international conference outside the US in Sydney. Bernardi was one of its keynote speakers. The conference was part of a talkfest called the Pacific Rim Exchange co-sponsored by Grover Norquist's Americans for Tax Reform and Heartland. The Australian co-sponsor was Melbourne think tank the Institute of Public Affairs (IPA), where the Victorian Liberal Party powerbroker, Michael Kroger, sat on the board.

The Liberal Party was keen for Bernardi and his network to stir up the populist anger against Gillard and the carbon price, but Abbott and his office wanted him kept at arm's length. Some of the activists Bernardi and Andrews were in bed with were fringe players, with links that could embarrass the opposition leader. At one of the early rallies, Abbott and two senior female Liberal MPs were videoed in front of a giant placard reading 'JuLiar, Bob Brown's Bitch'. Events like that made Bernardi's relationship with Abbott fraught.

'I was cannon fodder,' said Bernardi bluntly. 'You want something said that needed to be said [but] we don't want any hands on it? Send Bernardi out. Absolutely happy to have the ability to tap into the resources and get some stuff, but they were happy to turn that tap on and off according to how politically expedient it was,' he recalled bitterly. The opposition wanted a 'people's revolt' against Gillard, but they didn't particularly want to know how it was organised or who was funding it.

By May 2011, the Liberal Party was riding the tiger of right-wing populism, hoping to destroy the emissions trading scheme and bring down the Gillard government. It was the same model the Republicans were embracing in Washington against the Democrats. It was a model that was pushing the conservative parties further to the right, especially on climate change. The 'people's revolt' against Gillard would fundamentally fracture conservative politics in Australia, fostering splinter parties and deepening divisions in the Liberals. It would destroy any chance of uniting the major political parties to face the enormous challenge of climate change.

One key player who understood and applauded this tectonic shift in conservative politics was the ageing godfather of the right in Australia, Hugh Morgan. In 2010 he had taken on the presidency of the climate-sceptic Lavoisier Group, after his old factotum, Ray Evans, was struck down by illness.

In a rousing address to the Lavoisier faithful that November, Morgan credited the Tea Party movement with the Democrats' stunning losses in that month's US mid-term elections. Barack Obama was still in the White House, but the Republicans had regained control of the House of Representatives. As Morgan gleefully noted, the moderate Republicans who supported the push for an emissions trading scheme were defeated by insurgent Tea Party candidates. 'We will see in the US, from now on, a rapid withdrawal from all the follies of the global warming scam,' Morgan crowed.

Morgan's influence in the Victorian Liberal Party was still pervasive. He was now chairman of its cash cow, the Cormack Foundation. He warned Australian business against supporting any new attempt by Gillard to put a price on carbon pollution, flagging that it would be reversed by the next Liberal government. 'Those rent-seekers in Australia, who have been staking out positions for favourable treatment under a Gillard decarbonisation policy should take note,' he said.

By 2011 Morgan had joined forces with mining magnate Gina Rinehart, who had launched her own Australian anti-tax lobby group, crusading against a price on carbon and Labor's mining tax. This lobby, the Australians for Northern Development and Economic Vision, or ANDEV, had the rather astonishing demand that northern Australia, home to much of the country's mineral wealth, should be carved off into a special economic zone favoured with low taxes and looser government regulation, to help develop mining and agriculture.

Hugh Morgan signed up as an early member of the group, together with an eclectic mix of climate sceptics, politicians and right-wing businessmen, including an old ally of Rinehart's father called Ron Manners, a Western Australian miner. Manners was a sponsor of the 2009 climate-sceptic conference of the

US Heartland Institute. Rinehart also enlisted her own favourite climate-sceptic scientist, Professor Ian Plimer, who she later made a director of two of her key companies. Plimer was an advocate of the line that global warming had paused. 'There is no problem with global warming. It stopped in 1998,' he told *The Spectator*. 'The last two years of global cooling have erased nearly 30 years of temperature increase.'

Rinehart was an admirer of Plimer, as her spokesperson explained to *Four Corners* when she was asked about her support for the sceptic scientist. 'Mrs Rinehart suggests that the media should also permit to be published that climate change has been occurring naturally since the earth began, not just the views of the climate extremists,' she said.

Rinehart invited Tony Abbott to dine with her and Plimer and began quoting the opposition leader in her speeches. She was angry about the proposed carbon price and about Labor's mining tax, known as the Minerals Resource Rent Tax (MRRT), on iron ore and coal. She wanted them axed and she was happy to spur on the people's revolt against both taxes. 'The sooner we can, as a majority, let our politicians know that Australians are fed up with wasting taxpayers' money and do not want a carbon tax or MRRT on thermal coal (at least), the better,' she said.

Gina Rinehart was all for a people's revolt if it coincided with her interests, and was going to use her own loud voice to help the cause along.

CHAPTER ELEVEN

THE CARBON WAR

WAYNE SWAN WAS standing at the window of his electorate office in suburban Brisbane when his eyes lit on something unusual behind the local Nundah Village Shopping Centre. 'It was absolutely surreal,' the Labor treasurer recalled. 'I start to see these limousines come around the corner—one limousine, two, three. They all pull in. I think, hell, what's going on here?'

Gina Rinehart had arrived. Swan had just returned to Brisbane after a gruelling week delivering the budget and had reluctantly agreed to a short meeting with the billionaire.

Swan's relationship with Rinehart was still raw after her role a year earlier in the campaign to kill Labor's mining tax. Back then she stood on a flat-bed truck in Perth, replete in pearls and heels, screaming, 'Axe the tax' to a crowd of miners, businessmen and Liberal Party politicians. Swan was not thrilled to see her, which is why he had chosen to meet her in Labor's heartland rather than his ministerial office in Canberra. But Rinehart clearly had her own agenda.

'The entry room starts to fill up with all these people, all these Indian guests,' Swan recalled. 'Suddenly there's not enough seats to fit all these Indian guests!'

Rinehart and her entourage crammed into Swan's modest electorate office, where Rinehart did the introductions. 'These are my commercial partners,' she told him. The group of executives worked for the Indian billionaire G.V. Krishna Reddy, head of India's GVK Group, which owned a fleet of power plants back home. Rinehart was hatching a deal with Reddy for him to buy up the majority stake in her Galilee Basin coal assets in Central Queensland. The planned mines didn't even have approval yet, but she wanted to discuss tax relief from the federal government.

Swan was completely stunned. 'I'm sorry, I'm the treasurer, I can't do this, I can't go any further,' he told her. Swan didn't have any expert advisers with him in Brisbane who could even discuss her mega-coal project. After some awkward small talk, Rinehart got up to leave. On the way out, one of the Indian executives handed the treasurer a large box. Inside was an invitation to the wedding of Reddy's granddaughter in Hyderabad.

Swan declined the invitation, but he later discovered two senior opposition members had happily accepted. The National Party's Senate leader, Barnaby Joyce, and the Liberals' deputy leader, Julie Bishop, flew off in Gina Rinehart's private jet for GVK's big Indian wedding. Both Joyce and Bishop were already ardent campaigners to abolish the mining tax and oppose the carbon price.

Three months later, Rinehart sold her majority stake in the Galilee Basin coal assets for $1.26 billion to Reddy's Indian conglomerate. The mines still had no final approvals and no rail line to move the coal to a port. Despite this, the Indian conglomerate paid Rinehart's company $500 million up-front.

This coal deal would help make Gina Rinehart the world's richest woman. Coal prices were surging and Australian coal exports were on the rise. To Rinehart, to her Indian backers and to her political allies in the opposition, the carbon club looked unassailable. But a deep crack was opening up in the club's

foundations. Some of its biggest members were beginning to seriously question the long-term viability of coal-fired power.

———————

When Julia Gillard first announced that her new minority government would put a price on carbon, Marius Kloppers, the head
of the resources giant BHP Billiton, decided it was time to talk
about climate change. In a groundbreaking speech to the Australian British Chamber of Commerce in September 2010, Kloppers
lined up on the other side of the battlelines to Gina Rinehart.
BHP Billiton, Kloppers proclaimed, 'acknowledges that the mainstream climate science is correct and that we need to stabilise (and
eventually reduce) the carbon concentration in the atmosphere'.

Kloppers went on to warn that Australia's coal-fired power
system had to change. 'Australia will need to look beyond just
coal towards the full spectrum of available energy solutions,' he
said. 'Failure to do so will place us at a competitive disadvantage
in a future where carbon is priced globally.' The resources boss
wrapped up his speech by calling for a price signal to cut carbon
emissions. He suggested a carbon tax or an emissions trading
scheme for coal-fired power stations. His speech pretty much
punched a big hole in Tony Abbott's climate policy.

Kloppers quickly came under friendly fire, and with some
justification. Reading between the lines, it was clear that, if
the Gillard government wanted to put a price on carbon, BHP
would be knocking on the door wanting free permits to pollute
or compensation.

'He came out much to everybody's surprise and made a big
speech saying there should be a carbon price. And then—almost
in small letters—but don't include BHP in there,' said Richard
McIndoe, who was still TRUenergy's chief executive and running
the big Yallourn brown-coal generator in Victoria. 'I just took

that as him promoting his gas business. He was just trying to talk up the gas business, and down the coal business basically.'

There was some truth in this, but Kloppers was also finally turning the page of history for BHP. Despite the setbacks at the Copenhagen climate summit, the global giant could read the writing on the wall. Climate science was a reality that couldn't be ignored by public companies that were answerable to shareholders, banks, insurance companies, stock markets and the courts. The legal risks were too great. For at least a decade, BHP had been toying with carbon pricing. Its chief executive back then, Paul Anderson, had tried to push it, but he'd met with resistance from other resources companies. Now, Kloppers believed, the time had come to make a noise publicly.

While BHP Billiton had big coalmines in Australia and South Africa, it was also a serious player in gas, and Kloppers would soon bet US$12 billion on a shale gas deal in America. Like many global executives in the fossil fuel business, Kloppers believed the shift away from coal power was inevitable, even if it took a few decades. The Golden Age of Gas had arrived. He was banking on gas, which had lower greenhouse gas emissions than coal, as being the boom fuel that would bridge the gap between coal-fired power and renewable energy. There were many in the power business who thought the same thing.

The brown-coal-fired generators in Victoria, which had gone to war with Rudd and Penny Wong, were also shifting position. Richard McIndoe's TRUenergy still wanted coal-fired power plants, but it was also looking to gas and renewables. 'We took the view that an efficient emissions trading scheme was something that we could live with, especially given the more realistic transition period being proposed,' said McIndoe.

So when Greg Combet started serious negotiations with the big greenhouse gas emitters in 2011, he discovered a new mood for change. 'It was going to happen, so let's just get the best deal we can,' was how Combet read their reaction.

The climate change minister got the same feedback from the new owners of Victoria's giant Hazelwood brown-coal power station, the French gas company GDF SUEZ had merged with International Power. Its old local boss, Tony Concannon, was on the sidelines. GDF SUEZ's French chief executive, Gérard Mestrallet, had no problem with a carbon price. 'We understand this, we are in the European emissions trading scheme,' Combet recalled him saying.

There was, of course, a big catch. 'They were all in favour of a carbon price, until they had to pay one,' said Swan bitterly. If they had to pay, the big polluters wanted big compensation and, the higher the carbon price, the more compensation they wanted. The large aluminium companies like Alcoa, with union-ised plants, went about squeezing the government as hard as they could. And they weren't the only ones.

'I used to comment to my staff, probably unfairly, that the American executives, when they do an MBA, they do it in rent-seeking,' Combet recalled. 'They'd come and, quite nicely actually, just threaten you that they were going to shut down: "Too bad about the seat of Corangamite in Victoria—a thousand jobs lost there, plus the supply chain."'

———

The level of compensation demanded by the companies almost blew up the deal between Labor and the Greens. In the middle of 2011, a brawl over the amount being demanded by brown-coal power generators came to a head at a meeting in Parliament House between Gillard, Combet and the Greens deputy leader, Christine Milne. It was the winter break and the mood was grim. There was no place open even to buy a coffee.

Milne was upset to see the former head of the Department of Resources and Energy, John Pierce, at the meeting. Labor's energy minister, Martin Ferguson, had appointed Pierce as chairman of the Australian Energy Market Commission, the powerful rule-maker for the whole electricity system. Gillard wanted him and the government's energy security adviser to lay out the case for paying the generators $5.5 billion over five years to keep the lights on. Without that money, they argued, their power companies would be at risk. After hours of wrangling, Pierce was adamant the compensation was 'essential for the management of risks to energy security'.

Gillard told Milne the government couldn't ignore their warnings. 'They are our formal advisers on these matters and, if we don't take their advice, we will be liable in the event blackouts occur.'

Milne was devastated. She was the Greens deputy leader and about to give the nod to paying billions to the most polluting coal-fired power generators in the country. She asked herself, 'Should we just pull the plug on the whole thing?' She turned to her adviser, Ross Garnaut, who had tried to fight the generators during the Rudd years and failed. He urged Milne to play the long game. 'Look, Christine, it's only money,' she recalled him saying. 'Yes, it's a hideous waste of money, but you are getting what we want and that is the policy framework that will bring down emissions.'

Still angry, Milne demanded the regulator, Pierce, put his advice in writing. 'People in the future are not going to believe that the official advisory body to the prime minister advised paying compensation to coal-fired generators in an age of climate change,' she said.

Pierce did put his advice in writing. He also recommended paying the compensation in permits to pollute; but in the end they were paid upfront in cash for the first two years of the emissions trading scheme. The energy executives were amazed.

'The decision to pay it all up-front, or to front-load it, was actually a big surprise,' said TRUenergy's Richard McIndoe. He thought the government was simply trying to bulletproof the emissions trading scheme against the opposition. 'There's a school of thought that says once you've paid someone a load of money, it's very difficult to repeal and unpick those payments.'

Milne did score one win in the gruelling negotiations with Gillard and Combet. It was for a project that would have a lasting impact on Australia's efforts to make the big switch from coal-fired power to renewable energy. Oddly, it was born out of another defeat.

Both sides had to agree on a figure for the carbon price for the first three years of the emissions trading scheme. This was the price businesses would pay for permits to pollute. The Greens wanted a high price, around $40 to $60 per tonne of greenhouse gas emissions, in the hope it would drive a rapid shift to renewable energy. Labor wanted a price that wouldn't throw the economy into a tailspin. In the end the Greens leader, Bob Brown, and Gillard settled on a compromise of $23 per tonne.

As part of that compromise, Swan and Combet offered to fund a scheme that had long been of interest to Milne. They agreed to set up a $10 billion financing agency, inspired by Britain's 'green bank' that could lend money to big, innovative, renewable energy projects. The Clean Energy Finance Corporation was the first of its kind in Australia. A former senior executive with Bankers Trust, Jillian Broadbent, was appointed as chair. Its first chief executive was a former senior executive with Macquarie Group, Oliver Yates. Despite the heavyweight business credentials of its managers, the opposition forever tagged it 'Bob Brown's bank'.

The government also set up the Australian Renewable Energy Agency (ARENA) to fund research into renewable energy. These two new agencies, combined with the Rudd government's

20 per cent Renewable Energy Target, would spark a surge of investment in clean energy.

Gillard, Swan and Combet unveiled the Clean Energy Package in July 2011. It was a compromised bundle of reforms that promised big polluters $9.2 billion to protect jobs and industries over three years, but it also raised a lot of money from the polluters. There was money and tax breaks to protect households from the impact on electricity prices, and a plan to begin cutting back on electricity from heavily polluting brown-coal power stations. It was, in the end, the most sweeping reform of Australia's carbon-heavy economy the country had yet seen.

But the fight to get this far had taken an enormous toll on the government and on Combet in particular. He was physically exhausted and punch-drunk from the opposition's attacks. 'The fact that the politics had gone so badly in an area of policy that I was responsible for ate away at me,' he said later. 'I was in about as bad shape as I had ever been.'

Bad shape or not, there was still a fraught battle ahead for the minority government to get the climate policy through parliament. While some big greenhouse players were happy to sign up to the emissions trading scheme, others were not, and they were lining up with the opposition to fight it every inch of the way.

The day the policy was released, the reaction from Abbott, the coal lobby and the Murdoch media's highest profile columnist was savage. The Australian Coal Association took direct aim at Combet and Labor's voters in the coal seats. 'We will see ordinary people from Blackwater to Moranbah in Queensland, right through to the Hunter and the Illawarra in NSW, lose their jobs because of this tax,' the coal lobby's chief, Ralph Hillman, warned.

The *Herald Sun*'s Andrew Bolt was ferocious: 'Julia Gillard's carbon dioxide tax is the most brazen fraud to be perpetrated by an Australian government.'

And Tony Abbott kept hammering his old line. 'It's a tax increase,' he said. 'It's socialism masquerading as environmentalism.'

They were all revving up. The full-frontal assault in the carbon war was about to begin.

Tony Abbott jumped up onto the makeshift stage in front of a giant banner touting www.stopcarbonlies.com. The opposition leader was shouting to be heard above the people's revolt. 'This is the statement that has reverberated around this country,' Abbott yelled. "There will be no carbon tax under the government I lead."' The crowd waved their placards for the cameras, 'No Carbon Tax, Election Now'. On the fringes of the rally the chant went up, 'Ditch the witch! Ditch the witch!'

Abbott was making his big pitch to bring down Julia Gillard's minority government and kill the climate policy. Cory Bernardi strode through the crowd, surveying the 6000-odd pensioners, retirees, small businesspeople and right-wing activists bussed in that morning for the big Tea Party-style rally on the lawns in front of Parliament House. It was a triumph for the campaign he and his Washington ally, Tim Andrews, had done so much to foster.

Word of the rally had spread through social media, helped along by old-school shock jocks from Sydney's radio 2GB. Gina Rinehart's good friend Alan Jones, 2GB's top presenter, had flogged the rally all week. Jones's efforts to bring down the Gillard government were now legion. For weeks he'd been cheering on Tony Abbott and encouraging his listeners' attacks on the prime minister. 'The woman's off her tree,' Jones told them, 'and quite frankly they should shove her and Bob Brown in a chaff bag and take them as far out to sea as they can and tell them to swim home.'

It was August 2011 and the political war over climate change was spinning into dangerous territory. The vote on carbon pricing in the lower house was just weeks away. Insults, abuse and threats filled the airwaves, newspapers and social media. The targets were not only politicians but climate scientists. At the Canberra rally that morning, speakers whipped up the crowd over the 'corruption' of science. Few were more hysterical than an obscure Queensland engineer, Malcolm Roberts, from the newly formed Galileo Movement.

'They are stealing your money and stealing your freedom,' Roberts shouted at the crowd. 'Carbon dioxide does not drive temperature; temperature determines CO_2 levels—a complete reversal of what these people are spreading! Their lies and deception! Brown and Gillard!'

In a few short months, Roberts and his friends had secured Alan Jones as a patron of their movement and claimed a long list of climate-sceptic scientists as 'independent advisers', including Professor Bob Carter from James Cook University (JCU), his close colleague Peter Ridd, and Ian Plimer from Adelaide. A string of high-profile US climate sceptics, including some from groups funded by the Koch brothers' donor network, were also listed on the Galileo Movement's website. With these names attached to his, Roberts had moved from the fringes of the internet to the heady world of the big players. He would ultimately become a senator for Pauline Hanson's far-right One Nation party.

The 'people's revolt' supercharged the attacks on mainstream climate science. Climate Change Minister Greg Combet had tried to forestall the attacks by setting up a Climate Commission of experts to tell the public about the real risks from climate change. Instead, the head of the commission, Dr Tim Flannery, became such a target of ridicule and abuse he was left reeling.

A highly regarded zoologist, popular science writer and former Australian of the Year, Flannery was overwhelmed by the media assault on him and the commission. 'We were constantly being pilloried by the gutter end of the press,' said Flannery.

The brunt of the attacks came from 2GB and Murdoch's News Limited columnist Andrew Bolt, who repeatedly castigated Flannery for warning that Australia's towns and cities in the future could face the prospect of running out of water because of climate change.

Back in 2004, when Perth was under pressure from steadily falling dam levels, Flannery said, 'I think there is a fair chance Perth will be the 21st century's first ghost metropolis.'

Flannery later qualified the comment and said it was conditional on governments not taking action. Perth did have a water crisis, but the state government had since built a desalination plant to supplement the city's supply. Flannery's remarks on the water shortages nevertheless prompted ridicule and attacks that jumped from the airwaves to public forums.

'We attended one meeting at Parramatta where there was a group of people who had an effigy of me that I think they were hanging or something,' Flannery said. 'People were saying "Come over here, you cunt, and I will smash your head in." We had to have security to get in and out of meetings.'

Head of the Australian National University's Climate Change Institute, Professor Will Steffen, who served with Flannery on the commission, saw the torrent of abusive emails flowing into their inboxes, some containing veiled threats. After one climate presentation Steffen was alarmed when a colleague told him about one participant who'd said he had a gun licence and 'he knew how to handle these idiots talking about climate science'. The ANU's security was tightened and Steffen's climate institute moved to more secure premises.

The debate was toxic. Prominent climate-sceptic scientists and politicians like Cory Bernardi were also being abused and getting their fair share of vile letters and emails. Australia's chief scientist, Professor Ian Chubb, was despairing. 'Every time I think it's reached a low, we then go on and reach a new low,' he said.

The brawls over climate science were a backdrop to the corporate lobbying campaign against the climate bill which grew louder as a vote on the bills came closer, and was amplified by a media in overdrive. Sydney's *Daily Telegraph* managed to find extraordinary new angles on the pain the carbon price would inflict on voters. 'Death tax: Grieving family's funeral carbon fee,' one headline screamed.

Abbott famously predicted the South Australian steel town of Whyalla would be 'wiped off the map' by the carbon tax—a warning first sounded by the local branch of the Australian Workers' Union. The coal lobby warned 'potentially' 25 mines would be shut down in the first four years of the scheme. This was despite the generous assistance packages for industry and booming investment in coalmines.

In September 2011, Gillard finally brought the climate bills to a vote in the lower house. It was a crucial test for her minority government and Abbott immediately branded it 'the longest suicide note in Australian history'.

Gillard tried to seize on the historic moment. In the age of climate change, Australia, the nation defined by its quarry vision, was finally putting a price on carbon pollution: 'We govern in a world of change, a world that is transforming. My task as Prime Minister, my party's task in government, this parliament's task as it meets today, is to lead our country through this transformation. Not to hide from change,' she said as she tried to look to the future. 'Because the final test is not: are you on the right side of the politics of the week or the polls of the year? The final test is this: are you on the right side of history? And in my experience, the judgement of history has a way of speaking sooner than we expect.'

With the backing of the independents, Tony Windsor and Rob Oakeshott, and the Greens, the climate policy passed that October. Gillard embraced Combet on the floor of the house, relieved it was finally done. She also managed to kiss Kevin Rudd.

One month later the bills passed the Senate and came into law. For Milne this was the high point of her entire political career. 'It was elation really,' said Milne, 'to have this path and know that we were actually going to seriously make way on the climate.'

———

They had won the battle, but would now lose the climate war. Even before the Senate vote, Australia's most influential figures were standing shoulder to shoulder with the climate sceptics and Tony Abbott. In a powerful signal to the country's Catholics, Australia's senior archbishop, Cardinal George Pell, arrived at Westminster Cathedral Hall in London to address Britain's top climate-sceptic think tank, the Global Warming Policy Foundation. The foundation's academic advisory council included some of Australia's most cited climate sceptics, Professor Bob Carter and Ian Plimer, and Pell drew heavily on their work.

He criticised the financial costs that climate change 'true-believers' were imposing on local economies and claimed global warming had 'stopped'. Carbon dioxide was, said Pell, 'not a pollutant, but part of the stuff of life', a fact never in dispute— the debate was only ever about the amount that should be in the planet's atmosphere.

'Animals would not notice a doubling of CO_2 and obviously plants would love it,' Pell told the audience. Drawing on the work of Ian Plimer and Bob Carter, Pell was emphatic there was no current global warming. 'In the 1990s we were warned of the "greenhouse effect" but in the first decade of the new millennium "global warming" stopped,' he said.

Shortly after Pell embraced the sceptics, so too did Liberal Party grandee John Howard. While insisting he was still an 'agnostic' on climate change, Howard lent his imprimatur to Professor Ian Plimer's latest book, which told schoolchildren and their parents to challenge the science of global warning. Howard launched *How to Get Expelled from School: A guide to climate change for pupils, parents and punters* at Sydney's Tattersalls Club, an event hosted by his old friends from the Institute of Public Affairs (IPA) along with the Sydney Mining Club. 'I think Professor Plimer has done us all a service by bringing this material forward,' Mr Howard told the crowd of business executives.

These campaigns in the carbon war would drag on for another eighteen months, but by April 2013 a Coalition victory at the federal election due later in the year was pretty much inevitable. At the IPA's 70th anniversary dinner that month, Abbott looked like a winner. He had been asked to speak to the IPA's top-drawer supporters, including Hugh Morgan, Cardinal George Pell and— in pride of place, seated between Rupert Murdoch and his senior News executive Robert Thomson—Gina Rinehart, now one of the IPA's most generous donors. Rinehart had just been awarded the IPA's prestigious Free Enterprise Leadership Award after the IPA had spent the last two years pushing her low-tax vision for northern Australia.

Andrew Bolt did the honour of introducing Abbott, who reassured the IPA's chief, John Roskam, he would fulfil part of the wish list Roskam had drawn up for the new government. 'The Coalition will indeed repeal the carbon tax, abolish the Department of Climate Change, abolish the Clean Energy Fund.' In a nod to Rinehart, Abbott also promised his new government would support the development of northern Australia and abolish Labor's mining tax.

Not long before the IPA dinner, Hugh Morgan and the Cormack Foundation had agreed to give $3 million to the federal Liberal Party to help bankroll the next election. By now Morgan's warnings to Australian business to resist the carbon price looked like a safe bet.

With Gillard's government struggling in the polls, the party's factional bosses turned against her, restoring Kevin Rudd as prime minister just in time to call the election for 7 September. Her government and her political career were over. Exhausted by the carbon war and the political infighting, Combet also quit.

A few weeks later a report from Combet's department showed that, one year after the carbon price began, Australia's greenhouse gas emissions from electricity generation in the National Electricity Market were down by a remarkable 7 per cent from a year earlier. Renewable energy generation was up. Thanks to both state and federal subsidies, the number of Australian households with solar rooftop panels had skyrocketed during the Rudd–Gillard years from 7000 to 1.3 million. Economic growth had not collapsed and industries had not closed down.

But these were small victories. Business was now betting on Abbott to win the election. 'Abbott was supreme,' said Combet. 'He was ascendant. Business follows politics and by that time it was absolutely obvious.'

Rudd's new climate change minister, Mark Butler, found business had turned on Labor's climate change policy with Abbott's victory assured. 'Industry had just got into his *Lord of the Flies* feeding frenzy, where they just wanted to kill the whole thing,' he said.

In a belated effort to defend Labor against the attacks, Rudd conceded the $23 three-year fixed carbon price Gillard had negotiated with the Greens was a 'carbon tax'. He promised to shift it to a floating price tied to the European Union's Emissions Trading Scheme. 'It allowed us to do a big headline of terminating the carbon tax,' said Butler. 'As a signal of a shift from old prime minister to new prime minister it worked.'

The problem was: it didn't work for the budget. The EU's carbon price had temporarily collapsed to less than $9 a tonne. That meant if Labor did get re-elected there would be much less money in the budget to fund the clean energy technology, industry assistance and support for low income earners. But, as Rudd and Butler knew, that was academic. Labor was going to lose.

On election night Tony Abbott could barely contain his excitement as the seats fell to the Coalition. He took the stage in front of a triumphant crowd at the swish Four Seasons Hotel in Sydney and claimed victory. 'The government of Australia has changed,' he told them. The chant went up: 'Tony, Tony, Tony.' When the new prime minister listed the promises his government would act on, the first was the abolition of the carbon tax.

Mining magnate Gina Rinehart, Australia's richest woman, celebrated the night with her good friend Barnaby Joyce, the National Party's most vocal climate sceptic. She joined Joyce and his wife, Natalie, at the West Tamworth Leagues Club in northern New South Wales where Joyce had won the seat of New England after resigning from the Senate. The National's leadership and deputy prime ministership were now firmly in his sights.

From that night on, climate change was no longer a defining challenge or even a political necessity. It was back on the sidelines of politics. Australia was going to double down on its bet on fossil fuels. Up in Queensland, the dredges were already working overtime, expanding ports to ship more coal, and now gas, out through the Great Barrier Reef.

CHAPTER TWELVE

THE STATE OF THE REEF II

PARIS IN SUMMER can be a lot of fun but it wasn't happening for Jon Day on this visit. He had come to UNESCO headquarters in the Place de Fontenoy, but not to see its dazzling art collection or its beautiful Japanese garden. Day was part of the Australian delegation attending the annual meeting of the World Heritage Committee, and he was bracing himself for a diplomatic slap-down. High on the agenda was the state of the Great Barrier Reef and the committee was not happy.

As the director for conservation at the Great Barrier Reef Marine Park Authority (GBRMPA), Day was one of the people responsible for safeguarding the World Heritage listing of the reef. For Day, the reef was the most important World Heritage area on the planet, and he was right across the World Heritage values that kept it on the list. But at that Paris meeting in June 2011, Day realised the committee was now asking whether Australia was putting those values in jeopardy. 'That was the *wow* moment,' Day recalled. 'This is really serious.'

The reason for the committee's concern could be found back home in Queensland. Bulldozers and dredges had begun work on a giant liquefied natural gas (LNG) hub on Curtis Island just inside the World Heritage area of the Great Barrier Reef. Curtis Island sits just off the industrial city of Gladstone, and the plans included a massive expansion of the port there. Not only had the Gillard government signed the approvals for the mega project, it had done so without advising the World Heritage Committee beforehand. That breached the spirit, if not the letter, of the World Heritage Convention and really upset the people in Paris.

By the end of the meeting at UNESCO headquarters, Australia had copped a stunning rebuke for its stewardship of the Great Barrier Reef. The World Heritage Committee formally noted its 'extreme concern' about the approval of the Curtis Island development. '"Extreme concern" is very tough language for the Committee,' Day said later. 'Gladstone was just the tipping point, a pretty major tipping point, but it wasn't just Gladstone.'

What particularly troubled the World Heritage Committee were plans for six new ports or port expansions, including Gladstone, up and down the Queensland coast. Despite all their talk about the danger of climate change, in 2011 Labor governments in Canberra and Brisbane were gearing up for the next fossil fuel export bonanza. Already the world's largest coal exporter, Australia was now hoping to become the world's largest LNG exporter. That meant bigger ports, and more of them, inshore from the Great Barrier Reef. These ports would need huge dredging operations and earthworks that would churn up the waterways, smother seagrasses, disturb fishing grounds, disrupt the habitats of dugongs and turtles, and see thousands of tonnes of dredge spoil dumped at sea.

The World Heritage Committee was putting Australia on notice that it was seriously worried about the threat these developments posed to the Great Barrier Reef. It wanted the government to urgently invite a mission from UNESCO headquarters to investigate the port expansions, and it wanted a detailed assessment of all the new developments on the Queensland coast that might threaten the Reef.

'If you look at everything individually you don't see the big picture,' warned Tim Badman, who as a director of the International Union for Conservation of Nature (IUCN) was then a key delegate to the World Heritage Committee. Badman also called for the Curtis Island mega project to be suspended, but that was never going to happen. There was too much at stake in Australia.

———

The Curtis Island project was on Environment Minister Peter Garrett's desk before the 2010 election. The plans covered a $16 billion LNG plant proposed by Australian oil and gas company Santos and its foreign partners, plus a $15 billion plant proposed by British Gas. These were just two of four LNG plants planned for Curtis Island; Origin Energy with partner Conoco Phillips, and Shell with a Chinese partner, were pushing ahead with proposals as well.

Queensland's Labor government was backing the Curtis Island project to the hilt and had already signed most of the state approvals. But Garrett 'hit the pause button', as he put it, on the massive port expansion at Gladstone. 'I suspected it was difficult, if not impossible, to manage the impacts of large-scale dredging on the waters adjacent to the Great Barrier Reef Marine Park,' he said later.

Garrett was being asked to tick off on the biggest dredging operation ever attempted inshore from the Reef. The Gladstone Ports Corporation originally planned to move up to 42 million cubic metres of dredge spoil, much of it from the seabed in Gladstone Harbour or from around Curtis Island. The bulk of it was to be retained behind an 8 kilometre wall in the harbour right opposite the island. Excess spoil was going to be dumped at sea, just 400 metres from the Great Barrier Reef Marine Park, potentially affecting seabirds, dolphins, dugongs and a host of migratory species.

The chairman of the Reef's Marine Park Authority, Russell Reichelt, had already flagged his worries about the scale of the dredging and sea dumping. 'We raised extreme concerns with it,' Dr Reichelt said. 'We regarded it as not acceptable in the World Heritage property and gave that advice to the Department [of Environment] at the time.'

But the Curtis Island project had big supporters in federal cabinet. Labor's treasurer, Wayne Swan, and the resources and energy minister, Martin Ferguson, were pushing it because it meant jobs and investment for Queensland. Ferguson had already touted the first major export contract from Curtis Island even before the approvals were signed.

After the 2010 election, Gillard moved Garrett out of Environment and into Education. His replacement, Tony Burke, approved the first Curtis Island project in just six weeks. Swan and Ferguson joined Cath Tanna, the Australian head of British Gas, at a press conference a few days later to talk up the great benefits of the planned LNG hub. Only then did the government officially inform the World Heritage Committee in Paris of the approvals.

Burke later explained his swift approval by pointing out that Curtis Island sat opposite the industrial Gladstone Port, which already had a big coal export terminal. This was true. The coal terminal was in fact named after Cath Tanna's father, the late Reg Tanna, who once ran the port. Clearly Curtis Island was not sitting in pristine Reef waters. There was also a nearby alumina refinery. But given the industrial activity around the port, it was all the more likely that kicking off a massive dredging operation could have serious environmental consequences for the World

Heritage area. Some of the dredge spoil had a good chance of being contaminated with heavy metals.

Burke imposed scores of new environmental conditions on the Curtis Island project that reassured the GBRMPA. These included regular testing of the water quality in Gladstone Harbour and other areas affected by the dredging and dumping. 'We made sure that each of the changes, particularly on issues relating to water quality that the Great Barrier Reef Marine Park Authority wanted, were included in the final approvals,' he said. But these conditions, along with those imposed by the Queensland government, would end up looking much better on paper than in practice.

In May 2011, with the dredging operation well under way, then prime minister Julia Gillard and Queensland's Labor premier, Anna Bligh, arrived on Curtis Island and donned high-vis vests and hard hats to launch the first LNG plant construction site. Somewhat undiplomatically, this big public-relations event came just weeks before the Australian delegation was due to meet the World Heritage Committee in Paris.

Standing on the Santos site, Gillard talked up the employment benefits of the project for the cameras. 'It's going to mean 6000 jobs,' she declared. Bligh was even more enthusiastic, heralding the moment as a 'new gas age' for Queensland and trumpeting its environmental benefits. 'Gas is a cleaner and lower emissions fuel than coal and it will be an important part of the region's transition to clean energy,' said Bligh.

Selling the giant LNG project as good for the planet was a big call. Bligh was echoing the accepted wisdom that gas in the short to medium term could replace coal in the global transition to clean energy. Switching from coal to gas can cut greenhouse emissions between 30 and 50 per cent. But if the gas being exported was just adding to overall energy consumption, it wasn't going to bring down global emissions. At best it would slow their rate of growth.

There was also a downside to the LNG plants for Australia's own greenhouse footprint. Exporting the gas would not help Australia's transition away from coal-fired power. In fact, it would result in soaring gas prices in Australia, making it far more costly to replace coal. Also, importantly, producing LNG on Curtis Island would create a lot of emissions at home. Coal seam gas piped from fields in Queensland would have to be purified and cooled to minus 161 degrees Celsius at the plants before it was exported. That would take large amounts of energy and ultimately greatly increase Australia's own emissions.

But back in 2011, the greenhouse emissions from the Curtis Island project were not the issue troubling the World Heritage Committee. It was the industrial development and the giant dredging operation that concerned UNESCO headquarters in Paris and local environment groups in Queensland. And it didn't take long for those worries to escalate.

Just months after dredging started, local Gladstone fishermen began pulling in nets of barramundi covered with awful parasitic skin infections and exploding red eyes. Blue salmon, whiting and fingermark bream were also discoloured. Crabs were appearing with ulcers, and sharks showed bright-red skin lesions. Some commercial fishermen suffered similar skin lesions. By mid-September, the Queensland government stepped in and temporarily closed the waters around Gladstone to fishing.

The Gladstone Ports Corporation insisted the dredging was not seriously impacting the water quality or causing the fish disease. The state government suggested the giant floods that had broken the drought and deluged Queensland the previous summer were the problem. But no one could come up with a definitive explanation. The fishermen didn't believe any of them.

'They say it's naturally occurring, but we've never seen it and it all seems to coincide with the dredging starting,' said Mark McMillan, who had fished the Gladstone waters since he was a kid working for his father.

When the UNESCO fact-finding mission arrived in Australia in March 2012, the explosion of developments inshore from the Reef was a hot topic. Heading the mission were Dr Fanny Douvere, the coordinator of the Marine Programme at the World Heritage Centre, and Tim Badman from the International Union for Conservation of Nature. The pair met with the federal environment minister, the Queensland government, the LNG operators, mining companies, port authorities, environment groups and fishermen.

The result of the mission was a damning report for the World Heritage Committee that seriously questioned Australia's management of the Great Barrier Reef. It recommended an independent review of the environmental problems in Gladstone Harbour; took issue with the unprecedented scale of coastal development inshore from the Reef; and asked the federal government to stop any new port developments that would seriously impact the Reef's conservation.

When the World Heritage Committee met in June 2012 in St Petersburg, alarm bells were already ringing in Canberra. The meeting dealt an embarrassing blow to Australia, the custodian of the Reef. The committee concluded it would now consider inscribing the Great Barrier Reef 'on the List of World Heritage in Danger' unless Australia reported back within seven months that it had made real progress on the committee's concerns, especially about port development.

When the news broke, environment groups unloaded on both the Queensland and federal governments. Greenpeace branded the report 'a damning indictment on the threat posed to the Great

Barrier Reef by coal port developments and gas port developments in Queensland'.

Environment Minister Tony Burke sounded contrite. 'The Great Barrier Reef has been at a crossroads,' he told reporters, promising the government would do better. 'There is a way of having sustainable development,' he said. In words that would come back to haunt him, Burke declared, 'We're talking about the Reef. If you can't get your environmental protections and systems in place on this one, you may as well give up.'

Up in Queensland, the government was less fazed by the threat of the Reef being put on the 'in Danger' list. A new premier was in power. The Liberal National Party (LNP) had smashed Labor in a stunning election win in March that year. Premier Campbell Newman and his LNP colleagues had swept into power on the pro-growth slogan 'CanDoQLD'. Stopping port development inshore from the Reef was not an option, said Newman. 'We are in the coal business. If you want decent hospitals, schools and police on the beat, we all need to understand that.'

With those words, the new premier starkly summed up Australia's dilemma in 2012. Despite its attempts to curb its greenhouse gas emissions, Australia was still a huge beneficiary of the carbon-intensive global economy. The previous year, local and global investors had laid out some $116 billion for energy projects in Australia, mostly for fossil fuel. Apart from Curtis Island, BHP Billiton was busy expanding its coalmines, while Chevron, Shell and ExxonMobil were pressing ahead with the largest resources project in the nation's history, the giant Gorgon LNG project off Western Australia.

Jobs, corporate profits, export income and government revenue still strongly depended on Australia's big greenhouse-emitting industries. The nation's political leaders and most Australians were yet to seriously question the idea that the nation could keep exporting fossil fuel for decades to come—whatever the consequences for the Great Barrier Reef or the planet.

The new Queensland government certainly thought coal exports were going to boom in the future. Premier Newman immediately put a high priority on kickstarting coalmining in the Galilee Basin in Central Queensland. The Galilee promised to deliver jobs and a stream of state royalties from at least five planned mines there. All these planned projects were bigger than any coalmine then operating in Australia. The Galilee Basin held huge reserves of thermal coal the miners hoped would fire power stations in India and South-East Asia. But it would need even bigger ports inshore from the Great Barrier Reef to ship it out.

The Queensland Resources Council was right behind the Galilee development. 'These are some of the biggest mines we'll ever see in Australia,' boasted the council's head, Michael Roche. 'We don't have coal projects of this order and they are being designed at a scale that again will develop very good cost structures.'

The determined push to open up the giant Galilee coalfields was the final straw that galvanised a core group of environmentalists to mobilise a nationwide campaign against the coal industry in Australia. In early 2012, they decided it was time to mount a full-throated challenge to the coal industry's social licence to operate. The impact of climate change on the Reef was central to their argument. And while it got off to a bad start, the campaign would soon put the 'coal versus coral' debate squarely on the political agenda.

The activists' strategy was called 'Stopping the Australian Coal Export Boom'. It was largely written by two veteran environmentalists, John Hepburn from Greenpeace, and Bob Burton, a former Wilderness Society campaigner who worked with CoalSwarm, a US online database monitoring global coal exports. The strategy put a lot of coalmine expansions under the spotlight in New South Wales and Queensland, but 'The Battle of Galilee' was central to it. The activists were determined to stop the planned Galilee Basin coalmines.

'The Galilee projects, if they got up, would be opening up a huge new source of carbon emissions from Australia on top of all the other coalmine expansions,' Burton explained later. 'There was a wave of big new proposals and we thought, if we sit back and do nothing the consequences would be enormous.'

The anti-coal strategy called for a coordinated campaign to disrupt and delay building of rail lines, ports and key infrastructure for the new mines. It planned to create uncertainty over coalmine investment; challenge the coal companies' access to farmland and water rights; and lobby to remove subsidies for the industry. It proposed raising up to $6 million to fund court cases and campaigns.

The aim was to turn Australians' support for the coal industry on its head. 'Change the story of coal from being the backbone of our economy, to being a destructive industry that destroys the landscape and communities, corrupts our democracy, and threatens the global climate,' was how they summed up the campaign to withdraw the social licence from the coal industry.

When the strategy document leaked to the *Australian Financial Review* in March 2012, all hell broke loose. Federal Labor politicians lined up to condemn it. Labor's federal treasurer, Wayne Swan, called it deeply irresponsible: 'I think many people confuse the fact that whilst we need to reduce greenhouse gas emissions, we've got to keep the lights on and the power flowing.' Trade Minister Craig Emerson slammed the activists: 'They are deluding themselves if they think the world is just going to flick the switch to renewable energy,' he told the ABC. 'It would mean mass starvation and they ought to wake up to that instead of living in a fantasy land and organising these sorts of campaigns.' Resources Minister Martin Ferguson was appalled: 'Reports of elaborate strategies designed to destroy Australian industries and jobs are very disturbing.'

The coal companies were equally alarmed. Rio Tinto's spokesman called it 'a blueprint for economic vandalism'. BHP pointed out the jobs and taxes that would be lost if the campaign succeeded. Xstrata's chief executive, Peter Freyberg, just couldn't comprehend it: 'destroying one country's coal industry is not an appropriate response to concerns about the climate change impacts of coal,' he said.

The furious reaction overlooked one salient point: the activists' strategy was not aimed at existing coalmines; it was aimed at new coalmines and mine expansions. They argued that if nations were going to stick to their promise in Copenhagen to keep global temperature from rising more than 2 degrees Celsius, that meant the world was supposed to be headed for carbon neutrality by 2070. Doubling or tripling Australia's coal exports was going to make that a whole lot harder.

Despite being braced for a backlash, Burton was amazed by the coal industry's ferocious response to the document. 'The reaction reflected their attitude of, "How dare anyone challenge us?"' he said. 'They really believed they were the heroes of the country.'

The activists should not have been surprised. The coal industry and Australia's leaders of the major political parties were in lock step in supporting the expansion of the coal export business. Australia was not going to give up its share of fossil fuel revenue only to see another country get the jump on us and profit instead: 'If our coal is not supplied, all we do is deny ourselves the jobs, the investment, the tax revenues to fund services; but the coal will be supplied,' said the Queensland Resources Council chief, Roche.

The Galilee coal boom was going to happen. If that meant more port expansions inshore from the Great Barrier Reef were needed, they would also happen, even if there was a problem with the World Heritage Committee.

————

When members of the committee gathered for their annual meeting the following year in June 2013, Australia's management of the Great Barrier Reef was back on the agenda. The committee repeated its request to Australia that no new port development or infrastructure be permitted outside the existing major ports or port areas. It also repeated its warning that, unless there was more substantial progress on its concerns, the committee would continue to consider inscribing the Great Barrier Reef 'on the List of World Heritage in Danger' at its meeting in 2014.

Jon Day from GBRMPA listened to this with dismay. Once again he was on the Australian delegation at the World Heritage meeting. He knew back home the federal and Queensland governments had at that very moment plans on the table to push through another big port expansion inshore from the Reef. This time it was the coal port of Abbot Point on the North Queensland coast about two hours south of Townsville. It was earmarked as one of the export hubs for the Galilee Basin mines.

Technically, the Abbot Point plans were just an expansion of the existing port, not a new port. But the dredging operation was going to remove 3 million cubic metres of spoil from the seabed and dump it in the Great Barrier Reef Marine Park. Senior staff at the Marine Park Authority were strenuously opposed to the dumping, but the port expansion had strong political backers in Canberra and Brisbane.

As Day put it later, both governments should have been looking at alternatives, especially given that Australia was on notice from the World Heritage Committee at the time. 'Sure, they may cost more, but again we're dealing with a World Heritage Area, the most important World Heritage Area on the planet, a magnificent marine protected area that the world wants us to protect.'

In June 2013, the director of environmental assessment at the Marine Park Authority, Adam Smith, bluntly advised his superiors that the Abbot Point expansion plans should be opposed: 'The likely impact of the dredging and disposal in the Great Barrier Reef World Heritage Area currently proposed would be environmentally and socially unacceptable,' he warned.

His opinion was backed up by the majority of scientists in the authority and by coral reef scientists outside it. But the pressure from both the Queensland LNP government and the federal Labor government to support it just kept mounting.

––––––––

Queensland billionaire Clive Palmer was one powerful supporter of opening up the Galilee Basin. The former Gold Coast real estate broker turned mining entrepreneur held exploration permits over almost 17,000 square kilometres of the Galilee. In 2012 Palmer's Waratah Coal was hoping to get one, if not two, huge coalmines approved for development. He had potential buyers in China; but he needed financiers with deep pockets, a rail line to transport the coal over 500 kilometres to the coast, and a terminal to ship the coal from Abbot Point.

A giant of a man, Palmer enjoyed throwing his weight around. He had big ambitions and a mouth to match. A lifetime member of the Queensland LNP, Palmer had been a mover and shaker in state politics for years. He first worked for the party when the hard-nosed National Party premier Joh Bjelke-Petersen ruled in the 1970s and 1980s, before that government sank under the weight of scandals. When the Queensland National and Liberal parties merged in 2008 to form the LNP, Palmer became a generous donor to the party both in Queensland and federally. The Queensland party president, Bruce McIver, was one of his mates.

In March 2012, when the LNP crushed Labor and took office in Queensland, Palmer's ambitious coal plans in the Galilee looked like coming up trumps. At the election-night victory party, Palmer went over the top praising the new premier, Campbell Newman. 'He's the most successful political leader in the nation's history, no one's ever got this sort of vote in any state or federal election,' he enthused.

But within months of the state election, Palmer's Galilee coal plans were on hold and his relationship with the Newman government was poisonous. It emerged that Gina Rinehart and her Indian partner, GVK, with the help of some heavy lobbying, had got the jump on Palmer. Their giant Alpha coal project was seen as more advanced in its planning, and the deputy premier, Jeff Seeney, announced Rinehart and her partners had got the first approval for a Galilee mine. GVK's proposed rail line from its mine to Abbot Point port was also seen as a workable plan.

Palmer was seething, snarkily telling the ABC's *Four Corners*, 'I don't think it was anything against us personally. I think it was the fact that Mr Seeney and Mr Newman love GVK and loved India.'

In a second blow to Palmer, the Newman government's expansion plans for the Abbot Point coal port did not include extra space for his company. That meant GVK and Rinehart's Hancock Prospecting had secured terminal access at Abbot Point, but Palmer had not.

Another serious Indian player, the Adani Group, had also secured port access at Abbot Point by buying a 99-year lease over the whole terminal. Adani, founded by billionaire Indian patriarch Gautam Adani, was also exploring a mega-mine in the

Galilee. If that came off, Adani's Carmichael project, with its plans to build the largest coalmine in the southern hemisphere, would also leap ahead of Palmer.

Well known for his incendiary temper, Palmer declared war on the Newman government. 'God bless the backbench in Queensland and God bless democracy, because things are going to change because the public will not put up with this sort of rape on their state,' Palmer fulminated on the ABC's *Lateline*. Newman and Seeney hit back at Palmer, accusing him of trying to bully the government over his Galilee coal plans. 'This is all about a billionaire that hasn't been able to get his way with our government,' said Seeney.

The political battle between Palmer and the Newman government over the Galilee spun out of control and the LNP threatened to expel Palmer. The billionaire didn't blink. Instead he launched his own political party aimed at destabilising the LNP. His political friends weren't surprised. Independent MP Bob Katter saw the Newman government's decision to support Rinehart and GVK over Palmer as the tipping point: 'I think a lot of the politics of Australia over the next ten years will be written around that decision,' the old maverick politician predicted.

Katter was right. In 2013, the billionaire unveiled his Palmer United Party (PUP) just in time for the federal election that year. He would ultimately throw $15 million into it. The media scorned him as a rich buffoon, but his old sparring partner, the National Party's Barnaby Joyce, knew better. 'He comes across as a sort of effervescent, sometimes bumbling character and that is a ploy,' he warned. 'You're a fool if you think he's a fool.'

In the last week of the 2013 federal election campaign, using saturation television advertising, Palmer stripped votes from Labor in Queensland but also some from the LNP. On election night, he snatched the House of Representatives seat of Fairfax on Queensland's Sunshine Coast by a hair's breadth. But that was not the real prize. The Palmer United Party ultimately won three seats in the Senate, and this would result in it sharing the balance of power there. That ended any chance that Labor's carbon emissions trading scheme would survive.

Palmer had promised to kill off the emissions trading scheme, the 'carbon tax', as he called it. His interest wasn't just political he had an acute understanding of its effect on business. By his own calculation, Labor's carbon scheme cost his emissions-intensive Townsville refinery, Queensland Nickel, around $25 million over its first two years. If the scheme was scrapped, Palmer's company had a lot to gain. Perhaps not surprisingly, the PUP campaign was funded by Queensland Nickel.

Both Palmer and Tony Abbott agreed that Labor's emissions trading scheme had to go, but Palmer had his own demands about how this should be done. The Abbott government would now have to work with the billionaire mining entrepreneur to fashion Australia's new climate change policy. There were two things both Palmer and Abbott could agree on: Australia would have no price on carbon, at least for the foreseeable future. And the Galilee Basin should be opened up for coalmining.

In the dying days of the Labor government in Canberra in 2013, a huge report on the Great Barrier Reef landed in the Department of the Environment. It was part of Australia's response to the World Heritage Committee and its threat to put the Reef on the 'List of World Heritage in Danger'. Compiled by the Marine Park Authority, the purpose of the document was clear: how to improve Australia's management of the risks facing the Reef.

The thumping report looked at everything from port dredging to the crown-of-thorns coral-eating starfish. But down the back of the report, under the heading, 'The Future of the Great Barrier Reef', there was one unavoidable, grave warning: 'Climate change remains the most serious long-term risk facing the Reef and is likely to have far reaching consequences for the Region's environment,' it stated.

In blunt terms it spelt out the threat: 'The urgent need to limit global warming to two degrees Celsius above pre-industrial levels has been recognised by almost 200 nations. At present global emissions are not on track to achieve such a target and even a two-degree Celsius rise would be a very dangerous level of warming for coral reef ecosystems, including the Great Barrier Reef, and the people who derive benefits from them. To ensure the Reef remains a coral-dominated system, the latest science indicates global average temperature rise would have to be limited to 1.2 degrees Celsius.'

Fourteen years after the coral reef scientist Ove Hoegh-Guldberg was branded an 'alarmist' for his groundbreaking paper on the threat to the Reef from climate change, the report could not have been clearer. It rated the likelihood of warming sea waters from climate change over the few decades as 'almost certain', and assessed the consequences of this for life on the Reef if this happened as 'catastrophic'.

But despite the threat to the Reef from rising greenhouse gas emissions, in 2013 the major parties encouraged Australia to stay the course as the world's number one coal exporter and, in the not too distant future, become the number one LNG exporter. The 'quarry vision' was still strong on both sides of politics. Within months of the Labor government's defeat that year, Martin Ferguson, its long-time resources and energy minister, became chair of the advisory board of the country's main gas and oil lobby.

Down in Canberra, where the new Abbott government was about to take office, the threat of warming waters on the Great Barrier Reef barely caused a ripple.

CHAPTER THIRTEEN

THE PURGE

ON THE FIRST day of the new Abbott government, Australia's climate scientists got a pretty clear message. Within 24 hours of the swearing-in ceremony at Government House in Yarralumla, the new environment minister, Greg Hunt, called the head of the Climate Commission and sacked him.

'It was a short and courteous conversation,' Dr Tim Flannery recalled. 'I'm pretty sure that cabinet hadn't been convened when they did it. My very strong recollection is that it was their very first act in government.'

One of Flannery's colleagues on the commission, Professor Will Steffen from the ANU's Climate Change Institute, was also sacked, along with every other commission member. 'I think we were the first definitive action of the Abbott government,' Steffen recalled. 'They got rid of us and you could probably measure it in hours rather than days.'

What surprised the scientists most was not their hasty sacking on 19 September 2013, but how quickly the government obliterated their work. 'The website that we'd spent a lot of time building was taken down with absolutely no justification as far as I could see,' said Flannery. 'It was giving basic information that was being used by many, many people—teachers and others—just to gain a better understanding of what climate science was actually about.'

The commission had been set up by the Gillard government as an independent source for the public to understand climate change and its impacts. The aim was to 'help build the consensus required to move to a clean energy future'. But the commission and its members had been pilloried as 'alarmist' by sceptical columnists in the Murdoch media and by radio shock jocks from the beginning. Flannery was expecting the commission to be disbanded, but the decision to kill its website hurt.

'The fact that they would close a website that costs many tens, if not hundreds of thousands of dollars to put together and was heavily patronised and was entirely apolitical told me how far they would go. And that was the line they crossed with a lot of the commissioners. They were outraged.'

Other commissioners—Professor Lesley Hughes, a Macquarie University biologist and IPCC author; Gerry Hueston, a former Australasian BP president; and Roger Beale, the former head of the Department of the Environment during the Howard years— were all sacked by letter. Beale would wear the sacking as a badge of honour. 'I was almost proud to be sacked—for the first time in my life—from that post by Greg Hunt,' he said later.

Despite the attempt to erase them from history, a group of the commissioners were determined to push on with their work. Within days of their sacking, they launched a publicly funded website under a new name, the Climate Council. It raised almost $1 million in less than a week and kept pumping out reports on climate science for the public to read.

Hunt argued shutting down the commission was just part of the government's plans to 'avoid duplication of services'. His own department, he said, could give independent advice on climate change. But Hunt, like everyone in the Abbott government from senior cabinet members down, knew the prime minister was a climate sceptic who questioned a lot of what the commission had reported. Abbott had made it clear he was dubious about mainstream climate science, opposed putting a price on carbon emissions, and was hostile to renewable energy targets.

'I think that the proposition that climate is changing drastically, that man-made carbon dioxide emissions are the cause, and that therefore we must drastically reduce, almost at any cost, our carbon dioxide emissions—I think that proposition is not well founded,' Abbott said repeatedly.

Hunt liked to distance himself from Abbott on climate change by professing his own belief in the science. He struck an agreement with Abbott that the government would not challenge climate science publicly. But nevertheless Hunt worked hard on Abbott's strategy to dismantle or review almost every major climate change policy put in place by Labor. That meant a purge. Unknown to Flannery and his colleagues, other sackings were already taking place. Hunt was not the only executioner; other senior ministers had to take up the axe as well.

———

The Saturday after Abbott's election, Treasury secretary Martin Parkinson was pushing his shopping trolley through Coles in the upmarket Canberra suburb of Manuka when his mobile rang. On the line was Australia's most senior bureaucrat, Ian Watt, head of the Department of Prime Minister and Cabinet. He told Parkinson to come to his office that afternoon before abruptly hanging up. 'No pleasantries, no nothing,' Parkinson recalled. His wife Heather Smith, also a senior public servant, was in the shopping aisle with him. 'I think this is going to end badly,' he told her.

More than any other public servant in Canberra, Parkinson

had a long history with climate change. He had worked for three governments on schemes to put a price on carbon emissions. He had been on the Shergold inquiry for the Howard government when he was a deputy secretary at Treasury. Then he set up the first Department of Climate Change under Rudd, to put in place the Carbon Pollution Reduction Scheme. When the Gillard government promoted him to Treasury secretary he still worked closely with Blair Comley, who had succeeded him at the department, to help get up the emissions trading scheme, the 'carbon tax' that Abbott was so keen to dismantle.

Until now, no one in the Abbott government had suggested Parkinson's past would be a problem. After all, no incoming government had ever removed a Treasury secretary as far as anyone could remember. Parkinson was a respected public servant who had worked for both sides of politics. Since the election a week earlier, he'd been talking with the new treasurer, Joe Hockey. He even got a call from the ebullient Hockey on the Friday night. 'Great first week—really think we're going to work well together,' Hockey told him.

But when Parkinson arrived at Ian Watt's office that Saturday afternoon he found him waiting anxiously with the Public Service commissioner, Steve Sedgwick. Watt couldn't look his old colleague in the face. Instead he read out some legal words saying, effectively, the Abbott government didn't have confidence in Parkinson's ability to run the Treasury. Oddly, they then asked him to stay on to help sort out Abbott's promised Commission of Audit on government spending and deliver the first budget. If Parkinson was willing to do that, they might find him an overseas posting at the International Monetary Fund.

The Treasury secretary was not impressed. 'I looked at Ian and I said, "I'm not going anywhere. If they want to sack me, they can sack me."' After some difficult words back and forth, it was understood

Parkinson would stay on for the budget. The two bureaucrats then gave him a hug before he left. He was totally stunned.

Parkinson heard nothing from his minister, Joe Hockey. So on Monday, he rang Hockey's chief of staff, Grant Lovett. It was soon clear that Lovett was not in the loop: he asked the Treasury secretary how his weekend went. 'Shithouse,' Parkinson replied. 'You bastards sacked me.'

A few minutes later Hockey was on the line. 'He had no idea this was going to happen,' Parkinson recalled. 'He was clearly upset. In fact it was a combination of upset and furious. And he went in to bat for me.'

Abbott hadn't told Hockey he was purging his most senior, experienced public servant. The prime minister's only explanation was that he wanted the new government to 'stamp its authority' on Treasury and he wanted everyone in the system to work enthusiastically, 'as we reshape our country'. Hockey bowed to Abbott and sucked it up.

The purge that targeted Parkinson became known in Canberra as 'The night of the short knives'. Watt also had to remove the secretary of the Resources and Energy Department, Blair Comley, who had run the Department of Climate Change; and Don Russell, the head of the Industry Department, where many of the climate change programs ended up. The departure of Russell, a one-time chief of staff to former Labor prime minister Paul Keating, was at least not unexpected.

'People were sent hell, west and crooked,' said Allan Behm who had worked for Labor's climate change minister Greg Combet. 'I couldn't find any work. Nobody would touch us.'

––––––––

Abbott was just getting started. He was determined to completely overhaul Australia's climate change policy. He had already put the wheels in motion more than a month before the election. While still opposition leader, a confident Abbott had written to Watt to tell him the bureaucrats should begin drafting laws to repeal Gillard's Clean Energy Act as soon as the election was over. 'It is my clear intention that legislation to repeal the carbon tax and emissions trading scheme will be the first bills introduced into the new 44th Parliament,' Abbott had written.

That same day, Abbott had sent a warning letter to the chair of the Clean Energy Finance Corporation (CEFC), Jillian Broadbent. During the caretaker government period before the election, he wanted her to make sure the corporation 'immediately ceases to assess or make any further approvals or payments'. The incoming treasurer, he told her, would notify the board to suspend the corporation's operations as soon as they won the election. Hunt and then Shadow Finance Minister Andrew Robb also wrote to the CEFC's board members before the election with the same message: we're going to disband you, so don't enter into any new contracts.

But when one opposition frontbencher contacted Oliver Yates, the corporation's feisty chief executive, Yates told him to back off. 'I said I don't know exactly what the law is here but knock your socks off and tell me I shouldn't be pursuing my job whilst you're not my minister,' Yates recalled telling him. 'As soon as you are my minister, as I understand it, you can't ask me not to pursue my obligations. Under my role as a statutory officer, my power comes from Parliament.'

Yates was absolutely right. The former Macquarie executive was lacking in some of the social niceties, but he knew how to read legal advice. Not Hunt, Hockey or Abbott had the power to shut down the CEFC until they got a new law through parliament, and that included getting it through the Senate.

Despite this, straight after the election, Yates and his chair, Broadbent, were summoned to a meeting with Hockey and the new finance minister, Mathias Cormann. Broadbent was deeply worried that the government, in its haste to kill off the corporation, would have a fire sale of all its investments in renewable energy projects. That would trash the value of the assets, something she really wanted to avoid.

She told Yates to go easy in the meeting, so they could get that message through to the government: 'We were meeting with our new responsible ministers. I didn't want Oliver opening with a sales pitch on the merits of the CEFC. They had just been appointed, we were congratulating them on being appointed, and we were asking them not to destroy value when they dismantle us or when they get the approval to dismantle us.'

Broadbent told both ministers the corporation's investments were good and they would be able to sell them at 100 per cent of their value if that became necessary. Hockey was genial, but Broadbent quickly realised that neither he nor Cormann had a clue what the corporation did—that it was working in syndicates with big private investors to get large renewable projects built. 'It was part of their platform to unwind it, but they had little understanding of what we were doing. So in the end, the Treasurer said, "What exactly do you do there, Jillian?"'

To the Liberal ministers, the CEFC was just 'Bob Brown's Bank', a Labor–Greens scheme that had to go. 'They didn't want to really know about the CEFC,' Broadbent recalled. 'It wasn't their idea and their policy platform was to close it, and they almost convinced themselves it was a great big Green Bank, or whatever jargon they used to describe it. And they had low appetite to know anything about it.'

Despite Broadbent's warning to the ministers, Treasury officials soon followed up and advised her they would be appointing one of the top accounting firms to liquidate the corporation's assets. But Yates had already been talking to the new Senate crossbenchers and was convinced Abbott and Hockey would not be able to shut the corporation down. They were going to be blocked in the Senate. 'The numbers were always in our favour,' said Yates, 'so it didn't worry me particularly from day one that they would ever get it through the Senate.'

Abbott was determined to press ahead anyway. When the new parliament sat in November 2013, Hockey and Hunt introduced a raft of legislation to shut down everything from the emissions trading scheme and the CEFC to the Climate Change Authority. Abbott was doubling down on his divisive strategy that had been so successful in opposition. His time in government would now entrench the endless war in Australia over the politics, economics and science of climate change. And a cheer squad of powerful supporters were on hand to encourage him on his course.

That November, the old sceptic warhorse Hugh Morgan launched a timely attack on the IPCC's climate scientists, shortly before the legislation to kill off Labor's climate policies was brought in. Morgan compared them to 'Chicken Little', alarmists who kept warning the sky would fall in. Not long before this, the Cormack Foundation, which Morgan chaired, had stumped up another $300,000 for the Institute of Public Affairs. The think tank's latest publication, *Climate Change: The Facts 2014*, was soon in the works with essays from Australia's best known climate sceptics, Andrew Bolt and Professor Bob Carter, along with the global names now so familiar in Australia: MIT's Professor Richard Lindzen, Dr Pat Michaels from the Koch-funded Cato Institute, and former UK chancellor Nigel Lawson.

At the same time as Morgan's attack on the IPCC, Lawson's UK think tank, the Global Warming Policy Foundation, welcomed Abbott's old mentor, John Howard, to deliver its annual lecture. The former Liberal prime minister used the opportunity to enthusiastically back Abbott's plans, in a speech he called 'One Religion is Enough'.

'The Australian public has now elected a government that has a pragmatic approach to the issue of global warming and a determination to treat our mining industry as a prized asset,' Howard told the London crowd. He confidently predicted serious concerns about climate change were on the decline: 'The high level of public support for overzealous action on global warming has now passed. My suspicion is that most people in Australia, on this issue, have settled into a state of sustained agnosticism.'

In his deeply pragmatic way, Howard argued that Australians were more interested in their rising electricity bills than global warming. Like Abbott, he was keen to shift the blame for those rising prices onto the carbon tax and renewable energy. Howard's old antagonism to the Renewable Energy Target was still festering. He told the UK think tank that in Australia there was 'a growing consciousness that large subsidies are being paid to the production of renewable energy, and that this is having an increasingly heavy effect on low-income earners'.

Howard's view was shared by Abbott. The Renewable Energy Target was another climate policy the new prime minister wanted dumped.

One very vocal opponent of renewable energy already had Abbott's ear—Maurice Newman, the man he had hand-picked as his new chair of the Prime Minister's Business Advisory Council. A prominent climate change denier, Newman was well known for his ferocious opposition to wind farms. He had threatened to join a legal action to sue a local farmer who'd planned to install a wind turbine near his country estate outside the picturesque town of Crookwell, three hours south-west of Sydney.

Newman was a well-connected conservative networker, a former chairman of the ABC and the Australian Stock Exchange.

He applauded Abbott's hostility to the Renewable Energy Target, blaming it, along with the emissions trading scheme, for destroying Australia's industrial competitiveness. Newman was furious that Australia, in his view, had become 'hostage to climate-change madness'.

'The scientific delusion, the religion behind the climate crusade, is crumbling,' Newman told *The Australian* shortly after his appointment by Abbott. 'Global temperatures have gone nowhere for 17 years. Now, credible German scientists claim that "the global temperature will drop until 2100 to a value corresponding to the little ice age of 1870".'

The ice age prediction seemed unlikely to pan out. But one thing Newman could bet on was an inquiry into Australia's Renewable Energy Target, a move Abbott flagged before the election. The government picked another high-profile climate-sceptic businessman for the job, Dick Warburton, a former chairman of the Caltex Australia oil company. Warburton defended his appointment, saying, 'I am not a denier of climate change. I am a sceptic that man-made carbon dioxide is creating global warming.'

The Warburton review was a challenge for Hunt, who'd been a supporter of the Renewable Energy Target. When Hunt, along with Industry and Science Minister Ian Macfarlane, announced the review, they both knew it would be politically fraught. Abbott had gone to the election promising to keep the Renewable Energy Target. Now he wanted to backtrack on that promise, and the review could be a way to do it. 'I would've liked to have drastically reduced or abolished the Renewable Energy Target in early 2015,' Abbott said later. 'As we know, the Renewable Energy Target has done untold damage to our power system.'

News of Warburton's appointment was a blow to the renewables industry, already reeling from the U-turns in Canberra's climate policy. Investment was dropping like a stone. Over at the CEFC, Oliver Yates watched it collapse around this time. 'It absolutely froze it, it chilled it,' said Yates. 'At that stage renewables were much more expensive than they are today. You needed an incentive so they could compete in the electricity market.'

Abbott would look back at this collapse with pride. 'Good,' he said later. 'I mean, subsidised renewable energy has done so much damage. It's driven prices up, it's driven reliability down. I mean all these rent-seekers love it because it's a licence to print money. But it's printing money at the expense of consumers and the jobs of people in heavy industry.'

Abbott belonged to a school of thought that believed the Renewable Energy Target was an unnecessary subsidy that burdened the economy and electricity users. He did not buy the idea that, if Australia was going to make the long and painful journey to clean energy, it had to be kickstarted with subsidies—especially if there was no price on carbon emissions to drive investment.

Put simply, the Renewable Energy Target worked by making the big buyers of wholesale electricity, like electricity retailers, purchase renewable energy certificates to cover a fixed quota of renewable energy. These certificates were generated from two different sources: large renewable energy power stations and small-scale systems like solar rooftop panels or solar hot-water units bundled together, usually by an installer.

When Labor raised the Renewable Energy Target to at least 20 per cent of electricity generation by 2020, its aim had been to supercharge the shift to renewables, including wind and solar. Such clean energy sources, until then, had been dominated by hydro. Nonetheless, Labor had made sure that large greenhouse-emitting industries, like steel and aluminium, which exported to the world market, were cushioned from the full costs of the target.

By 2014, the number of certificates from small-scale units had blown out as Australians embraced solar energy with a passion. Over 2 million small units had been installed for homes, small businesses and community groups; well over half of these were rooftop solar panels.

Abbott and some other ministers were up in arms about the cost of the Renewable Energy Target to household electricity bills. Overall, it added somewhere between $15 and $50 a year to an average bill. But the Renewable Energy Target was also predicted to lower electricity prices as time went on, and, for many house-holders, having solar panels already reduced their bills.

Thanks to the Renewable Energy Target, large-scale renewable generation had also been on the rise until Abbott's election, but that too upset many in the government. Renewables accounted for around 13 per cent of electricity generation, but because overall electricity demand was falling, wind farms in particular were now threatening to push out some coal-fired power. That was seen as bad for coal, and it also would make the electricity grid more difficult to operate.

A chorus of Coalition politicians joined the attacks on renewable energy, particularly wind farms. Hockey complained to Alan Jones on radio 2GB, 'I drive to Canberra to go to parliament and I must say I find those wind turbines around Lake George to be utterly offensive. I think they're a blight on the landscape.'

The new MP for Hume in regional New South Wales, Angus Taylor, also went on Jones's program attacking wind farms. Taylor had gone to war with the ACT government over the expansion of wind turbines in his electorate. Another NSW Liberal back-bencher, Craig Kelly, echoed these complaints.

Not surprisingly, when the Warburton review came down in August 2014, it recommended winding back the Renewable Energy Target. It argued that the target, which was set at 41,000 gigawatt hours (GWh) of generation in 2009, could now end up higher than the mandated 20 per cent by 2020, because of falling electricity demand.

But renewable energy was incredibly popular with voters, and Warburton's report was attacked by all sides of politics, including by Coalition members and senators in regional Australia. There was no way slashing the Renewable Energy Target would get through the Senate. Then South Australian Labor premier Jay Weatherill saw the backlash hit the Abbott government. 'It was a jobs issue,' he explained. 'And it became a regional jobs issue because the over-whelming majority of investment in renewable energy is occurring in the regions, so this became a political issue for them.'

After a protracted fight inside the government, Hunt and Macfarlane finally managed to persuade Abbott to compromise with Labor over the Renewable Energy Target, but that took eight months to lock in. The deal cut the target from 41,000 GWh to 33,000 GWh—estimated at around 23.5 per cent of electricity generation by 2020.

The compromise with Labor infuriated opponents of the Renewable Energy Target, including the prime minister's senior business adviser, Maurice Newman. The day the compromise deal was announced, Newman managed to exact some revenge. He wrote one of his inflammatory columns in *The Australian*, in which he virtually accused the UN of a global conspiracy for wanting to reduce greenhouse gas emissions.

'It's a well-kept secret, but 95 per cent of the climate models we are told prove the link between human CO_2 emissions and catastrophic global warming have been found, after nearly two decades of temperature stasis, to be in error,' Newman wrote. 'Why then, with such little evidence, does the UN insist the world spend hundreds of billions of dollars a year on futile climate change policies?' The reason, wrote Newman, was simple: 'This is not about facts or logic. It's about a new world order under the control of the UN.'

Newman's attack was aimed at the top UN climate official, Christiana Figueres, who had met with Hunt in Australia that week. Hunt was forced to brush off Newman's UN conspiracy theory at his press conference, saying delicately, 'individuals are entitled to their views'. At this point, Hunt was still trying to make sure the Renewable Energy Target deal would make it through his own party room. 'It's a sensible package, it's a sensible balance,' he told reporters. 'I think and hope that we can just get on with the job of producing energy from here on.'

But as far as Abbott was concerned, most of that energy was still going to come from fossil fuels. And he wasn't taking advice from anyone who seriously questioned that.

———

At least one man in the Canberra bureaucracy believed he had a duty to advise the Abbott government on climate change, whether it listened or not. Former governor of the Reserve Bank Bernie Fraser was chair of the Climate Change Authority, set up by Labor's Julia Gillard to give independent, expert advice to the government on climate change policy including Australia's 2020 target to cut greenhouse gas emissions in line with its UN commitments.

Unlike the purging of Tim Flannery and the Climate Commission, Hunt didn't have the power to sack Fraser and the Climate Change Authority; it was set up as an independent statutory agency. The Abbott government saw it as stacked with Labor and Green appointments. But Hunt needed a new law to get rid of it and, once again, he had to get that law through the Senate, where the government didn't have the numbers.

In February 2014, the authority released a report that put climate science front and centre of its advice to the Abbott government. It called for Australia to make a much bigger effort on climate change. It recommended Australia strengthen its proposed 2020 target to cut greenhouse gas emissions from 5 to 19 per cent, based on 2000 levels. It also proposed that Australia would need to look at emissions cuts of between 40 and 60 per cent by 2030. Not surprisingly, the environment minister, Greg Hunt, didn't want to hear this.

'He wasn't interested in what the authority did or what I did,' Fraser recalled. 'He was very much under the thumb of Abbott and he really wasn't brave enough to challenge Abbott. Anything that Abbott said was no-go; he wasn't going to argue with that.'

Hunt would dispute he was under Abbott's thumb, but not that he ignored the Climate Change Authority's advice. Hunt wanted to shut down the authority and get rid of its entire board, which included Fraser and Australia's chief scientist, Professor Ian Chubb.

'Every time Abbott or Hunt spoke about climate change or the authority, it was all about abandoning or terminating the authority,' said Fraser. 'That was a major plank of their policy.'

While the government didn't have the numbers in the Senate, the Climate Change Authority had a temporary reprieve and Fraser decided to use it. At 73, Fraser had spent almost his entire adult life in public service and knew how to give frank and fearless advice. He had been Treasury secretary for the Hawke Labor Government before being appointed governor of the Reserve Bank. If Abbott and Hunt weren't going to listen, he would make the case to the public.

So in March 2014, Fraser gave an extraordinary speech to the National Press Club, laying out the authority's tough advice on climate change. He started with the Copenhagen Accord and the understanding that the world needed to keep global temperatures from rising more than 2 degrees to avoid dangerous climate change. 'We are halfway towards this increase of two degrees,' he warned his audience. 'We're halfway there now.'

On the current climate modelling, Fraser explained, if the world wanted a reasonable chance of keeping to the 2 degree goal, it needed to 'budget' the amount of greenhouse gas emissions it put into the atmosphere from here on in. In other words, there was a global carbon budget and each country had a share of that budget. 'Our fair share is calculated,' Fraser said, 'at about one per cent of this global emissions budget.' If Australia was going to do its fair share in meeting that budget, it needed to make the kind of deep cuts to its emissions that the Climate Change Authority was canvassing. He put up a range of practical proposals for the government to reach these ambitious targets: everything from setting emissions standards for cars to creating a fund to buy carbon offsets overseas.

Fraser told the Press Club he understood that many Australian businesses would argue tooth and nail against the authority's advice. And he bluntly concluded, 'It seems clear to me that in the area of climate change policy, the Government is backing in business interests and big business interests for the most part ahead of community interests.'

Fraser never regretted giving that speech. 'I was trying to make a case without scaring the pants off everybody about climate change,' Fraser recalled later. 'I was trying to persuade people that this was real.'

While the Abbott government ignored the authority's advice, some in the energy business were listening. At the CEFC, Oliver Yates would repeatedly raise the authority's work on the emissions targets with investors. In his view it was 'the most influential piece of policy' on climate change that came out in the Abbott years.

In long-term investment decisions on energy, Yates believed, the 'macro outlook' had to be dictated by the climate science.

And the science said greenhouse gas emissions in Australia had to fall. That meant Australia's coal-fired power stations were going to have to be shut down sooner rather than later.

'The Climate Change Authority in 2014 put out their final report, stating that the government needed to reduce its emissions between 40 and 60 per cent by 2030 to stay on the 2 degree path. They wrote down the schedule of coal-fired power station closures,' Yates said. 'If you look at that and say "Am I going to defeat science?" the answer is no, I am not going to defeat science.'

For the Abbott government, Bernie Fraser and the Climate Change Authority were just more Labor–Green climate advisers it wanted to purge. So in the May 2014 budget, Treasurer Joe Hockey announced the authority's funding was being stripped back to zero. The budget also slashed funds for the Australian Renewable Energy Agency (ARENA) and cuts were made to the country's top science agency, the CSIRO. All this was expected and barely made the news. What did hit the headlines were Hockey's unpopular budget cuts to health, education, welfare, the ABC and SBS. The anger over those cuts rocked the government to its core.

Abbott and Hockey suddenly found themselves accused of breaking their election promises—just what Julia Gillard had been accused of three years earlier. Abbott had promised, before the election there would be no surprise cuts in the first budget. 'I trust everyone actually listened to what Joe Hockey has said last week and again this week,' Abbott told SBS News in his last week as opposition leader. 'No cuts to education. No cuts to health. No change to pensions. No change to the GST, and no cuts to the ABC or SBS.'

But on budget night Hockey wielded the axe and dropped a bombshell proposal to make Australians pay a $7 Medicare co-payment for a visit to their GP. Within weeks, Abbott's approval rating crashed, along with his political capital.

All this was excellent news for Clive Palmer, the new MP for Fairfax. In July, his PUP party was also set to share the balance of power in the Senate and he knew Abbott would now need him more than ever. Palmer and his senators were being wooed by a growing number of anxious lobbyists fighting program cuts going up to the Senate.

One of the more surprising people wooing the billionaire was the former head of the Australian Conservation Foundation, Don Henry. The veteran environmentalist was trying to save as many of Labor's climate programs as he could. He wasn't alone. Bob Brown's former chief of staff, Ben Oquist, was also working hard on the PUP leader. But Don Henry had one big lure to attract Palmer—the former US vice-president Al Gore, the climate change hero of *An Inconvenient Truth*.

Palmer, it turned out, was a fan of the Kennedy Democrats. He was not only a donor to the Kennedy Library in Boston, he'd read the speeches of Al Gore's father in the US Senate. When he heard Don Henry was on the board of Al Gore's Climate Reality Project, Palmer was keen to talk to him.

Henry soon found himself as the go-between for Palmer and the former vice-president. Gore was coming to Australia that June and Palmer asked Henry 'if Mr Gore would consider supporting or partially supporting or providing some encouragement for me if I took certain positions?' Henry recalled.

Palmer agreed PUP could help save the Climate Change Authority, ARENA, the CEFC and the Renewable Energy Target. This was great news for Henry and Gore. But Palmer drew the line at 'the carbon tax'. He had gone to the election promising to axe it and there was no way he would renege on that. Besides, while Palmer had stood aside from running his Queensland Nickel company when he became an MP, his family was still involved with the business. The carbon tax had hit the company like it had others.

In the end, Gore faced a tough decision, said Henry. 'Do you stand up supporting various pretty significant pieces of climate action but still see the loss of cap and trade [the emissions trading scheme]? Or do you just let that opportunity pass?'

Both Gore and Henry agreed they needed to save what climate programs they could. So on 25 June, Gore boarded Palmer's private jet from Melbourne and came to Canberra for a joint press conference with Palmer in the Great Hall of Parliament House.

'We came to the agreement, then had dinner that night,' Palmer recalled. His colleague Glenn Lazarus and independent MP Cathy McGowan joined Palmer and Gore in the Parliament House dining room. 'Al Gore was very practical,' said Palmer. 'The Greens were very disappointed and very angry.'

That was an understatement. Former Greens leader Christine Milne slammed Gore for being part of the Palmer deal. 'His credibility in Australia was all but destroyed in a single day,' she pronounced.

A few weeks later, Palmer supported Abbott and Hunt to abolish Labor's emissions trading scheme, 'the carbon tax', in the Senate. Abbott, Hunt and the front bench celebrated their victory. They had finally killed off the scheme to put a price on Australia's greenhouse gas emissions. This bitterly contested policy had convulsed both Coalition and Labor governments for over a decade; it helped bring down two prime ministers and one opposition leader. Now Abbott and Hunt had finally driven a stake through its heart.

Among the biggest winners from this decision were the brown-coal power generators in Victoria. As the Climate Change

Authority later reported, the country's ten main coal generators had received some $2 billion from the Gillard government to compensate them for the 'loss of value' in the assets on their balance sheets caused by the introduction of the emissions trading scheme. The heaviest polluting plants had pocketed about half a billion dollars each. When Abbott dumped the scheme, it was never repaid.

'So we effectively got two years of compensation upfront and then the scheme was, post-election, repealed,' said Richard McIndoe the former head of TRUenergy, owner of Yallourn brown-coal plant in Victoria. Much to his surprise, no one from the Abbott government asked for the money back. 'No. No one ever raised it,' said McIndoe. 'It was never raised at all.'

After the government killed off the emissions trading scheme, Hunt pushed through his replacement plan to cut Australia's greenhouse gases. His Emissions Reduction Fund passed the Senate with Palmer's support. It worked on the idea of paying companies and farmers to reduce emissions voluntarily. Its cost was estimated at $2.55 billion over four years and it was widely criticised as an expensive, inefficient way to cut greenhouse gas emissions. While it promised to remove about 14 million tonnes of carbon emissions a year, one study found this cut was equivalent to just one-third of the annual emissions of the Australian energy company AGL—owner of some of the biggest emitting coal-fired power stations in the country.

Palmer's quid pro quo for supporting Hunt's plan was that Abbott agree to keep Bernie Fraser's Climate Change Authority. Palmer wanted the authority to examine his own ideas for an emissions trading scheme in the future. 'They agreed to that. They really needed Palmer's vote,' Fraser recalled. But Hunt made it clear the Abbott government would not support any new emissions trading scheme and it would still abolish the Climate Change Authority as soon as it got the chance. 'It was a lifeline from Palmer,' said Fraser. 'We grabbed the line, but it was a pretty flimsy line.'

Abbott wasn't happy with the Palmer deal, but he was delighted the carbon price was gone at last. 'With the help of Greg Hunt, I think we came up with a very sensible climate change policy,' Abbott said later. 'From then to the end of my time as leader there was no real dissension.'

———

Abbott was convinced he could brush aside the science of climate change; but that didn't make it go away. In November 2014, the IPCC released its latest advice, warning that greenhouse gas emissions caused by human activity were the highest in history. It predicted that continued emission of greenhouse gases would cause long-lasting changes for people and the planet, increasing the likelihood of 'severe, pervasive and irreversible impacts'.

That same year, Australia's top science agencies, the CSIRO and the Bureau of Meteorology, spelt out what this meant for the country. Australia's climate, they reported, had already warmed by 0.9 degrees Celsius since 1910. They explained the frequency of extreme weather was increasing; rainfall had declined since 1970 in the south-west of the country, while winter and autumn rains had also declined since the mid-1990s in the south-east. And they detailed how extreme fire weather had increased since the 1970s and the fire season had lengthened across large parts of Australia.

When the agencies looked to the future, they found the projections were even more serious. The number of extreme fire-weather days in southern and eastern Australia was expected to keep increasing along with temperatures. That meant an even longer fire season, further declines in rainfall in southern Australia and an increase in the frequency and severity of droughts.

When the rains did come, extreme daily rainfalls could increase, along with the risk of flooding.

Despite all this scientific advice, Abbott wasn't convinced climate change was an urgent issue for his government. On the contrary, his office backed demands for an investigation into the Bureau of Meteorology, which co-authored the climate reports.

Some of these demands came from high-profile climate sceptics in a wide-ranging attack on the bureau reported by *The Australian*'s environment editor, Graham Lloyd. He detailed accusations that the bureau had been 'manipulating historic temperature records to fit a predetermined view of global warming'. The claims were based largely on the work of Jennifer Marohasy, a former environment researcher with the Institute of Public Affairs. She was urged on in her attacks by Maurice Newman, chair of the prime minister's Business Advisory Council. Newman savaged the bureau in his column in *The Australian*, chiding it for its membership of the World Meteorological Organization, which was, he claimed, 'an anthropogenic warming propagandist'.

The battle over the weather bureau went on for weeks at the highest levels of the Abbott government. By this stage there were concerns inside the government that the bureau's reputation, both at home and abroad, could be damaged. After pushback from Hunt and his parliamentary secretary, Simon Birmingham, Abbott agreed to limit the investigation to an expert review that would work with the bureau.

Even so, by the end of Abbott's first year in office, some of the Australia's leading climate scientists felt under siege in their own country. The prime minister had made it clear he didn't believe most climate scientists, even those who worked for his government. But while Abbott's climate scepticism could shape policy in Canberra, it would soon put him in conflict with the most powerful leader in the world.

CHAPTER FOURTEEN

THE TARGET

THE SUMPTUOUS STATE banquet in the Great Hall of the People was quickly overshadowed when the host and the guest of honour stood, locked eyes, and raised their glasses. With that simple gesture, Barack Obama and Xi Jinping sent a message around the world. A new global climate agreement was possible.

Outside, in Tiananmen Square, Beijing's choking smog haze had lifted a little, thanks to the Chinese officials who had shut down factories and kept some cars off the streets in the weeks leading up to the Asia-Pacific Economic Cooperation (APEC) summit in November 2014. The summit was a non-event but the historic meeting on the sidelines between President Obama and President Xi kickstarted climate talks that many had thought were going nowhere.

The world's top two greenhouse gas polluters had agreed to seriously tackle their nations' emissions over the next sixteen years. It was a bold promise, aimed at pushing the rest of the world to sign up to an ambitious new UN climate agreement in Paris in 2015.

Standing side by side with the Chinese president, Obama ramped up the Beijing deal: 'By making this announcement today, together, we hope to encourage all major economies to be ambitious—all countries, developing and developed—to work across some of the old divides so we can conclude a strong global climate agreement next year.'

The US had agreed to cut its emissions by between 26 and 28 per cent by 2025, just over a decade away. The cut was from a baseline of 2005 but was still more ambitious than many expected. China in turn agreed on a date when its own emissions would finally peak—2030—then begin to decline. It even suggested it could reach that target earlier. As part of its plan, China flagged it wanted to get 20 per cent of its energy from clean sources by 2030.

The Beijing deal had been the best part of a year in the making. Nine months earlier, Secretary of State John Kerry and his top climate envoy, Todd Stern, had flown to China for the first talks. But the aim was not just a top-down deal with China. They wanted to create the momentum for a UN agreement at the 2015 Paris climate summit that included every nation on the planet.

'I think it was a pivotal event. It sent a very powerful signal to countries all over the world that this Paris agreement that everyone had been working on could actually happen after all,' Stern recalled. 'The US and China were the historic antagonists, the biggest player in their respective blocks of developed and developing countries. But when you had Obama and Xi come to that agreement it was a big signal.'

For Obama, the Beijing deal was a game changer. It drew a line under the debacle of the 2009 Copenhagen summit, when he had been snubbed by the Chinese leadership and left to carry the can for a weak outcome. The US was determined this would not happen again in Paris. Obama was now in the last two years of his presidency, and tackling climate change was a legacy project.

Obama had hand-picked a tough political operator as his senior adviser to push the climate strategy. John Podesta had served as President Clinton's chief of staff, and Obama called him back in 2013 to take on the job. 'He said, "I need someone to drive this whole climate action plan including the international side. Come to the White House and do that for me,"' Podesta recalled.

The China–US climate deal in November 2014 took many in Tony Abbott's government by surprise. 'I think they had to be taken by surprise because we surprised everybody. That was the intent,' said Podesta. The whole purpose was to 'send a powerful signal that the two largest economies in the world were reorienting towards a clean-energy economy'.

It was not a signal Abbott wanted to pick up. This was a problem for Podesta, who had gone to Beijing with Obama when the president sealed the deal. A few days after the Beijing announcement, Obama and Xi were due to fly into Brisbane for a long-planned G20 meeting of the world's major economies hosted by Abbott. The White House wanted to carry the momentum from Beijing to Brisbane with climate change squarely on the G20 agenda. Abbott was resisting.

For Obama, the Brisbane meeting was crucial to the strategy. There he planned to announce a US$3 billion pledge to the UN's Green Climate Fund—the pot of money to help developing nations adapt to climate change and shift to clean energy. The money was key to winning broad support for a new climate agreement. Obama's aim was to put pressure on the other rich nations at the G20, including Australia, to make big pledges to the fund a few weeks later at a UN climate meeting in Lima.

'We needed to make that announcement,' Podesta explained. 'That was the one–two punch between the US–China statement and the Green Climate Fund statement. We wanted the G20 to take up and add momentum through the series of events the Australian government was hosting.' Unfortunately, said Podesta, 'they weren't that keen on that.'

Podesta could not have been surprised. He knew Abbott's history as a climate sceptic. 'He was the Donald Trump of the times,' was how Podesta later described the Australian prime minister. 'He was definitely rowing in the opposite direction.'

Two months earlier, in September, Podesta had hit out at Abbott's government during a closed-door meeting of Australian and US political insiders in Washington. 'I put aside diplomatic niceties. I gave a full and frank description of my views,' Podesta recalled. Australia, he said, needed to step up on climate change and do more. That was also a message Obama took to Brisbane.

But Abbott was a vocal opponent of the UN Green Climate Fund and didn't want to put money into it. For him it was 'a Bob Brown bank on an international scale'. One of the aims of the UN Green Climate Fund was to get poor countries to ditch heavily polluting coal-fired power in favour of cleaner energy.

The wrangling between Washington and Canberra over climate change and the UN Green Climate Fund went on until the world leaders flew into Brisbane. Australian officials knew climate change had to be discussed, but Abbott and his office were still resisting. Obama made it plain he would not be thrown off course. 'Obama insisted that climate be included in the G20 agenda. We had worked to elevate climate at the G20 for years,' said Ben Rhodes, Obama's former deputy national security adviser.

On the opening day of Brisbane's G20, Abbott gave a low-key speech to the world leaders about his budget woes, asylum seekers and his success at getting rid of Labor's 'carbon tax'. Grudgingly he also conceded climate change would be on the leaders' agenda. 'Obviously I would like this discussion to focus on the politics of economic reform, that's what I would like the discussion to do,' Abbott told the leaders. 'In the end, though, this is your retreat. It is open to any of you to raise any subject that you wish.'

The US president did just that. That afternoon he gave a speech at the University of Queensland that reverberated around Australia, annoying the hell out of Abbott and his office. In front of around 1000 students and local dignitaries, Obama unveiled America's $3 billion pledge to the UN's Green Climate Fund.

In his speech, Obama took a swipe at the mantra that coal was the key to lifting countries out of poverty, a line pushed by the Abbott government. While the Green Climate Fund could be used for 'cleaner' coal plants, Obama hoped it would drive the push towards all forms of clean energy. 'It allows us to help developing countries break out of this false choice between development and pollution,' said Obama. 'Let them leap-frog some of the dirty industries that powered our development—go straight to a clean energy economy that allows them to grow, create jobs, and at the same time reduce their carbon pollution.'

What particularly upset Abbott was that Obama linked rising greenhouse emissions to the state of the Reef. 'The incredible natural glory of the Great Barrier Reef is threatened,' Obama said starkly as he pressed the case for a new climate agreement. 'We can get this done. And it is necessary for us to get it done. Because I have not had time to go to the Great Barrier Reef and I want to come back, and I want my daughters to be able to come back, and I want them to be able to bring their daughters or sons to visit. And I want that there fifty years from now.'

By the time the applause died down, it was clear Obama had not only sent a message to the G20 leaders but to Abbott in particular. 'The speech was a public stake in the ground [at] a global meeting in which the chair was trying to avoid the subject,' said Podesta bluntly.

In Abbott's circle, Obama's speech was seen as payback. 'A very deliberate smash at us,' was how one insider put it. A few Australian advisers blamed Podesta, but the tough words were Obama's alone.

'What was interesting is that the [written] speech had a short section on climate, but then in delivery Obama ad-libbed in a much longer section completely on his own,' Ben Rhodes recalled. 'I think it had to do with several factors, which I discussed with him at the time: a very positive response from the audience, which was younger; his focus on climate at the time coming off the Beijing trip; and yes—his frustration with Tony Abbott.'

Abbott was furious about Obama's speech and its references to the Great Barrier Reef. He called on his foreign minister, Julie Bishop, to get out and defend the government. 'Tony asked that I make a public defence of Australia's response to challenges regarding the Great Barrier Reef,' Bishop recalled. 'I sought to be constructive in that response, as I also felt that President Obama had not been as fully informed as we had assumed, prior to making his remarks.'

In the end, despite all this aggravation, Abbott had to put his name to the final G20 leaders' communiqué that supported 'strong and effective action to address climate change'. The communiqué also called on countries to set their 2030 targets to cut emissions for a new climate agreement in Paris, and backed financing for developing countries to tackle climate change through bodies 'such as the Green Climate Fund'.

'All of us support strong and effective action to address climate change,' Abbott said dismissively in his final G20 press conference. But he dodged a question on whether Australia would put money into the Green Climate Fund. Instead, he launched a passionate defence of coal-fired power. 'Coal is going to be, now and for the foreseeable future, a very important part of the world's energy needs,' Abbott insisted. 'It has to be because, if it's not, we are never going to provide energy to the 1.3 billion people who don't have it.' He did manage to add a proviso that coal-fired power needed to be more efficient.

Soon after the G20 meeting, Australia's delegation left for the UN climate talks in Lima, Peru, where the UN's Green Climate Fund was now well and truly on the agenda. Julie Bishop led the delegation, but Abbott asked Andrew Robb, the trade minister well known for his climate-sceptic views, to go with her. As it turned out, Robb proved to be an ally rather than a chaperone when Bishop pressed Abbott from Lima on the need to make a pledge to the fund. Japan, Canada and a host of other rich countries had already put up their hands. 'There was regular consultation with the prime minister and others about the rationale, including a call to cabinet, from memory,' Bishop recalled. 'Part of my argument was our standing with the nations of the South Pacific, who regarded this fund as essential to their future prospects.'

Abbott was still dubious, but he finally agreed to put up $200 million over four years, using money pulled out of the foreign aid budget. Even under an Abbott government, Australia could not afford to be completely out of step with its allies over climate change.

But despite Australia's pledge to the Green Climate Fund, Abbott, like many of his cabinet colleagues, was convinced Australia's coal industry would keep powering the world indefinitely. The big coal companies had his government's resolute support, and they knew it.

———

In the lead-up to the G20 summit in Brisbane, some of the coal giants had been keen to capitalise on Australia's support. One in particular, America's Peabody Energy, was closely in tune with the government's thinking. Earlier that year, Peabody's chief executive, Greg Boyce, had launched a worldwide public-relations campaign pushing back against renewable energy and promoting coal-fired power plants. The company's big message was that energy poverty, not climate change, was 'the world's number one human and environmental crisis'.

Peabody's 'Advanced Energy for Life' campaign was designed in part by the global lobbying firm Burson-Marsteller. It pushed the alarming statistic that 3.5 billion people in the world were 'without adequate access to energy'. Images of desperate children and babies pressed the case that coal could end 'energy poverty' using today's 'advanced clean coal technologies'. The problem was, of course, no technology was actually delivering commercial clean coal at this point but a small footnote in the glossy brochures had a very broad definition of clean coal that included plants that were simply more efficient. Coal power could replace highly polluting wood stoves, but even the most efficient new coal plants would still produce a lot more greenhouse gas emissions than renewables.

Despite this, by the time of the G20 meeting in Brisbane, Peabody was claiming 500,000 people around the world had joined its social media campaign to lobby world leaders to end 'energy poverty'. At the end of the G20, Abbott was echoing Peabody's messaging.

Peabody's line on energy poverty wasn't new. It had pushed it since 2011. But Advanced Energy for Life rebooted the message at a time when the company needed to keep coal on top in the energy mix. Prices had slumped and climate change was posing an existential threat to coal's long-term future. Peabody was also under big financial pressure. In the past, the company had aggressively attacked climate science and funded climate-sceptic groups, but this was becoming less and less effective. The campaign to end energy poverty was a far easier sell for its political allies. Peabody's critics believed it was designed in part to fight Obama's new regulations in the US restricting carbon dioxide emissions from coal-fired power stations. But it was also pitched at the Indian and Asian markets, where coal imports were rising.

Peabody was a big coalminer in Australia with a board seat on the peak mining lobby, the Minerals Council of Australia, alongside BHP, Rio Tinto and Glencore. It had been part of the campaign against Gillard's emissions trading scheme run by the old coal lobby, the Australian Coal Association. That lobby had now been absorbed into the Minerals Council. The council's new chief executive, Brendan Pearson, had just finished a two-year stint at Peabody's Australian arm.

Peabody's energy poverty message in Australia was a bit disingenuous. While Australia's thermal coal exports were powering electricity plants in Asia, its lucrative coking coal exports, largely from Queensland, were fuelling steelmaking in Japan, China, Korea and Taiwan, not exactly lifting the world out of grinding poverty.

But Abbott embraced the cause of 'energy poverty' with a passion, and so did his key ministers. In May 2014 the prime minister had made a rousing defence of Australian coal to the annual minerals industry parliamentary dinner, telling the mining executives that it was 'our destiny in this country to bring affordable energy to the world'.

The prime minister managed to weave the crusade against energy poverty together with Australia's ambition to retain the crown as the world's biggest coal exporter. 'It's particularly important that we do not demonise the coal industry,' Abbott told the executives, 'and if there was one fundamental problem, above all else, with the carbon tax, it was that it said to our people, it said to the wider world, that a commodity which in many years is our biggest single export, somehow should be left in the ground and not sold. Well really and truly, I can think of few things more damaging to our future.'

Abbott's mantra became 'coal is good for humanity' and it was harnessed to justify opening up the vast thermal coal reserves in Queensland's Galilee Basin. By mid-2014, Gina Rinehart and her Indian partner GVK were making little headway in the Galilee; their costs had risen, the price of coal had fallen and some approvals remained outstanding. Both the state and federal governments were now pinning their hopes on the Indian billionaire Gautam Adani to get the first Galilee mine up and running.

The environment minister, Greg Hunt, signed the federal approvals for Adani's Carmichael mine in July that year. On paper it was largest coal project in Australia's history. If it went ahead as planned, the Galilee coal burnt in power plants in India and elsewhere would emit as much greenhouse pollution each year as the whole of Queensland. But Hunt brushed that problem aside, instead pushing the moral case to fight energy poverty.

'Our task now is to ensure that the breakthroughs in technology and the reduction in emissions—which can be done through better technology—occurs,' he said. 'But if you're asking us to stop 100 million people in India having for the first time electrification or having significantly extended electricity—the great goal of bringing humanity out of deep, grinding poverty—we're not a) able to stop that as a country and b) we shouldn't be condemning people to poverty.'

The go-ahead for the Adani mine dramatically shifted the politics of climate change in Australia. It supercharged the campaign by climate activists to strip the social licence from coalmining in Australia. And it put the 'coal versus coral' debate at the heart of that fight.

It also turned the spotlight back on Abbot Point, the coal port inshore from the Great Barrier Reef that was earmarked to ship much of the Galilee coal. Both the Queensland and federal governments supported the port's threefold expansion. Hunt had gone ahead and approved plans to dredge 3 million cubic metres of spoil from the seabed at Abbot Point and dump it in the Great

Barrier Reef Marine Park. Hunt said he did this only after 'a very careful and deep review' and with conditions that managed to win over the Marine Park Authority's chairman, Russell Reichelt. But leaked internal emails showed that the authority's scientists and officials had fought against the approval down to the wire.

Hunt's Abbot Point dumping decision turned out to be a strategic blunder. It derailed Australia's effort to stop the World Heritage Committee listing the Great Barrier Reef as 'in Danger'. The committee's annual meeting that year was held in Qatar, and by the time the Reef debate came on, the WWF had briefed committee delegates on the internal dissent in the Great Barrier Reef Marine Park Authority over Abbot Point. The committee's critical view of Australia and its delegation hardened.

'It just didn't go well for them,' said Richard Leck, who was marine spokesman for WWF Australia and was at the meeting. 'At that point the whole delegation stood up and walked out, which, again, was not a good look.'

In the end, the World Heritage Committee sharply noted its 'regrets' at the Abbot Point dumping decision and flagged it was possible the Great Barrier Reef could be inscribed on the 'List of World Heritage in Danger' by 2015. Australia's delegation was humiliated.

That debacle elevated the crisis over the Great Barrier Reef to the Prime Minister's Office. In the discussions that followed, Bishop and Hunt persuaded Abbott to mount a massive diplomatic campaign to prevent the Reef being put on the 'in Danger' list. Curiously, one of the most powerful arguments put to the government was that an 'in Danger' listing of the Reef could create big problems for Queensland's coal industry.

If the Reef was on the endangered list, everything from shipping coal through Reef waters to financing new coalmines inshore from the Reef could cost more. 'If you've got an endangered listing on something, the international banking community is going to be much more reluctant to lend capital around major project developments,' one government figure explained. That would include mines, railway lines and ports.

This was confirmed by the late head of the Queensland Resources Council, Michael Roche, in an interview he did with Professor Tiffany Morrison from James Cook University. Roche explained that the 'in Danger' listing would make it harder for governments to sign off on developments inshore from the Reef. 'The biggest developer they were worried about was the mining industry,' said Professor Morrison. 'They were really worried that having a World Heritage Area "in Danger" on the edge of all these developments would have an effect on the reputation of the mining industry, their social licence to operate and governmental appetite for regulation.'

By November 2014, the mounting global concerns over climate change were suddenly making Australia's coal industry look vulnerable. But the government's response was to double down on its support. This was clear just two days after the G20 meeting finished in Brisbane, when Abbott welcomed the Indian prime minister, Narendra Modi, to address federal parliament.

Abbott used the opportunity to spruik the proposed Adani coalmine in the Galilee and its big ambitions to feed new coal-fired power plants in India. 'If all goes to plan,' he said, 'next year an Indian company will begin Australia's largest ever coal development, which will light the lives of 100 million Indians for the next half-century.'

But when Modi spoke, he was not so bullish. He cheered on Australia as a partner in India's development, but when it came to powering his country, Modi called for 'energy that does not cause our glaciers to melt—clean coal and gas, renewable energy and fuel for nuclear power'.

Even energy-hungry India, it seemed, couldn't quite share Abbott's unchecked enthusiasm for coal-fired power. India, along with China and the other big developing countries, were by now responsible for more greenhouse gas emissions going into the atmosphere than the developed world. The Indian prime minister could read the scientific advice and the writing on the wall. If they kept polluting at the rate they were going, the planet would be unable to avoid dangerous climate change.

———

At the end of the summer holidays, in January 2015, Peter Costello, the former federal treasurer, was enjoying a barbecue with friends when news came through that immediately put his political antennae on high alert. In the Australia Day honours that morning, the prime minister had awarded the nation's top gong to Prince Philip. Abbott had made the 93-year-old husband of the Queen a Knight of the Order of Australia.

'You talk about a figurative barbecue stopper,' said Costello. 'This was a literal barbecue stopper. You know, I think everyone saw what was on the end of everybody's knife and fork.'

In retrospect it was hard to believe, but Tony Abbott's downfall began with this self-inflicted wound. The diehard monarchist suddenly looked completely out of touch with modern Australia. Worse still was his dreadful political timing. The Queensland premier, Campbell Newman, was in the final days of a bruising state election campaign. The following Saturday, Newman and his LNP government suffered a brutal swing against them, which put a minority state Labor government into office.

When federal Coalition MPs returned to Canberra that February, the polls were still bad for the Abbott government. They had never recovered from the previous year's horror budget. The discontent over Abbott's leadership exploded and angry backbenchers pushed for a leadership spill. Some went public, attacking both Abbott and his deeply unpopular, omnipresent chief of staff, Peta Credlin.

Just seventeen months into his prime ministership, Abbott suddenly found his leadership in play. Speculation about a challenger swirled around Julie Bishop and Abbott's erstwhile rival, Malcolm Turnbull, now the communications minister. Neither put up their hand. Instead Abbott faced an 'empty chair' challenge. Even without a rival, the partyroom vote against him was a shock: 39 to 61, in support.

The vote shook Abbott to the core. 'We have decided that we are not going to go down the Labor Party path of a damaged, divided and dysfunctional government,' he tried to persuade reporters after the ballot. But the political animal in him knew his enemies could smell blood. He described the challenge as his 'near-death political experience' and promised his colleagues that things would change within six months.

But Abbott now had a target on his back. If the polls didn't improve, the Liberal MPs would turn on him. 'They will stick with a prime minister who can win and they'll cut one off who can't,' said one of Howard's old ministers, Amanda Vanstone.

Abbott knew Turnbull would be stalking him. Abbott's first big scalp in the climate wars would be unlikely to pass up the chance for vengeance. 'I lost the prime ministership because Turnbull desperately wanted it,' Abbott said later. 'From the day he decided to remain in the parliament he was with greater or lesser degrees of intensity working to get back into the leadership.'

With his own leadership weakened, Abbott now had to wrap his head around another target. The UN climate summit in Paris was looming at the end of the year and Abbott had to decide on the 2030 target to cut Australia's greenhouse gas emissions. It threatened to be one of the most divisive issues to come before the

Coalition party room, and Abbott, the climate sceptic, would have to find a target they could all agree on. He didn't want it to be him.

———

Over at the Climate Change Authority, Chair Bernie Fraser had some advice for the government on Australia's Paris target. The authority was still hard at work, thanks to the reprieve granted to them by Clive Palmer and the crossbenchers in the Senate. In July 2015, Fraser released the authority's recommendations for Paris. They were based on three principles: what the climate science was saying, what other developed countries were doing, and what was in the interests of all Australians, including future generations.

The authority urged the Abbott government to lift its ambition to tackle climate change. It recommended that Australia cut its greenhouse emissions 30 per cent by 2025 from 2000 levels. It didn't give a specific target for 2030 but advised looking at an emissions cut of between 40 and 60 per cent from 2000 levels, depending on how quickly technology moved, the efforts of other countries and a range of other factors. The authority said these targets would put Australia in the middle of the pack with other developed countries and be a fair contribution towards keeping global temperatures from rising more than 2 degrees Celsius.

While the targets were challenging, the authority said, 'they are no more so than the targets many other developed countries have been pursuing in recent years and are committing to in the post-2020 period.'

Again, Hunt didn't want to hear this advice. For months he'd been wrangling with the Prime Minister's Office to get anywhere close to a credible target for Paris, let alone an ambitious target. Abbott had initially flagged an embarrassingly low target—a 13 per cent cut in emissions by 2030. That would have left Australia virtually isolated.

Hunt had another pressing reason to get a more plausible target from Abbott. He, along with Bishop, Australia's Ambassador for the Environment, Peter Woolcott, and the new Queensland Labor government, had just managed to pull off a diplomatic coup and keep the Great Barrier Reef from being listed on the World Heritage Committee's 'in Danger' list that year. In the lead-up to the committee's annual meeting, Hunt announced that the government would ban the dumping of dredge spoil in the Great Barrier Reef Marine Park, including the dredge spoil from the Abbot Point port expansion.

In light of Australia's victory at the committee meeting, Hunt and Bishop agreed Australia needed a Paris target that at least recognised the Reef was at risk from climate change, and showed that Australia supported the goal of keeping the rise in global temperatures within the 2 degree limit.

Inside the government, debate over the Paris target dragged on until August that year. Abbott's office went back and forth with Hunt and Bishop. Bishop was going to lead the Australian delegation to the Paris talks, but Hunt had to argue the case in cabinet for the target and, more importantly, explain how Australia would meet it.

Right from the start, Abbott wanted to stick to John Howard's old policy of 'No Regrets' that Australia had taken to Kyoto. He wanted a target that would inflict minimal cost on Australia's economy, and, in particular, on its big greenhouse-emitting industries.

Hunt and Bishop used the pledges by other developed countries to push up the target Abbott wanted. Finally they settled on a weak emissions cut of 26–28 per cent by 2030, on the higher baseline of 2005. It looked like a copy of the US target that Obama had announced, but it was far less ambitious—the US planned to meet its target by 2025. The EU target was also higher, a 34 per cent cut by 2030 off a baseline of 2005.

After detailed briefings from Foreign Affairs, the Attorney-General's Department and Treasury, Hunt convinced a sceptical Abbott that the 26–28 per cent target could be met without new policies or much pain. They would partly rely on Hunt's new Emissions Reduction Fund, which was already paying business and farmers to cut emissions. It was an exhausting and exhaustive effort by Hunt and Bishop.

Abbott reluctantly accepted the Paris target, believing it would be only a small cost on the budget. 'The advice I had with Greg Hunt was, at the most, we would be able to get there by a modest replenishment of the Emissions Reduction Fund,' Abbott recalled.

But many in the mining lobby were not convinced this would be enough to reach even the weak target agreed to by Abbott. They wanted the government to leave open the option of buying international carbon credits in case Australia's emissions blew out, something both Bishop and Hunt also thought might be necessary. Abbott was completely opposed to this. 'Greg was always keen on international carbon credits, and I was always profoundly hostile to them,' Abbott remembered. 'And if he'd said to me that we were going to need international credits to get there I would've said, "That's bullshit, we can't do it." If we had to do it with international carbon credits I would've said, "Well sorry, the targets are beyond us."'

As part of the decision on the Paris target, Abbott, almost unnoticed, quietly agreed to ratify Australia's second Kyoto target—a 5 per cent cut in emissions by 2020. This was done for a good reason—it would save the Treasury a fortune.

It worked like this. By ratifying the 2020 target, Australia would reap another windfall gain from the unambitious climate target the Howard government had negotiated at Kyoto back in 1997. Australia was one of the very few developed countries allowed to increase its greenhouse emissions under its 2012

Kyoto target—up to 108 per cent of a 1990 baseline. Australia had then easily met that target, coming in at 104 per cent. As a result, Canberra now had a surplus of carbon credits under Kyoto—128 million tonnes' worth.

Environment groups scorned these surplus credits as a 'hot air loophole'. The UK, Germany, Denmark and other rich countries all had surplus credits under Kyoto, but they were prepared to allow them to lapse. Canberra would not. The Abbott government wanted them rolled over to help meet its weak 2020 Kyoto target. The rollover credits would save Hunt's Emissions Reduction Fund from having to buy more carbon credits from business and allowed the minister to say Australia would meet its 2020 target.

———

Despite all the angst inside the Abbott government about the 2030 Paris target, the revolt in the Coalition party room didn't happen. The prime minister and his conservative allies were by now under siege on too many other fronts. Abbott unveiled the government's Paris target on 11 August 2015, selling the low-ball target as a practical solution that wouldn't put up electricity bills. 'We are committed to tackling climate change without a carbon tax or an emissions trading scheme that will hike up power bills for families, pensioners and businesses,' the prime minister said in a joint statement with Hunt and Bishop.

Hunt realised the toughest criticism of the Paris target could come from outside the party. Conveniently, the day before he and Abbott released the target, Murdoch's *Daily Telegraph* splashed a story on the hair-raising costs of the opposition's supposed Paris targets: 'ALP's $600b Carbon Bill'. The *Tele's* national political editor, Simon Benson, took aim at Labor's leader, Bill Shorten, and claimed 'Labor's ambitious plan to cut carbon emissions by 40 to 60 per cent by 2030 would deliver a devastating blow to the economy'; it would also 'cost tens of thousands of jobs' and 'would likely lead to the closure of all 37 coal-fired power stations in Australia'.

The story implied Shorten was going to support the Paris targets recommended by Bernie Fraser's Climate Change Authority— even though Shorten had yet to make a decision and a specific 2030 target had not been recommended by the authority. The story cited Treasury and Department of Industry modelling as the basis for the 'shocking predictions'.

When Bernie Fraser saw the story, he was furious. He issued a statement rejecting the claims about the Treasury modelling given to the authority. He also said the Abbott government's Paris target 'would put Australia at or near the bottom of the group of countries we generally compare ourselves with'.

Hunt returned fire by backing the *Tele* hit job to the hilt, saying, 'Labor is now trying to deny their own commitment after learning that the modelling they commissioned showed how devastating it would be for Australian families and the economy.'

Relations between Fraser, the veteran public servant, and Hunt hit rock bottom. In a phone call with Hunt, Fraser bluntly told the minister his claims were 'bullshit'. Fraser remembers his daughter being in the room when he said it. 'Her eyes were popping out of her head.' A few weeks later Fraser announced his resignation as chairman of the Climate Change Authority, telling his fellow board members the atmosphere with the government was 'poisonous'.

The *Tele*'s story on Labor did not head off criticism from the environment groups about the government's weak Paris target. The Climate Institute warned that Australia's target 'fails tests of both scientific credibility and economic responsibility'. It would leave Australia, by 2030, still with the highest per capita emissions in the developed world.

As the Paris climate summit drew closer, Australia's target also became a problem with its South Pacific neighbours. Along with other low-lying island nations, they were now pushing for the Paris summit to limit the global temperature rise to 1.5 degrees Celsius, a far more ambitious aim than 2 degrees.

As far as Abbott was concerned, this was never on the table and he thought the idea was ridiculous. 'I always regarded this as hocus pocus essentially,' he said later. 'This idea that somehow we could turn the thermostat up or down by adjustments to carbon dioxide, I mean I absolutely accept that climate change happens, mankind makes a contribution, we have to take prudent sensible measures to deal with it. But there are so many factors in climate, quite apart from carbon dioxide, the idea that you could make micro changes to the globe's thermostat by having a few million more or less tons of carbon dioxide was always, I thought, fairly dubious.'

Australia's Pacific Island neighbours would disagree. They believed the difference between 1.5 and 2 degrees could be the difference between their survival and their annihilation. Perhaps fortunately for them, Abbott would not be in power when the Paris climate summit opened.

After months of speculation, Malcolm Turnbull finally launched his leadership challenge against Tony Abbott in September 2015. Significantly, Turnbull barely mentioned the words climate change in his pitch to his Liberal colleagues. His appeal was purely pragmatic. He could win the next election and Abbott could not. Abbott had lost 30 Newspolls in a row and that was his undoing.

'The polls meant the bush was tinder dry and it was just waiting for a spark,' was how Costello put it.

Some of the very MPs who had torn Turnbull down over climate change now resurrected him. But not everyone had forgotten the bitter words he had thrown at his colleagues back then over the party's climate policy: 'Any policy that is announced will simply be a con, an environmental fig leaf to cover a determination to do nothing.'

But the new pragmatic Turnbull, at his first press conference as the leader, went out of his way to praise the Coalition's climate policy under Abbott, particularly the Paris targets cobbled together by Hunt and Bishop. 'It was a very, very good piece of work,' he said.

More than anyone, Turnbull knew under Abbott there had been a fragile truce in the Liberal Party's civil war over climate change, but the divisions were still just below the surface. They could easily blow apart his leadership again.

Abbott, for now, appeared to accept the finality of his defeat. He made an unexpected promise that he would not destabilise Turnbull. 'There will be no wrecking, no undermining, and no sniping,' he said. But he soon made it clear he was going to stay in the parliament. That guaranteed the fragile truce over climate change would remain just that—fragile. Abbott would resume hostilities at the first sign of Turnbull faltering.

CHAPTER FIFTEEN

HOSTAGE

Paris was still mourning its dead when Malcolm Turnbull flew into the shattered city and made his way to the Bataclan Theatre, where scores of young concertgoers had been gunned down just weeks earlier. Laying a wreath at the memorial outside, the Australian prime minister offered words of sympathy: 'We are all together. We are with France. We are with the people of Paris, we are with all people committed to freedom in this battle against terrorism, against violence, against violent extremism.'

The terrorist attacks in Paris in November 2015 had taken 130 lives, injured many more and rocked the government of François Hollande. They also nearly derailed the long-planned December climate summit, the UN's 21st Conference of the Parties (COP21). But world leaders rallied and so did France, allowing Turnbull to take his place at the podium of the Paris summit. In characteristic style, he struck a bullishly optimistic note about the climate crisis that threatened the planet.

'We are not daunted by our challenge. It inspires us. It energises us,' Turnbull told his fellow leaders. 'We do not doubt the implications of the science or the scale of the challenge. But above all we do not doubt the capacity of humanity to meet it—with imagination, innovation and the prudence that befits those, like us, who make decisions that will affect not just our own children and grandchildren but generations yet unborn.'

It was left to other leaders to hammer home the urgency of the climate crisis and the need for an ambitious deal. Barack Obama did just that. Fourteen of the fifteen warmest years on record had been since 2000, he reminded them, and 2015 would be the warmest of all. That summer, said Obama, he had seen how the sea had swallowed villages in Alaska, where the permafrost was thawing, fires were burning and glaciers were melting.

The leaders flew out of Paris before the tough negotiations were done, fearing a repeat of the Copenhagen failure. The late-night wrangling was left to ministers and to the French who, against all odds, pulled off a success with the help of some White House lobbying. For the first time, the world said yes to a landmark agreement for all countries not only to keep the global temperature rise below 2 degrees Celsius but to aim to limit it to 1.5 degrees.

At the last minute, Foreign Minister Julie Bishop backed the 'High Ambition Coalition', the campaign supported by Pacific Island nations to work towards the 1.5 degrees goal. Bowing to pressure from Australia's allies and its Pacific neighbours to get the Paris deal done, Bishop tweeted that Australia was in.

In reality, the pledges in Paris to cut greenhouse emissions still put the world on a path to almost 3 degrees of warming by the end of the century—exposing the planet to dangerous climate change. But the key to the Paris Agreement was the commitment by nations to review and if necessary strengthen the targets every five years. The aim was to get to a carbon neutral world from 2050. Rich countries were again expected to take the lead.

For weary campaigners, Paris was the big deal. 'It is a historic agreement and one many, including myself, would have not thought possible as little as a month ago,' wrote Erwin Jackson, deputy CEO of Australia's Climate Institute, who had worked on the cause since he was a young activist in 1992. 'I feel numb. This is a mix of exhaustion and the implications of the outcome today not having sunk in. Time will tell but today I feel we turned a corner. The future is brighter than it was yesterday.'

For Australia, the implications of the Paris Agreement were profound. It sent a signal that coal-fired power was on the way out. Unless 'clean coal' could work economically—and that still looked unlikely—developed countries like Australia would need to phase out coal-fired power stations by around 2030, within fifteen years of the Paris Agreement. China would need to phase them out by 2040 and other developing countries by 2050.

For Australia's carbon-intensive economy and its lucrative coal export business, this was the fork in the road. If the Paris agreement held, state and federal governments would need to step up plans to retrain or retire thousands of workers in coal-fired power plants. It meant transforming Australia's energy market. It meant shrinking thermal coal exports and finding new jobs for miners if there was going to be a just transition. It meant transforming greenhouse-gas-intensive industries like aluminium, steel and cement. It meant setting serious vehicle emissions standards, switching to electric vehicles, and scaling back emissions from liquefied natural gas and livestock production.

Australia would also have to bump up its weak 2030 target of a 26–28 per cent cut in emissions. That target had Australia on track to remain the highest per capita greenhouse emitter in the developed world and was in line with 3–4 degrees of warming, not 1.5.

'The time for piecemeal, unstable and short-term policy is over. The real work for Australia starts now,' was how John

Connor, the head of the Climate Institute, summed up the challenge for Turnbull when the Paris Agreement was done.

———

Despite Turnbull's rallying speech in Paris about facing up to climate change, there was little sign of it in Canberra. Turnbull was hamstrung. When he brought down Abbott that September, Turnbull had done a deal with the National Party to get their backing for a Coalition government. National's leader Warren Truss and his deputy, Barnaby Joyce, were both climate sceptics. As part of that deal, the Nationals had insisted there would be 'no carbon tax', in effect there would be no price on carbon pollution. That meant the policy put in place by Abbott would be very hard to change.

Turnbull was also boxed in by the Coalition party room, where the climate sceptics effectively exercised a veto over policy. By rough estimates, at least one-third of Turnbull's cabinet and a third of partyroom members were climate sceptics. 'It sounds about right,' said Turnbull later, 'but it depends on what you mean by climate sceptics.'

Former western Sydney Liberal MP and Turnbull ally Craig Laundy explained there were two types of sceptics in the federal party room. There was, he said, 'The sceptic in that they don't believe it's real. And there's a sceptic that thinks it's dumb politics to do anything in the space.' Tony Abbott and his allies like Craig Kelly were climate-science sceptics, said Laundy. Other MPs lined up with the sceptics for political reasons. 'These are the people who want red meat for the base,' said Laundy frankly. 'The young conservatives, they want to have a fight on it against the Labor Party.'

Turnbull had two cautious ministers holding the government's climate policy together. He kept Hunt on as environment minister and chose the clever, ambitious Josh Frydenberg as his new resources and energy minister. A star of the Victorian right, Frydenberg would not blow up the party room. Inspired by the conservative 1950s prime minister Sir Robert Menzies, he saw himself as a future leader and even kept one of Menzies' old walking sticks in his office.

Turnbull also brought Martin Parkinson back in from the cold, making him Canberra's top bureaucrat in charge of Prime Minister and Cabinet. But real change to the climate policy could only come if Turnbull won the 2016 election. Even then, he would need a convincing win to take on the party room, and that was not going to be easy. The polls were tightening, the electorate was still angry with the Coalition over the first budget cuts, and Turnbull was losing moderate swing voters by stalling on climate change and marriage equality.

The federal party was also seriously short of cash for the 2016 campaign. So much so that Turnbull asked the old Victorian powerbroker Ron Walker for help. Despite being ill with cancer, Walker agreed to set up a meeting for Turnbull with Hugh Morgan and Charles Goode, the men controlling the party cash cow, the Cormack Foundation.

Turnbull seemed oblivious to Morgan's past anger with him for backing a carbon price when he was opposition leader. Turnbull could never fathom Morgan's climate scepticism, which he saw as wacky rather than a strategic defence of Australia's fossil fuel industry.

'There is a particular constituency, and they tend to be men of a certain age group who are ferociously opposed to the concept of anthropogenic climate change,' said Turnbull. 'Maurice Newman is a good example. And it's weird. You take a guy like Hugh Morgan, who either is an engineer or used to employ hundreds of engineers, and who understands making decisions on the best available science. Yet they've got this massive scientific consensus on climate and they say, "No, it's all wrong." I find it extraordinary. It's a real phenomenon.'

Despite their differences, in March 2016 Morgan, along with Goode and Walker, met for lunch at Melbourne's fusty Athenaeum Club. The Cormack Foundation had given $3 million to Abbott's 2013 campaign and it was still regularly funding the Institute of Public Affairs (IPA) and other right-wing think tanks. But the Cormack grandees were more cautious about funding Turnbull. Their priority, they said, was the upcoming Victorian election.

'I spoke to Hugh and Charles. They gave us some money, but they weren't that helpful,' Turnbull recalled. 'They said they didn't want to have to sell any shares. I said, "Well the reality is we're in the fight of our lives here, and there's a lot at stake."'

In the end the Cormack Foundation gave Turnbull's campaign around $1 million, rather than the $2 million or more it needed. But, unbeknown to Turnbull, they also decided to fund two high-profile climate-sceptic senators running against the Liberal Party: the far-right libertarian Senator David Leyonhjelm; and South Australian businessman Senator Bob Day, from the Christian-based Family First party. The two got $25,000 each. Day had been a close friend of Morgan's alter ego, Ray Evans, the kingpin of Australia's climate-denial movement. Day had delivered a eulogy at Evans' funeral two years earlier.

With the Cormack Foundation keeping a tight hold on the purse strings, Turnbull was forced to put up $1.75 million of his own money to fund the federal campaign in what turned out to be a knife-edge election. Labor had big union funding and its leader, Bill Shorten, campaigned better than many expected. Election night, 2 July, was a debacle for Turnbull. His prime ministership was left hanging by a thread. He ended up with a one-seat majority and no political capital to burn on climate change.

In the wake of the election, Turnbull faced a dangerously fractious Coalition party room; his government was dependent on the vote of every member, including Abbott and his supporters. With that in mind, a mutual friend decided by October it was time to broker a truce between Turnbull and his old nemesis, Tony Abbott.

The price for peace was high. Abbott wanted a cabinet post. Turnbull flatly refused. 'Malcolm made it absolutely crystal clear that, as far as he was concerned, I was never going back into the cabinet,' Abbott recalled. 'I said to Malcolm words to this effect: "Fair enough, you're the prime minister, you choose the cabinet, but you've got to understand that I'll do my own thing on policy, and you've got more to lose than I have."'

Turnbull insisted he was right to lock out Abbott. 'It was very obvious to me and everybody else in cabinet, almost without exception, that his agenda was to destabilise and blow up the government,' said Turnbull later. 'And I think everyone knows that. He had supporters in the media; he had fewer supporters in the party room. If he had been in the cabinet room he would have been able to do a lot more damage. There was not one Liberal Party member in the cabinet who thought he should be brought in, not one. The consensus view was that he was a hair shirt that we had to wear.'

With Abbott free to incite insurrection, reopening the Coalition's climate policy would be more difficult than ever. It would take an act of God to get it back on the agenda. And that's exactly what happened.

———

Shortly before 4 pm on the afternoon of 28 September 2016, South Australia's premier, Jay Weatherill, was sitting in his office in the state Parliament House when the lights suddenly flickered on and off. He didn't think much of it until a short time later, when his energy minister, Tom Koutsantonis, appeared in his office white-faced. 'We are system black,' he told the startled premier. The state had been completely cut off from the National Electricity Market (NEM) and 850,000 households, businesses and government offices were without power.

That afternoon an extraordinary burst of wild weather had ripped through South Australia. A supercell of thunderstorms—including tornadoes, some with wind gusts up to 260 kilometres per hour—had pummelled the state with wind, rain and giant hailstones. The storms uprooted trees, tossed farm sheds around paddocks and damaged or brought down 23 electricity transmission towers. Three major transmission lines were badly damaged, causing them to trip. In quick succession the faults on the power network multiplied. In the north of the state a group of wind farms shut down as their preset protection mechanisms kicked in.

After the last wind farm powered down, the big Victorian–South Australian interconnector, which linked the state to the NEM, suddenly boosted the flow of power into the network. That surge caused a shock to the system, shutting down the interconnector itself. This separated South Australia from the NEM, and the remaining generators in the state could not stay online. That was how South Australia went black.

With chaos engulfing his state, Weatherill called a crisis cabinet meeting. 'Our immediate concern was public safety,' he said, recalling that hectic night. Ministers and staff began contacting hospitals and emergency services. They needed to get more than 200,000 commuters home in peak hour in a city without a traffic light operating. Weatherill personally began calling radio stations, pumping out as much information to the public as possible.

Within hours of the blackout, Weatherill overheard an ABC

news story that claimed that wind farms had contributed to the blackout. 'Forty per cent of South Australia's power is wind-generated,' reported Chris Uhlmann from Canberra, 'and that has the problem of being intermittent—and what we understand at the moment is that those turbines aren't turning because the wind is blowing too fast.'

The premier was surprised. He knew that the wind turbines were likely to have shut down to protect themselves when the system tripped. Not long after, independent South Australian senator Nick Xenophon was on the airwaves and online ramping up the power crisis. 'I've been told that the Royal Adelaide Hospital was down,' he told one reporter, saying he feared for elderly and ill South Australians.

'So here they were, before anyone knew what was happening, they were climbing into us,' Weatherill recalled.

Power was restored to 80–90 per cent of the state by midnight, but the furious debate over the cause of the blackout had just begun. Federal ministers went public the next day, attacking Weatherill's Labor government for aggressively pushing renewable energy. 'It was Frydenberg. It was Turnbull. It was the deputy prime minister, Barnaby Joyce. They were all hopping into us,' said Weatherill.

Turnbull had put Frydenberg in charge of both Environment and Energy after the election, and now he and Turnbull hammered the line that the blackout was 'a wake-up call' for the Labor states for pushing renewable energy—even though Frydenberg acknowledged the blackout was triggered by the massive storms. Turnbull lectured Weatherill over the airwaves. 'We've got to recognise that energy security is the key priority and targeting lower emissions is very important, but it must be consistent with energy security.'

Weatherill was furious that Turnbull was using the blackout to mount a political attack before the crisis was even over. Parts of the Eyre Peninsula, a hub of mining and agriculture in South Australia, were still without power and remained so for up to eleven days. 'Not one offer of assistance from Malcolm Turnbull during this crisis,' he recalled bitterly. 'I rang him in the evening, within an hour of the blackouts, to inform him what was going on, and there was no communication from him throughout the whole thing. It was as though they were conducting themselves in relation to us as though we were some foreign belligerent nation, not a constituent element of the Commonwealth.'

In the many reports on the South Australian blackout that would follow, there was no doubt the intense storm was largely to blame. But it also exposed systemic problems in the NEM and its handling of renewables. The problem was not just South Australia's responsibility, it was the responsibility of the energy regulators and private energy generators, as renewable energy began to push out coal, making the grid less stable.

South Australia's blackout was a turning point for climate and energy policy for the Turnbull government. In the policy vacuum created since the federal Coalition had been elected, Weatherill had unveiled an ambitious target to make his state 50 per cent renewable by 2025. He had gone to the Paris climate summit pitching for Adelaide to become one of the world's first carbon-neutral cities.

Now South Australia became a political punching bag for Turnbull as the debate raged over renewable energy versus coal-fired power. Turnbull and Frydenberg used the blackout to wedge Labor. But, more importantly, they also used it to reopen the Coalition's own fraught climate and energy policy, which had been in limbo since Turnbull took office.

Just over a week after the blackout, Frydenberg and the state energy ministers held a crisis meeting in Melbourne, where

Frydenberg nominated Australia's chief scientist, Dr Alan Finkel, to prepare a major report on the future security of the NEM. At that point, as one senior government figure put it, 'we ended up in a situation where the federal government in effect owned the whole electricity issue'. Whatever the Finkel Review recommended, it was now certain the Liberal Party's fragile truce over climate and energy was doomed.

―――――

On the other side of the world, a superstorm of another kind was gathering in September 2016. Donald Trump suddenly looked as if he might have a chance of winning the White House in the race against Democrat Hillary Clinton. This was exciting news for America's climate change sceptics. After clinching the Republican nomination earlier that year, Trump had given an unusually well-prepared speech at an oil and gas conference in Bismarck, North Dakota. Trump promised he would pull the US out of the Paris Agreement on climate change if he was elected president.

Myron Ebell, the outspoken climate sceptic at the Competitive Enterprise Institute, had been following Trump's campaign closely. Ebell's influence had grown quite a bit since the days when he lent a helping hand to Australia's climate sceptics coming through Washington. Even so, he was surprised when he got a call that September from an adviser to Donald Trump's organisation. A group of Trump's people were setting up transition teams to make plans for the new administration if he won in November. The adviser asked Ebell if he would like to head the team for the US Environmental Protection Agency (EPA) that would put together a list of recruits for the top EPA jobs, among other things.

Under Obama, the EPA had spearheaded the Clean Power Plan, clamping down on greenhouse emissions from coal-fired power stations. Ebell had been a ferocious critic. He had branded the Clean Power Plan 'illegal', arguing that renewable energy producers would profit at the expense of coal and natural gas.

But even Ebell was stunned by the caller's offer. He was a political pusher and shover, he told the Trump adviser, not a lawyer or a bureaucrat. He had never actually served in any administration. As Ebell recalled the conversation, the Trump adviser said: '"Well, we've heard that you want to get us out of Paris." I said, "Yes."

'"And we've heard that you think that the Obama climate agenda has to go." Ebell replied, "Yes."

'"And Mr Trump has even said that he would like to abolish the EPA." And I said, "Well, you know, I agree with that."'

According to Ebell, the Trump adviser replied, 'Well, that's why I'm calling you.'

Ebell's appointment to the Trump transition team outraged environmentalists when the news leaked but it was cheered by the climate sceptics. Within weeks of Trump's victory, the new president appointed a high-profile sceptic, Scott Pruitt, to head the EPA. Pruitt wasn't Ebell's first choice but he was on Ebell's list. As a former attorney-general from Oklahoma, Pruitt had joined an action to sue the EPA to try to stop Obama's Clean Power Plan.

With Trump in the White House, climate sceptics were back at the centre of power. Ebell and his allies began lobbying to make sure Trump kept his promise to pull the US out of Paris. They knew if that happened it would punch a huge hole in the global effort to shift from fossil fuels to clean energy. With the US out, China would also be able to walk back from its Paris commitments.

Trump's election was a shot in the arm for climate sceptics around the world and especially in Australia. Just as Turnbull was about to reopen the Liberal Party's climate and energy policy, the sceptics were back on the march. And Tony Abbott would be joining them.

———

Shortly before the federal parliament opened for 2017, one of the worst heatwaves in Australian history began its sweep through New South Wales, Queensland and South Australia. Nearly a hundred bushfires burned in New South Wales, Sydney temperatures reached their hottest on record and Brisbane sweltered through a record run of 30 consecutive days above 30 degrees.

Climate and energy policy had forced its way onto the national agenda for the year. The Finkel review on the looming energy crisis was expected by June and the government would need to respond. So in early February, Turnbull decided to float some new ideas on energy in a speech to the National Press Club. But he did it by launching another partisan attack on Labor.

His government, said Turnbull, was drawing the battlelines with Labor on energy policy. He labelled Bill Shorten's plan to lift Australia's Renewable Energy Target to 50 per cent by 2030 and boost Australia's Paris target as 'a sure recipe to deliver much more expensive and much less reliable power'. He demanded governments and industry deliver affordable, reliable and secure power. While Labor dismissed the speech, it marked a critical turning point. Turnbull insisted he wanted the government's next energy policy to be 'technology agnostic', not favouring one form of energy over another.

Turnbull knew Australia had to tackle the vexed problem of getting more renewables into the energy market to replace coal-fired power if the country was going to meet even its weak Paris targets. This had to be done without jeopardising energy reliability. New forms of what was called 'firming power' to back up renewables had to be found for when the wind wasn't blowing or the sun wasn't shining.

Bold ideas were needed, and Turnbull had asked the Clean Energy Finance Corporation and the Australian Renewable Energy Agency to come up with them. Having been saved by Clive

Palmer from Abbott's purge, they were now useful to Turnbull. He asked them and the Finkel review to all look, in particular, at 'pumped hydro'. Put simply, the idea was to build plants that could pump water uphill to store it in an upper reservoir from the plant's main dam when surplus renewable energy was available. When firming power was needed, the water could then be quickly released through turbines to create power as the water flowed back into the downhill dam. That way pumped-hydro plants could essentially act as big batteries to help back up renewables and stabilise the power network.

Critically, Turnbull was acknowledging that Australia's ageing coal-fired power stations would be phased out and this couldn't be ignored. Already Hazelwood, known as Australia's dirtiest power station, was due to shut down the following month, taking a big slice of cheap coal-fired power out of the market. Its French owners, Engie, wouldn't prop it up anymore.

For Turnbull, energy policy was a toxic time bomb ticking away in the prime minister's desk drawer. As he put it, 'nobody was politically more at risk from it than me'. Australia was the world's largest coal exporter and he knew many in the Coalition believed clean energy had to include 'clean coal'.

Turnbull needed to keep coal in the policy mix. In his speech that day, Turnbull shocked many in the environment movement when he appeared to give support to building new high-efficiency low-emissions (HELE) coal-fired power stations in Australia. 'We have a vested interest,' he told the Press Club audience, 'in showing that we can provide both lower emissions and reliable baseload power with state-of-the-art clean coal-fired technology.'

Turnbull would later insist he wasn't backing a new coal plant. He was treading a careful line of being deliberately 'technology agnostic', knowing no one would actually invest in a new coal plant in Australia. 'Politically I could be agnostic,' recalled

Turnbull, 'confident in the knowledge that if economics and engineering were the criteria on which investment decisions were made, i.e. if the private sector made the decisions, there wouldn't be any coal-fired power stations built at all.'

But that February, Turnbull looked like he was trying to please all sides in the fractious Coalition party room as he steered them towards a new climate and energy policy. He rallied the right on his own side with increasingly shrill attacks on Labor's support for renewable energy. Inevitably, this backfired on Turnbull in one extraordinary episode that came to define the Liberal Party's climate change policy for years to come.

On 8 February, Adelaide was hit hard by a record heatwave as temperatures rose to 42 degrees Celsius. That evening the state suffered another blackout, albeit a limited one. Power was cut to 90,000 homes for 45 minutes and spot prices for electricity soared. The outage was caused by a series of failures involving to varying degrees the Australian Energy Market Operator (AEMO), an Engie- and Mitsui-owned gas generator (which was later sued by the Australian Energy Regulator) and a disastrous software glitch from the power network's private owner in South Australia.

But before the causes of the blackout were even known, Turnbull and his ministers decided to use the next day's question time in parliament to again ramp up their attacks on Labor's renewable energy policy, targeting Bill Shorten.

'Do you want to know how many Australian families will be worse off under a government led by this man?' Turnbull shouted across the parliament. 'Every single one. How many South Australians are worse off because of the Labor left ideological approach to power? I will tell you: every single one—every single one that wants to turn the lights on, that wants to put the air conditioner on, that wants to have a job and wants to have some investment.'

Turnbull didn't hold back. 'Yesterday it was 41 degrees—no power,' he roared. 'That was great! What a great achievement! Really, this is the triumph of the Labor Party.'

Frydenberg was next on his feet. 'This big experiment has failed the people of South Australia,' he warned, 'They pay on average more than 40 per cent more than the national average for electricity.'

Then, finally, it was the turn of the treasurer, Scott Morrison. With a flourish, Morrison held up a large lump of coal in his hand. 'This is coal,' he called across to the Labor benches as he raised it up. 'Do not be afraid. Do not be scared. It will not hurt you.'

Morrison had got the lump of coal from the Minerals Council office in Canberra, and now his passionate defence of coal echoed Abbott's two years earlier at the minerals industry dinner. Accusing Labor of having 'an ideological, pathological fear of coal' or 'coalophobia', Morrison outdid Turnbull in his zeal to attack Labor.

Turnbull grew stony-faced as Morrison hit full throttle and the lump of coal was passed around government backbench members. 'That was a crazy thing to do,' Turnbull said later. 'He'll be reminded of that every day. I've got no idea why he did it.'

It may have been crazy, but it sent a message to the entire country. Morrison was backing coal over renewable energy. That message was heard by every climate sceptic in Australia and in the Coalition. Unlike his leader, the treasurer was a staunch defender of coal-fired power.

The day after the bizarre theatrics, the summer heatwave hit Sydney and the NSW Coalition government faced its own power crisis. The big power company AGL notified the state's main aluminium smelter it was going to reduce its electricity supply so schools, businesses and homes across the state could stay connected. No one in the Turnbull government raised a word of criticism against their NSW colleagues over the crisis.

In Adelaide Weatherill was fuming. 'This was New South Wales—with the largest reliance on coal-fired generation of any system in the world—facing identical challenges,' he said later. 'By then the right had their hands around Malcolm's throat,' was Weatherill's take. 'He was in their thrall post the election result. He was weakened.'

The blackouts and the political fallout taught Weatherill a big lesson. He could not in future rely on the federal government or the energy regulators to stabilise the NEM and South Australia's power system. Instead, he and his energy minister began looking at a new energy plan that included firming power in the state. A month later, South Australia began talks with US billionaire Elon Musk, head of the Tesla electric car, battery and solar power business. Musk had offered to build a 100-megawatt grid-level battery for the state in 100 days to help stabilise the power network. It would be the largest lithium-ion battery in the world, capable of storing energy from renewables and helping to prevent load-shedding blackouts.

Not to be outdone, just days later Turnbull announced his own 'electricity game-changer'—an ambitious plan for firming power; a new Snowy Mountains pumped-hydro scheme. Called Snowy 2.0, the multibillion-dollar scheme promised to add 2000 megawatts of renewable energy to the NEM. The huge project was a pet scheme of Snowy Hydro's chief executive, Paul Broad; he had been waiting for an opportunity to pitch it to Turnbull, and the PM grabbed it with both hands. 'The unprecedented expansion will help make renewables reliable, filling in holes caused by intermittent supply and generator outages,' Turnbull said. 'It will enable greater energy efficiency and help stabilise electricity supply into the future.' But it would take years to build.

Scathingly, Morrison dismissed South Australia's 100-megawatt Tesla battery, calling it the 'Hollywood' solution, likening it to a tourist attraction like the Big Banana in the NSW holiday town of Coffs Harbour. 'I mean, honestly, by all means have the world's biggest battery, have the world's biggest banana, have the world's biggest prawn like we have on the roadside around the country, but that is not solving the problem.'

As it turned out, the Tesla battery was built by December that year and proved successful in preventing blackouts in South Australia.

Josh Frydenberg and the states' energy ministers released the highly anticipated report from the chief scientist Alan Finkel on Australia's energy market in June 2017. It was supposed to be a blueprint for rebuilding Australia's NEM into a world-class system to face the challenges of tomorrow—transforming Australia into a cleaner energy country.

With over 9.5 million customers dependent on the NEM, the Finkel review warned the country's energy system was in transition and there was no going back. 'We are at a critical turning point,' it said. The review ticked all the boxes, with recommendations for energy security, reliability, keeping prices down and lowering greenhouse gas emissions. The government wanted to make sure that, if coal-fired plants shut down, this didn't leave a gaping hole in supply that would allow electricity companies to price-gouge.

But the thorny question remained—how to do this without putting a price on carbon emissions? Why would investors put money into new clean energy without a price signal to say cheap coal power would be replaced? This was especially urgent, as the federal Renewable Energy Target was due to end in 2020.

Finkel's solution was a Clean Energy Target linked to Australia's Paris target of a 26–28 per cent cut in emissions across the economy by 2030. To meet even that weak target, a lot of emissions cuts would have to come from the electricity sector. Modelling for the Finkel review estimated that, by 2030, around 42 per cent of electricity demand would be met by renewables—not far off Labor's 50 per cent target.

Finkel's Clean Energy Target for the electricity sector would work a bit like the Renewable Energy Target, providing incentives for new generators of clean energy. The incentives wouldn't exclude coal, but it would need to be 'clean coal'. It looked like new HELE coal plants could find it hard to qualify as clean energy.

For Tony Abbott and the Nationals in the Coalition party room, this was not acceptable. For them it meant coal-fired power stations were on the way out. When Frydenberg tried to sell Finkel's Clean Energy Target to them, he hit a wall of resistance.

'The Finkel presentation that Josh made to the party room was hailed as a wonderful presentation, but essentially it was the death of coal,' said Abbott later. 'There were a lot of speakers. I think essentially there were about 30 speakers and twenty of them would be deeply sceptical. And many of them were saying, "Look, we can't just get rid of the coal industry, because it's a critical part of our power system and because it's such a massive part of our export earnings."'

Abbott's crucial ally in the partyroom debate on the Clean Energy Target was Angus Taylor, Turnbull's Assistant Minister for Cities and a rising star on the right. Taylor, a long-time critic of the renewable energy targets, was a hero in climate-sceptic circles, even though he insisted he was not a climate sceptic.

By September, the dissent in the Coalition party room was impossible to overcome. The Canberra press gallery was running stories that Abbott and six other backbenchers would cross the floor to defeat the Clean Energy Target if Frydenberg tried to introduce it. Turnbull could not afford a revolt. Frydenberg did not want to deep-six the chief scientist, but he and Turnbull had little choice other than to abandon the Clean Energy Target and the first real attempt to overhaul the climate and energy policy put in place by Abbott.

––––––

While Abbott and leading National Party players Barnaby Joyce and Resources Minister Matt Canavan were lobbying Turnbull intensely over coal, another more covert plan to defend coal-fired power was underway in Australia. It was funded by Glencore, the Swiss mining giant and Australia's biggest coalminer after it merged with Xstrata a few years earlier. Glencore's chief executive, Ivan Glasenberg, was a coal man; despite the Paris Agreement, he was still bullish on coal, especially thermal coal, and he needed the world to keep buying it.

In February 2017, around the time Morrison held up his lump of coal in parliament, Glencore ticked off on a campaign, called Project Caesar, to defend coal power in Australia and other markets worldwide. The company would later describe Project Caesar as 'a proactive communication campaign on coal'. The idea, it said, was partly 'to counter misinformation from environmental activists'.

Glencore was a notoriously aggressive operator, so it wasn't surprising it wanted a heavyweight lobbying firm to run the campaign. What was surprising was its choice of firm—Crosby Textor, the Liberal Party's long-time polling and political advice shop that had worked on Turnbull's 2016 election campaign.

Best known for running conservative political campaigns in the UK and Australia, Crosby Textor also ran campaigns for big corporates. As Sir Lynton Crosby reportedly put it, 'we represent clients of high net worth, companies and countries—pretty much anything beginning with the letter "c".' Glencore certainly fitted that bill.

The Glencore campaign boiled down to a no-holds-barred attack on renewable energy while lobbying Australian politicians and the public to build new HELE coal plants in Australia. As climate change science became harder to ignore, HELE plants became the coal industry's response to clean energy. But even on the industry's best estimates, the emissions from HELE coal plants would only be 30 per cent less than the emissions from an average NSW coal plant. HELE plants were still big greenhouse emitters, and they were expensive to build.

An investigation by *The Guardian* discovered that Glencore's campaign used a front group called Energy in Australia. Among the provocative stories it posted online was a story that the French government and French power company Engie were about to desecrate the unmarked graves of Australian soldiers who had died on the Western Front by erecting a wind farm near the Digger memorial in Bullecourt. As it turned out, that didn't happen.

Glencore refused to confirm or deny its campaign used the Energy in Australia front group. But videos and posts from Energy in Australia's website and Facebook page were authorised by a former Queensland LNP backbencher, Matt McEachan, and circulated to prominent climate-sceptic federal MPs, who shared them with voters. George Christensen, the federal LNP member for the Queensland mining seat of Dawson, was among the MPs who ran the posts.

By September 2017, as the Coalition partyroom fight over coal and the Finkel Clean Energy Target reached a climax, the lobbying campaign to support HELE plants went into overdrive. Crosby Textor polled voters on behalf of the NSW Minerals Council and found 81 per cent of NSW Coalition supporters in the state backed the push to build a new HELE plant, a story given a run by *The Australian*.

Glencore was by no means the only player lobbying in the fraught debate. All sides, including the renewable industry, were doing it—but not covertly. When the peak mining lobby, the Minerals Council of Australia, stepped up its pro-coal campaign, however, it opened up a deep split in its own ranks. BHP's senior executive, Mike Henry, came out publicly backing the Finkel Clean Energy Target, putting the mining giant at loggerheads with the Minerals Council's chief executive, Brendan Pearson.

BHP was a global company answerable to its shareholders. It had supported the Paris Agreement; it had gas interests that could profit from being a transition fuel, as well as other metals critical to renewables. It also could not ignore the science of climate change, growing shareholder pressure, or the number of lawsuits proliferating in the US against fossil fuel companies over this issue.

As well, BHP was increasingly worried that some in the Minerals Council were, with Pearson's support, lobbying to strip the charitable tax status from key environment groups campaigning on climate change. In late September, BHP essentially gave the council notice. If Pearson didn't go, BHP would review its membership of the council.

Pearson's decision to exit the Minerals Council, under pressure from BHP, appeared at first to herald a seismic shift in Australia on climate change and energy. But the stand by 'The Big Australian' changed little in the Coalition party room. Instead, it marked the moment when the climate sceptics in the Coalition stood their ground against big businesses that supported the move away from coal-fired power.

Right-wing columnist Janet Albrechtsen, a board member of the IPA, summed up that defiance in her regular column in *The*

Australian. Lamenting Pearson's departure from the Minerals Council, she savaged BHP for what she called its adoption of 'green politics in a way that works against Australia's national interest in having secure, stable and cheap energy'.

The calls for a return to cheap coal-fired power ignored reality. Big investors were moving away from coal. Money was going into renewables, which Australia's national energy market was still ill-equipped to deal with. The Turnbull government was stuck with an increasingly unstable energy grid, high greenhouse gas emissions and rising energy costs, including savagely high domestic gas prices.

But the hard-core members of the Carbon Club were still digging in. A month earlier, the IPA had launched its latest publication, *Climate Change: The Facts 2017*. The think tank was still being partly bankrolled by mining billionaire Gina Rinehart and other wealthy climate sceptics, and it was still attacking mainstream climate science. The new IPA tome was edited by Jennifer Marohasy, now a senior fellow at the Institute. She was again demanding an investigation into Australia's Bureau of Meteorology for supposedly manipulating temperature readings that showed Australia was warming due to climate change. This was despite an expert panel appointed by the government, made up of statisticians and scientists, finding 'no reason to question the accuracy of the dataset' maintained by the bureau.

Marohasy and her claims were nevertheless given star billing on Murdoch's Sky TV by shock jock Alan Jones. 'Governments don't want to know the lies that have been told about temperature,' Jones told the Sky audience. 'This is terrifying.'

Less than two years after the Paris Agreement had set hopes soaring that the world could avoid dangerous global warming, the climate wars in Australia were back in full swing.

CHAPTER SIXTEEN

SACRIFICING GOATS

Tony Abbott arrived in London pretty pleased with himself. After helping to kill off the Clean Energy Target, he was now being welcomed as an honoured guest of the Global Warming Policy Foundation. The UK climate-sceptic think tank founded by former chancellor, Nigel Lawson, had invited Abbott to give its annual lecture in October 2017.

Abbott was following in the footsteps of his heroes John Howard and Cardinal George Pell, who had both stood at this podium before him. His lecture had the cautious title 'Daring to doubt', but even he admitted it would best be remembered as his 'Sacrificing goats' speech.

In a provocative attack on climate policies, Abbott compared efforts to reduce greenhouse gas emissions to slow climate change as akin to animal sacrifice by ancient priests. 'Primitive people once killed goats to appease the volcano gods,' he told the audience. 'We're more sophisticated now but are still sacrificing our industries and our living standards to the climate gods to little more effect.' Abbott went on to roundly criticise the scientific consensus on climate science. 'Beware the pronouncement "the science is settled",' he warned. 'It's the spirit of the Inquisition, the thought-police down the ages.'

But beneath the hyperbole, Abbott used his UK lecture for a more strategic purpose. With the death of the Clean Energy Target, Frydenberg and Turnbull were working on a new policy to put to the Coalition party room. Abbott was spelling out the demands of climate-sceptic MPs that he wanted met. First, he said, no subsidies for renewable energy. Second, he wanted the federal government to fund a new HELE coal plant. 'Given the nervousness of private investors, there must be a government-built coal-fired power station to overcome political risk,' Abbott demanded—and he had support from the Nationals.

This was politically impossible for Turnbull. If the prime minister agreed to build a new coal-fired power station in Australia, it would destroy his political credibility. But Abbott would press this demand and rally his allies against Turnbull.

Six months after Abbott's goats speech, Malcolm Turnbull arrived in the French village of Villers-Bretonneux on the eve of Anzac Day 2018, to pay homage to Australia's World War I general Sir John Monash. With French prime minister Édouard Philippe, Turnbull was opening a new museum dedicated to the brilliant Australian. Among the dignitaries in the crowd was Tony Abbott; the Monash Centre had been his project, conceived during his short reign, and it was appropriate for him to be there.

But it was also awkward. Just weeks before, Abbott and his allies had launched a new pro-coal lobby group for Coalition backbenchers and named it after the general. The Monash Forum had the backing of around twenty Coalition MPs. The flyer that called them to arms praised Monash's peacetime legacy as an engineer who built coal-fired power stations. Among other things, the forum was petitioning the Turnbull government to build a new coal-fired power station, possibly on the site of Victoria's old Hazelwood Power Station in Victoria.

Abbott's allies, Kevin Andrews and Eric Abetz, were founding members of the group and they were joined by two of the Coalition's most vocal climate sceptics, Craig Kelly, chair of the Coalition's backbench committee on environment and energy, and Queensland LNP federal member George Christensen. The former deputy prime minister, Barnaby Joyce, also lent his support. Joyce had been forced onto the backbench a few months earlier after news broke that he had left his wife for a new partner who was pregnant with his child. No longer in cabinet, Joyce was another loose cannon.

Tellingly, the launch of the Monash Forum was promoted on Sky News by Abbott's old chief of staff, Peta Credlin. Abbott insisted its purpose was simply to back coal-fired power over renewables, not to destabilise Turnbull. 'It was about ensuring energy policy was about keeping the lights on, keeping prices low, or as low as possible,' said Abbott. 'I guess underlying all that was this instinct that you don't keep the lights on by eliminating the cheapest form of baseload power.'

Fatefully, at first Turnbull didn't see the Monash Forum as a threat to his prime ministership. He saw it as an absurdity. 'Coal became a kind of leitmotif for these guys and my response was we had to be technology agnostic. We wanted to be the "affordable, reliable energy party" not the "coal party" or indeed "the renewable party",' he said later.

'The thing that has destroyed their economic credibility, other than in their own echo chamber, is the fact that nobody in the industry believes that you could rationally build a new coal-fired power station. Literally nobody. The Tony Abbott, Craig Kelly, Barnaby Joyce agenda of building fleets of new coal-fired power stations—there is no private-sector dollar available to go into that. And the reason for that is the economics have changed.' But not everyone in business thought so. One of the lobbies posting its support for the Monash Forum was Energy in Australia, the social media front reportedly working with Crosby Textor's pro-coal campaign for Glencore, the largest coalminer in Australia.

Turnbull and his allies dismissed the Monash Forum as 'just the usual suspects' attacking the government. But its launch in April 2018 was well timed. It coincided with the 30th consecutive Newspoll that put the Turnbull government behind Labor. This was the fatal number of bad polls that Turnbull had used as his licence to bring down Abbott. Just as the government's latest climate and energy policy was due back in the Coalition party room, Turnbull's political capital was fast running out.

———

Josh Frydenberg unveiled the draft of the National Energy Guarantee to Coalition MPs in June 2018. The policy had been refined for months by the Energy Security Board, the energy regulators working under independent chair Dr Kerry Schott, a long-time bureaucratic fixer. The board was established by the state and federal energy ministers originally to implement the Finkel Review recommendations. The NEG, as it was dubbed, drew on many of Finkel's recommendations, but in the end it was based on two simple principles: energy reliability and reducing greenhouse gas emissions. It was also supposed to get power prices down. The new design barely lasted two months before it was ripped to shreds.

At first it looked like Frydenberg could make it happen. The idea behind the NEG was to avoid both a carbon tax and a subsidy for renewables. But it would require electricity retailers to provide reliability for customers while also meeting a greenhouse gas emissions target. Big power users also had to meet the emissions targets so they would have to buy some of their power from clean energy generators, creating an incentive to invest in new power plants.

To get support for the NEG, Frydenberg rounded up the major business lobbies, including the Business Council of Australia and the Australian Industry Group (AIG). They briefed the Coalition backbench MPs, including the climate sceptics, to explain why they needed to back it. Innes Willox, the AIG chief, warned them, 'To kill this off really would be wrecking-ball politics.' Manufacturers and small businesses were crying out for cheaper power prices.

Big business also desperately wanted certainty. It feared Labor could win the next election and push for much higher targets to cut greenhouse gas emissions. The head of the Business Council of Australia, Jennifer Westacott, described Labor's targets as 'economy wrecking'. The business lobbies were dead keen to get the NEG through.

Abbott was not impressed, but at this point he couldn't raise much opposition to it: 'The party room must've been much more thoroughly prepped for the National Energy Guarantee because it was like the place had been anaesthetised,' he said later.

Instead, Frydenberg's big problem at this stage was overcoming objections from the state energy ministers, who were crucial to making the NEG work. The Labor states were not just playing politics. Some were objecting because the federal government's proposed emissions target under the NEG was too weak.

The target to cut emissions in the electricity sector was to be just 26 per cent by 2030, but with a provision to review it. Given the existing Renewable Energy Target was expected to deliver most of these cuts by 2020, Labor and the Greens argued the NEG was unlikely to drive much future investment in clean energy. More importantly, the electricity sector needed to make much bigger emissions cuts if the whole of the economy was supposed to cut emissions by 26–28 per cent to meet Australia's Paris targets.

This was certainly the view of the ACT's energy minister, the Greens' Shane Rattenbury, who believed the NEG target was too low to reach even Australia's weak Paris target. 'We can't possibly get there if we only do that much in the electricity sector,' he said later. 'Because there is no way the agricultural sector was going to accept a 26–28 per cent cut. If you're going to take that approach and get 26 or 28 per cent out of every sector and everyone does it equally, well fine. But that was never, politically, never going to be the case.'

His views echoed objections from the main environment groups, including the Australian Conservation Foundation, whose supporters protested against the NEG outside parliament. They thought the Paris targets were too weak anyway, and the NEG would 'slam the brakes on clean energy and prop up dirty coal'.

In July, Abbott turned up the pressure. Echoing Trump, he called on the government to pull out of the Paris Agreement. He was soon backed up by the IPA and its sceptic stalwarts, Gina Rinehart and Hugh Morgan. But that month delivered an even bigger problem for Turnbull and Frydenberg. On Saturday 28 July, five federal by-elections were held across the country and, in a huge setback for Turnbull, the government failed to win even one of them, including the crucial Queensland swing seat of Longman.

The losses entrenched the view in the Queensland LNP that Turnbull was no longer a winner. If the result was repeated at the next election, the Coalition would be wiped out in that state. At this point, Turnbull lost the support of his influential home affairs minister, Peter Dutton, who had little interest in climate policy, but did hold a marginal seat in Queensland.

For Turnbull's ally, Craig Laundy, the by-election losses were the beginning of the end. Enough numbers in the party room were moving to put Turnbull's leadership in play: 'When you add all the Liberal Party members and senators in Queensland who have been spooked to the rump of the Monash Forum, you get to mid- to high twenties without blinking. Then what happens? Sky News kicks on inside Parliament House and they are just obsessed and infatuated with two agendas. One, tear down Malcolm. And two, use climate change to do it.'

The NEG quickly turned into the vehicle to destabilise Turnbull. Reports appeared that Abbott and the climate-sceptic rebels would cross the floor if Frydenberg tried to bring the NEG into parliament for a vote. Turnbull and Frydenberg at first tried to stare the rebels down, arguing the policy had been signed off by cabinet and supported in the party room.

'Abbott was trenchantly opposed, but certainly the feedback that Josh was giving us was that, apart from Tony, we probably wouldn't have any floor-crossers. That was his view,' Turnbull recalled. 'The opposition that blew up in the last week or so of my prime ministership was well beyond anything that Josh had anticipated, or at least in so far as I was aware. And it was being whipped up and supported in the media by the usual suspects in the traditional way.'

By this stage, numbers men for Dutton and Treasurer Scott Morrison were already head-counting. In a series of humiliating backdowns to appease the rebels, Turnbull and Frydenberg finally ended up ditching the emissions target from the NEG legislation altogether, stunning the government's business allies and state energy ministers. That week, Abbott watched Turnbull buckling and saw his leadership bleeding away.

'There was the first major adjustment to the NEG,' said Abbott, recalling Turnbull's retreat. 'He'd told the party room that the emissions reduction targets were going to be put into legislation to stop the Labor Party increasing them and then he suddenly announced, for God knows what reason, on the Friday night that was no longer going to be the case. [On Monday] Malcolm went into the primeministerial courtyard and dumped the whole policy because, while he might have been able to get it through the parliament with the Labor Party's support, he couldn't get the support of the Coalition party room for it.

'The NEG was fatal to his leadership,' Abbott said bluntly. 'Dropping it was fatal. But keeping it would have been fatal too.'

Jay Weatherill, who had lost his premiership five months earlier, looked in despair at the triumph, yet again, of the climate sceptics in the Coalition party room: 'Everyone has been dancing in circles around a bunch of climate deniers. But until you actually stand in the middle of Martin Place and say "I am a climate denier" they are not going to be satisfied,' he said later.

'It was never about energy policy, it was about killing Malcolm. Malcolm didn't realise he was dealing with political terrorists. They were never interested in negotiating a result. There was only one result they were interested in and that was killing him. And the way they killed him was to destroy his credibility.'

In a final attempt to save his job, Turnbull moved a surprise leadership spill on Tuesday 21 August to block a challenge from Dutton. Abbott was sitting next to Dutton: 'I said to Dutts, "Someone's got to run," and Dutts said, "Well I'm going to."'

Dutton lost, 35 votes to 48, but Turnbull was mortally wounded. The next day his top cabinet ministers began to defect. Mathias Cormann, the finance minister, was counting Dutton's numbers, and Morrison's backers were also hard at work.

In his last desperate days as prime minister, Turnbull called Rupert Murdoch to accuse his media empire of conspiring to bring him down. 'Rupert, this insurgency has been going on for a long time,' Turnbull recalled. 'We've talked about it many times. It's being fired up in your papers and on Sky at night.' Murdoch was in Australia, in part to attend the 75th anniversary dinner of the IPA with John Howard. Turnbull was fobbed off by the old mogul.

Turnbull also berated the senior cabinet ministers who were backing his ouster. He remembered railing at his finance minister, Mathias Cormann, asking him, 'Mathias why are you supporting it? This is just terrorism.'

Turnbull couldn't comprehend how climate and energy policy had again destroyed him when this one had been supported by every cabinet minister. 'Mathias was strongly in favour of the NEG, as was Scott. No dissent in cabinet. There was no one running interference in cabinet,' Turnbull recalled. 'Mathias and I worked closely together. His conduct was never explained by him.'

In the final ballot for the Liberal leadership, Morrison emerged the victor after key Turnbull supporters swung his way to block Dutton. At his last press conference at Parliament House, Turnbull blamed those he called 'the wreckers' for bringing him down. 'So, insofar as there has been chaos this week, it has been created by the wreckers,' he told reporters. 'I have done everything I can to maintain the stability of government and the stability of the Party. But, of course, if people are determined to wreck, then we know they will continue to do so.' With his family at his side, Turnbull walked away from Parliament House. The Liberal Party leader who had argued long and loud for climate science was gone. Whether he fought the climate sceptics head-on, as he did as opposition leader, or tried to appease them, as he did as prime minister, he could never defeat them. Climate and energy policy would remain hostage to the Coalition's sceptics and their allies for now. After all, the new prime minister had famously held aloft a lump of coal in parliament.

But the same year Scott Morrison replaced Turnbull, Australia's top scientific agencies, the CSIRO and the Bureau of Meteorology, released their annual 'State of the Climate' report. Its warnings should have grabbed the attention of the new government. It made it clear Australians would need to adapt to climate change coming down the line. The country had already warmed just over 1 degree Celsius since 1910, leading to 'an increase in the frequency of extreme heat events', the report said.

It laid out the climate projections for the coming decades. Australia would experience higher temperatures and more extremely hot days. There would be less cool-season rainfall across many parts of the south, and Australia would spend more time in drought. Alarmingly, it again projected more 'high fire weather danger days' and longer fire seasons for southern and eastern Australia.

In the ocean there would be more frequent, extensive, intense and longer lasting marine heatwaves. These foreshadowed more frequent and severe coral bleaching on the Great Barrier Reef. As the climate scientists knew all too well, that was already happening.

CHAPTER SEVENTEEN

THE STATE OF THE REEF III

AUSTRALIA'S BEST KNOWN coral reef scientist was sitting in a raucous meeting in Parliament House when he suddenly had an overwhelming urge to laugh. It was October 2015 and Ove Hoegh-Guldberg was trying to brief a group of Coalition MPs in Canberra on the latest research into global warming. Instead he and his two colleagues were under sustained attack.

'We just got savaged,' Hoegh-Guldberg remembered. 'And at one point I was sitting there and I just got the giggles. I thought. "This is just ridiculous."'

Hoegh-Guldberg, Mark Howden and John Church were now regarded as three of Australia's leading climate scientists. Until February that year, Howden had been a chief research scientist at CSIRO Agriculture; Church was the CSIRO's expert on sea-level rise. All had worked on the latest report by the Intergovernmental Panel on Climate Change (IPCC) and its findings were blunt. Human influence on the climate was clear. Human-induced emissions of greenhouse gases were the highest in history, and recent climate changes were already having widespread impacts on humans and the planet. 'The atmosphere and ocean have warmed, the amounts of snow and ice have diminished, and sea level has risen,' the report found.

The head of the Coalition's backbench committee on environment and energy, Craig Kelly, had asked the scientists to meet the MPs in the lead-up to the UN's Paris climate summit that December. But hours before the meeting, Hoegh-Guldberg learned Kelly had also invited the Institute of Public Affairs (IPA) to send along their preferred experts as well, including its emeritus fellow, Bob Carter, at the time Australia's leading climate-sceptic voice. Joining Carter was his fellow climate sceptic Jennifer Marohasy, who was still at war with the Australian Bureau of Meteorology over its temperature data. Rounding out the IPA team was researcher Brett Hogan, who had just written a new report titled 'The Life Saving Potential of Coal'. Hogan's report backed the opening up of new coalmines in the Galilee Basin, specifically the new mega-mines owned by Adani and the Gina Rinehart–GVK partnership.

As the meeting heated up, the IPCC authors were stunned by the ferocity of the attacks on them from the climate sceptics. 'We were all surprised at being treated like that and the unprofessional betrayal of trust that was involved in essentially staging an ambush,' Howden said later.

When Hoegh-Guldberg reacted with the giggles, one of the MPs demanded to know what he was laughing at. Hoegh-Guldberg replied, 'You have got to look at this from our perspective. Mark has come here, he is one of the most excellent agricultural scientists we have. John Church here is a world-leading ocean scientist. We came here to help you to understand this issue because it's so important for Australia. And all you can say is that it's a conspiracy.'

The October 2015 briefing of the Coalition backbenchers came to symbolise the deep split in the new Turnbull government over climate change that would dog it until the end. Carter and the IPA's executive director, John Roskam, were delighted with the showdown and the publicity it generated. 'What amused Bob and I at the time was how upset *The Guardian* newspaper was that Bob, Jennifer and Brett were giving politicians the alternative viewpoint to the "accepted consensus",' Roskam wrote later.

The showdown was the clearest signal yet to the scientists that under the new Turnbull government the climate wars were still in full swing, despite the latest IPCC warnings. That year, 2015, was the warmest year on record. Hoegh-Guldberg had spent the last two years writing, thinking and talking about the impact of global warming on the oceans and marine life, including on the Great Barrier Reef. While he was in Canberra, his fellow reef scientists were preparing for what they believed would be a new, potentially disastrous, outbreak of coral bleaching on the Great Barrier Reef caused by a rise in seawater temperatures expected that summer. Perhaps fortunately for Hoegh-Guldberg, this time he was not driving the effort to report on the imminent bleaching. Distracted by his workload with the IPCC, he was letting others take the lead.

———

Getting an Irishman to organise a volunteer brigade in the North Queensland summer might sound a bit crazy, but Terry Hughes was more than up for the mission. By October 2015 the fair-skinned, steely, marine biology professor had harnessed several hundred scientists, students and officials across ten organisations, from universities to federal and state agencies, to cooperate in charting the expected mass coral bleaching on the Reef.

Despite being born in Dublin, Hughes had dedicated his life to studying coral reefs, first in the Caribbean and then in North Queensland, where he was director of the Australian Research Council Centre of Excellence for Coral Reef Studies based at

Townsville's James Cook University (JCU). As 2016 loomed, Hughes and his colleagues became increasingly convinced that a marine heatwave would reach the waters of the Great Barrier Reef that summer, triggering a mass bleaching.

A few months earlier, in June 2015, the US National Oceanic and Atmospheric Administration (NOAA) had issued a warning about extreme sea-surface temperatures, which had resulted in significant coral bleaching in the northern hemisphere. As the months went on, it looked ominous for Australia. 'NOAA's forecasts were particularly extreme,' Hughes remembered. 'It's going to bleach very, very badly.' At the time, Australia was also enduring a strong El Niño event that brought warmer, drier weather and more clear blue-sky days. All this made the reef scientists even more nervous.

The National Coral Bleaching Taskforce put together by Hughes in October 2015 wasn't a formal body, more a network of researchers designed to share scarce resources and standardise the collection of data. Two key government agencies, the Great Barrier Reef Marine Park Authority (GBRMPA) and the Australian Institute of Marine Science (AIMS), were on board, along with Hoegh-Guldberg and colleagues at the University of Queensland and researchers at the University of Sydney, the University of Western Australia, the Western Australian Department of Parks and Wildlife, and Queensland Parks and Wildlife. Coral bleaching was expected to hit Western Australia's reefs as well. The universities and AIMS were already partners in Hughes's Coral Reef Studies centre at JCU.

At GBRMPA headquarters in Townsville, David Wachenfeld, then director of Reef recovery, and his colleague Rachel Pears were doing their own annual risk assessment of the Reef, and organising divers and boats for in-water surveys. The last coral bleaching had been back in 2006, but the two scientists also feared this summer would be a shocker and were keen to be part of Hughes's taskforce.

The taskforce initially planned underwater surveys on selected coral reefs over about two-thirds of the Reef, stretching from Rockhampton off the Central Queensland coast to Cape York in the north. While the surveying was supposed to start in February, in January 2016 reports came in from the far north that some corals were already starting to bleach around Lizard Island. Off Cooktown some corals were dying.

———

In a strange twist of fate, Bob Carter, the scientist who had done so much to cast doubt on the reality of global warming, was silenced just as the bleaching on the Great Barrier Reef first appeared. Carter died of a heart attack in Townsville on 19 January 2016. His death at 73 was a shock not only to his family and friends but to the global sceptic movement that lauded him as a hero. Tributes poured in from the US, Britain, Canada and New Zealand, praising the man who had been such a trenchant critic of the IPCC and Australia's climate scientists.

One of the first people to pay their respects was the then president of the US Heartland Institute, Joe Bast, who called Carter's death 'almost unspeakably sad'. In 2016 Heartland was still a serious force in America's climate-sceptic movement. The right-wing free-market think tank was supported in part by donations from the hedge-fund billionaire Robert Mercer, through his Mercer Family Foundation. That year Mercer also became one of the big bankrollers of Donald Trump's election campaign.

The Heartland Institute described Carter as one of its long-time policy advisers, and the year before his death had bestowed its Lifetime Achievement Award on him at its tenth international conference of climate sceptics in Washington. Also paying tribute to Carter that January was Myron Ebell from the Competitive Enterprise Institute in Washington. Ebell knew Carter well through the Heartland Institute and Ebell's Cooler Heads Coalition. They were often on the speakers list together. In December, just a month before Carter's death, he and Ebell had been in Paris for a 'counter-conference' to the UN climate summit. The climate-sceptic gathering was co-sponsored by the Heartland Institute, and Carter had used his Paris lecture to deliver another boots-and-all attack on the IPCC. In his tribute to Carter, Ebell praised the late palaeontologist as an outstanding scientist. 'I and many, many people around the world have lost a good friend. The Cooler Heads Coalition has lost an esteemed ally.'

Australian climate sceptics knew they too had lost a powerful ally. They had not forgotten Carter's role in barnstorming the bush with the National Party's Barnaby Joyce, whipping up opposition to Kevin Rudd's emissions trading scheme back in 2009. With unshakeable certainty, Carter had punched out his message that global warming had paused. Murdoch columnist Andrew Bolt praised the late sceptic for his 'guts' in identifying 'the pause in the warming' and commended Carter: 'He followed the evidence and not the crowd'. But few regretted Carter's passing more than the IPA. As its emeritus fellow, Carter had been its sceptic-in-chief. 'Bob was an immensely valued colleague and friend to us all at the IPA,' Roskam wrote in his eulogy for Carter. 'He will be very sadly missed.'

A few weeks after Carter's death, in February 2016, average sea-surface temperatures on the Great Barrier Reef were the hottest on record, 1.0–1.3 degrees Celsius above the average for that time of the year. In some spots off Lizard Island, the temperatures went up to 2 degrees above normal. The record monthly temperatures would continue right through until June.

Marine scientist Morgan Pratchett was on board the *James Kirby*, the JCU research vessel. Pratchett was part of Terry Hughes's National Coral Bleaching Taskforce and ended up surveying coral reefs as far north as Princess Charlotte Bay, a stunning waterway off the Cape York Peninsula. The bay is known for its extraordinary marine life, including large numbers of dugongs, and Pratchett was checking on the shallow-water corals. What he saw was devastating. Some of the corals were already dead or dying.

'You can tell if the coral is dead because the tissue starts coming away from the skeleton,' Pratchett explained. 'We were seeing that the temperatures were so hot the corals were just dying there and then from the hot-water exposure.'

Corals usually tolerate only a small variation in temperature, so when their normal summer maximum exceeds even 1 degree Celsius they can suffer heat stress. Most corals have tiny algae (zooxanthellae) living inside them that convert sunlight into food and give them their colour. If the heat stress is mild, the corals can recover from bleaching. But if the water temperature doesn't go down and the heat stress continues, the corals can starve to death over weeks or months. But, as Terry Hughes explained, the summer of 2016 was different. 'When we started to measure mortality we were very surprised to discover about half of the corals that died in the north died within about two weeks. They didn't just die slowly of starvation over many months; many of them actually cooked because the temperature was so hot.'

Up in Princess Charlotte Bay, some of the corals escaped bleaching entirely, but in follow-up surveys, others were dead. Pratchett found himself weirdly elated at simply seeing corals that were still alive. "You became almost immune to seeing the bleaching and the death and the destruction and then you got to a reef which had good coral cover and you were like—wow! Look, this reef is great!' he recalled.

In late February, as the bleaching grew worse in the far north, an unexpected intervention saved the southern half of the Great Barrier Reef. The biggest cyclone ever to make landfall in the southern hemisphere, Cyclone Winston, slammed into Fiji on 20 February, killing 44 people and tearing up the islands. By the time it reached Central Queensland days later, Winston had lost its energy but hung around as rain. Along with another down-graded cyclone, Tatiana, it cooled the southern Reef waters. Corals there that had shown early signs of bleaching were able to recover. That was a godsend for the Australian tourism industry.

Despite the saving rains, by March it looked like the 2016 coral bleaching was going to be the worst on record for the Great Barrier Reef. To get a handle on the full scale of the bleaching, Hughes organised a plane and chopper for aerial surveys.

'That's the only way you can really get the big picture. It takes eight days to crisscross the Barrier Reef and that tells us where the bleaching is most severe,' Hughes explained later. Bleached corals are incredibly visible from the air; but if you go up too late, many of them have already died and dead corals aren't visible. 'You can't measure mortality from the air,' said Hughes. 'That's why the vast majority of our work was underwater. So the whole time I was in the air, people were underwater—ground-truthing the accuracy of what I was seeing.' The underwater researchers would also go back to the same reefs eight months later.

Rachel Pears from GBRMPA joined Hughes on some of the aerial surveys. GBRMPA, AIMS and other outfits had boats and divers in the water as far north as Cape Grenville on Cape York, almost 300 kilometres north of Princess Charlotte Bay. One discovery further south hit the divers particularly hard: the stunning reefs around Lizard Island, the home of the Australian

Museum's Research Station, were at the epicentre of the catastrophe. 'On reefs and sites around Lizard Island, bleaching has approached 80–100 per cent, and many corals are already dying,' Dr Anne Hoggett, the co-director of the research station reported that April. 'This is by far the worst bleaching crisis we've seen, eclipsing the 2002 event.'

The aerial surveys and the early results of the underwater surveys shocked Hughes, and in late March he went public with the first findings. 'For me, personally, it was devastating to look out of the chopper window and see reef after reef destroyed by bleaching,' he told the ABC's 7.30 program. 'But really the emotion is not so much sadness as anger.'

Hughes laid the blame for the mass bleaching on climate change and went on to criticise successive Australian governments for not doing enough to tackle global warming. 'The government has not been listening to us for the past 20 years,' he said. 'It has been inevitable that this bleaching event would happen, and now it has. We need to join the global community in reducing greenhouse gas emissions.'

The coral bleaching of the Great Barrier Reef in 2016 finally forced the 'coal versus coral' debate into the political mainstream. The bleaching had hit just a few months after the Paris Agreement, when climate change was back in the headlines. In Paris, the Turnbull government had stuck to Abbott's weak targets to cut Australia's greenhouse gas emissions by just 26 to 28 per cent by 2030.

In the lead-up to Paris, the then environment minister, Greg Hunt, had also signed the final approvals for the huge Adani coalmine in Queensland's Galilee Basin. In making his decision, Hunt rejected the arguments from environmentalists that the greenhouse gas emissions from Adani mine would have a serious impact on the Great Barrier Reef because of their contribution to climate change. As Hunt put it in the approval decision, 'after giving consideration to the greenhouse gas emissions from mining operations and the burning of the mined coal, I found that the proposed action will not have an unacceptable impact on the world heritage values of the Great Barrier Reef Heritage Area'.

The massive coral bleaching now prompted scores of Australia's climate scientists to call publicly for a halt to any new coalmines. The scientists, including Hoegh-Guldberg and JCU's Morgan Pratchett, signed a full-page advertisement in Brisbane's *Courier-Mail* on 21 April. Headed 'Climate Change Is Destroying Our Reefs', it also called for Australia to rapidly phase out its ageing coal-fired power stations and stop building any new coal plants. 'The Great Barrier Reef is at a crisis point,' the advertisement read. 'Its future depends on how much and how quickly the world, including Australia, can reduce greenhouse gas emissions and limit ocean warming.'

The bleaching created a big dilemma for Hunt. He had just spent the past year fighting to keep the Reef off the World Heritage 'in Danger' list. The bleaching again threatened to put it on there. He had flown over the Reef in the early stages of the bleaching and kept emphasising that the southern and central sections of the Reef had escaped relatively well, citing the GBRMPA surveys to back this up.

But at this stage, GBRMPA wasn't playing down the bleaching and was blunt about its impact. In early May, the GBRMPA's chairman, Russell Reichelt, appeared at a Senate hearing in Canberra where he laid out the details. After the aerial surveys and 2000 in-water surveys by divers on 163 reefs, he said it appeared that 'virtually all reefs of the Barrier Reef have experienced some bleaching in the period between 1 February and today'. He described coral bleaching in the far northern section of the Reef, from Cape York to Lizard Island, as 'severe'; the bleaching in the northern to central area as ranging from 'severe to minor'; and the bleaching in the southern area as 'minor'. The critical surveys on how much of the bleached coral had died were still going on, he told the senators. But of the 900 reefs surveyed in the far north, GBRMPA thought the mortality was likely to be above 50 or 60 per cent. Reichelt also told the senators the bleaching 'is very strongly linked to global warming'.

Three days later, Turnbull called the 2016 election. Not long afterwards, tensions between the government agencies and Terry Hughes on the National Coral Bleaching Taskforce began emerging publicly. In the detailed reports on the bleaching released until then, there had been a large amount of agreement on the findings, but how the findings were being represented now became contentious. Hughes was determined to highlight the unprecedented scale of the damage.

At the end of May, Hughes released the latest figures from the Coral Reef Studies centre at JCU. It put the toll of dead and dying coral on average at 35 per cent on 84 reefs surveyed between Townsville and Papua New Guinea. Most of the losses, Hughes stressed again, were in the north. South of Cairns, the average mortality was around 5 per cent and south of Townville there was none. But both the key government agencies—GBRMPA and AIMS—decided to pull their support from that release.

A few days later Hughes was feeling the heat. 'There was a definite effort by the government to hose down the significance of the bleaching event,' as Hughes saw it. After the election was called, the Barrier Reef bleaching had become a weapon for Labor and the Greens to attack the government over its weak climate change policy.

The strains in the taskforce burst into public on 3 June, when GBRMPA and AIMS put out their own statement, 'The Facts on Great Barrier Reef Coral Mortality'. *The Australian* headlined its story on the new statement with the claim 'Great Barrier Reef: scientists "exaggerated" coral bleaching'. Written by environment editor Graham Lloyd, it alleged that, 'Activist scientists and lobby groups have distorted surveys, maps and data to misrepresent the extent and impact of coral bleaching on the Great Barrier Reef, according to the chairman of the Great Barrier Reef Marine Park Authority, Russell Reichelt.'

The report was careful not to accuse Hughes of distortion, but it said, 'A full survey of the reef released yesterday by the Authority and the Australian Institute of Marine Science said 75 per cent of the reef would escape unscathed.' Reichelt was quoted as saying, 'This is a frightening enough story with the facts, you don't need to dress them up.'

It looked like a serious division among the scientists over the impact of the mass bleaching. The initial survey results from GBRMPA put the overall coral mortality from the bleachings across the whole Reef at 22 per cent, which sounded a lot lower than Hughes's figure of 35 per cent but he was looking just at the northern two-thirds of the Reef. There were also other important qualifications. The results were preliminary, as the release said, covering the early losses. Corals kept dying as the weeks went on.

After more surveys later that year, GBRMPA's mortality figure jumped. The authority's final report on the bleaching put the mortality at 29 per cent of the shallow-water coral cover across the Great Barrier Reef Marine Park, with the big losses in the far north. This was not radically different from Hughes's final findings of 30 per cent mortality for the whole Reef, looking at survey results inside and outside the Marine Park boundary.

Despite their divisions, scientists from both the government agencies and the universities all agreed the mass bleaching and subsequent coral mortality was devastating to the Reef and that climate change had played a key role. The unusually high water temperatures, especially in the north of the Reef, told the story, Will Steffen from the Climate Council said later. 'When you looked at the pattern of reef bleaching and the fine-scale temperature pattern, they were virtually identical. That was a smoking gun that nobody could deny. Nobody.'

Well, not exactly nobody. Many of Australia's climate sceptics were not at all convinced that global warming had much to do with the worst coral bleaching on the Reef on record. And they had a loud voice in Canberra and in the media.

The divisions between the government agencies and Hughes's team fuelled stories on the climate-sceptic blogs and in the Murdoch media that the severity of the bleaching on the Reef was exaggerated. With Bob Carter dead, his friend and old JCU colleague Professor Peter Ridd made the running on the Reef bleaching. Ridd, a marine geophysicist, had been arguing for years that a lot of the research from Australia's coral reef scientists should be treated with scepticism and that the Reef was in good shape.

By 2017, Ridd and the IPA were pushing the case with gusto. That year Ridd wrote an essay for the IPA's latest collection *Climate Change: The Facts 2017*; he called it 'The extraordinary resilience of Great Barrier Reef corals, and problems with policy science'. In it he argued that corals were remarkably adaptable and 'perhaps the least endangered of any ecosystem to future climate change—natural or man-made'. The essay and its author got a big wrap on Sky News.

But both GBRMPA and AIMS took a different view. Coral reefs are resilient and can recover their coral cover with spawning from other corals that survive. But this can take years and the recovery depends on the corals not being hit by another catastrophic event in the meantime, like another bleaching.

Before the IPA's collection of essays was even launched, a second mass bleaching hit the Great Barrier Reef in March 2017. Once again, the sea-surface temperatures in Reef waters rose above average. The back-to-back bleaching was deeply worrying to all the government agencies and the academics and they put aside their differences. The scale of bleaching this time was smaller but still severe in parts. Much of the damage was surveyed from the air, so how much coral died was harder to quantify. But GBRMPA later found, 'Given the severity of bleaching observed, it is certain that the 2017 bleaching event caused a further decline in coral cover across the northern two-thirds of the Marine Park.'

The second bleaching was another big blow to the Reef, and Hughes believed it 'killed about 20 per cent of the corals'. Unlike the 2016 bleaching, the worst coral mortality was not just in the far north but down as far as Townsville. 'So between the two events, about one in every two corals died across a range of shallow-water habitats,' Hughes estimated.

After the back-to-back bleachings, Hughes argued the rising temperatures in the Reef waters shouldn't be referred to as just marine heatwaves but as extreme events. 'They are record-breaking temperatures,' he said. 'What we should be referring to is climate extremes that are damaging whole ecosystems to an extent that has never happened before.'

Marine scientists analysing the bleachings tried to look at both the positive and negative lessons that might be learned. One positive was that the hardier corals survived and appeared to be more resistant to heat stress in 2017. This meant there was more chance of coral cover eventually being restored. Some reefs in the south that managed to escape the multiple hits of bleachings, cyclones and crown-of-thorns starfish outbreaks were also in better condition. On the negative side, both GBRMPA scientists and Hughes reported that 'coral recruitment', new corals settling and growing on reefs, had crashed by an average of 89 per cent in 2018 in the aftermath of the bleachings. The hope was these shocking figures would improve after more spawnings—if the Reef wasn't hit again by a bleaching or a run of cyclones.

For fish that feed on coral the prospects didn't look great either. JCU's Morgan Pratchett had surveyed the dazzling butter-flyfish around Lizard Island since 1995. He saw a big decline in numbers for particular species that eat nothing but coral. Their decline came in the wake of the bleachings and an earlier run of intense cyclones that hit the local reefs. 'Normally we count something like seventeen species on those surveys and last time we did it we found only nine,' said Pratchett. 'And the densities of even the ones that we are still seeing are so much lower than they were four or five years ago.'

The impact of rising temperatures was not just seen in the water, but on land too. Over a decade before, GBRMPA's first climate change report had flagged that rising temperatures on the beaches of the northern Reef were causing the feminisation of green turtle populations. Because the gender of the turtles is determined in the nest, the higher temperatures where the turtles lay their eggs were skewing the sex ratio to female. By 2019, one study of green turtles cited by GBRMPA found only 1 per cent of the juveniles were male.

'The latest work is saying an unbelievably high percentage—99 per cent of juvenile green turtles in the northern Great Barrier Reef are female, which is the feminisation effect,' said GBRMPA's David Wachenfeld. 'That 99 per cent is the result of looking at juvenile turtles in the wild, it's not the output of a model, it is actually looking at turtles.' And it was not just green turtles. The hawksbill and flatback turtles in the north were also demonstrating the feminisation effect, reducing their ability to reproduce.

The role of GBRMPA is to work with both the federal and

Queensland governments to increase the resilience of the Reef. Big dollars have been poured into improving water quality, protecting the Reef from overfishing and controlling outbreaks of the crown-of-thorns starfish that prey on coral. The task is not just to keep the Reef on the World Heritage List. The Great Barrier Reef is a massive plus for the Australian economy, contributing over $6.4 billion a year, mainly in tourism revenue from the 2 million annual visitors who come from around the world and from around Australia.

But all the government resilience programs can only work up to a point, said Wachenfeld, who became GBRMPA's chief scientist in 2017. Like many of his colleagues, he argues that slowing global warming is critical for the Reef's future. 'Our best estimate is that up to 1.5 degrees [Celsius] we can manage to maintain something that looks like a functional Great Barrier Reef that continues to provide a tourism industry and a fishing industry and everything else. After that, all bets are off. As you start to get close to 2 degrees it is very unlikely that we will be able to do much to protect the Reef from that kind of climate change.'

In August 2018, after the back-to-back bleachings on the Great Barrier Reef, Ove Hoegh-Guldberg sat down in Brisbane to finalise a lengthy document he had written about coral reefs and climate change. It was not a paper for a scientific journal or for the IPCC; it was an affidavit for a court case. Hoegh-Guldberg had been an expert witness on climate change before, but this affidavit was very different.

He had agreed to give evidence in a landmark lawsuit against the president of the United States. The plaintiffs were 21 children and young adults, and the case was designed to test whether the American courts could play a role in getting the US government to act on climate change. It was called *Juliana v United States*, and the issue it raised was whether children had a constitutional right to a stable and safe climate.

The case was filed in Oregon state in 2015 when President Obama was in office, but under President Trump it was getting traction. 'It's basically on the denial of a future for kids,' Hoegh-Guldberg explained. In his affidavit for *Juliana v United States*, Hoegh-Guldberg laid out his opinion as a marine scientist. He argued that unless swift action was taken to eliminate the use of fossil fuels for energy, coral reefs will disappear or degrade beyond recognition. This would deprive both today's children and future generations, including the plaintiffs, not only of the splendour of coral reefs but the benefits that flow from them, including food and income.

'The fact that the U.S. Government has known of this problem yet has not taken proportionate steps to reduce greenhouse gas emissions from the fossil fuels that it exploits and promotes, is irrational and runs at odds with its constitutional responsibility to look after the security of its citizens, born and unborn.'

'We have to say something,' said Hoegh-Guldberg, explaining his decision to be part of the case. 'We cannot just hide in the shadows and be agnostic on the issue, given what the risks are.'

For more than two decades Hoegh-Guldberg had written some 300 peer-reviewed publications on oceans and coral reefs. He had been a lead author for the IPCC for its reports to inform world governments. Yet greenhouse gas emissions were still rising and so were the risks of dangerous climate change. For him and many other scientists, the courts were increasingly becoming a critical avenue to pressure governments and corporations into action.

The children's case would later be thrown out by a US federal appeals court. Writing the majority decision, Judge Andrew Hurwitz said that while the children had made a compelling case,

'The panel reluctantly concluded that the plaintiffs' case must be made to the political branches or to the electorate at large.' The lawyers for the children insisted they would appeal the decision but the ruling was clear: the children needed to get politicians and the public to listen to them, not the courts.

But in August 2018, despite the urgent warnings from climate scientists, neither the government in Washington nor in Canberra appeared to be listening. In Australia, the Liberal Party had dumped its energy and climate policy yet again, along with its leader. For the new prime minister, Scott Morrison, the risks from dangerous climate change just weren't a pressing issue for the majority of Australian voters or for their children.

Or so he thought.

CHAPTER EIGHTEEN

NO REGRETS

UNDER A CLOUDLESS blue sky, the new prime minister strode across the dry red dirt of Bunginderry Station wearing a baseball cap to protect him from the western Queensland sun. Scott Morrison had been in office only a couple of days when he arrived in the Channel Country to see first-hand the devastation of the big drought. The property where five generations of the Tully family had made their living was parched.

In August 2018, over half of Queensland was in drought, along with much of New South Wales, Victoria and South Australia. Families like the Tullys had endured this drought for the past five years. They had destocked some of their sheep and cattle and were hanging on, but costs were mounting and income was falling.

Morrison was offering hope and money. The government, he assured them, 'is one hundred per cent focused on what we need to do as a nation together to ensure there is not just relief from the drought but recovery from the drought'.

Australia's first Pentecostal prime minister was big on optimism. He had come with a phalanx of National Party heavies vital to the government's stability. He rattled off plans for dog-fencing, road programs and help with educating children, promising to work with the states and local government as the drought deepened. 'We have got to stay together as a country in our support,' he exhorted.

The prime minister's enthusiasm flagged only once on that day—when he was questioned about climate change. 'Can you tell me whether you believe that the prolonged dry period that people are experiencing out here is associated with human-induced climate change?' the reporter asked.

Morrison batted the question away: 'I don't think people out here care one way or the other whether it's that,' he replied quickly. But the reporter persisted. 'I'm asking you,' she said. 'Do you believe it is associated with human-induced climate change?'

Frustrated, Morrison stumbled before he found the words he wanted. 'Climate is changing, everybody knows that,' he said. 'I know what you're trying to ask, OK? I don't think that's part of this debate. That's my point. If people want to have a debate about that, fine. It's not a debate I've participated a lot in, in the past, because I'm practically interested in the policies that will address what is going on here right now.'

It was Morrison's first telling statement on climate change as prime minister and it would define his stand. He just didn't want to talk about it. And he didn't want to open up the Coalition's fraught internal debate on it.

But in Australia by 2018 climate change was part of the here and now. The cost of adapting to it was looming as one of the nation's greatest challenges in the 21st century. Climate change did not create a particular drought or a particular bushfire, but it was expected to make droughts and bushfires more intense, last longer, and exact a heavier toll on plants, animals, people and communities. That was the consistent advice from the government's own agencies.

Already this drought was shaping up to be worse than the Millennium Drought that had rocked Morrison's old mentor, John Howard, just over a decade earlier. Across eastern Australia the bushfire season had started unusually early. By August, the last winter month, there were 80 to 100 fires burning across New South Wales, with others in Queensland and Victoria. The year 2018 was going to be Australia's third-warmest on record and the warmest on record for New South Wales.

But climate change and the costs that would flow from it were not subjects Morrison was interested in discussing. 'I'm interested in getting people's electricity prices down,' he said as he stood in the hot sun. 'I'm going to leave that debate for another day.'

Six weeks later, in early October, the day for that debate came. The IPCC released its special report on the impacts of limiting global warming to 1.5 degrees Celsius above pre-industrial levels rather than 2 degrees. The report was commissioned as part of the Paris Agreement after many nations, including Australia's low-lying Pacific neighbours, warned 2 degrees of warming was too high. It reopened the global debate over whether world leaders should strengthen their responses to climate change.

But Morrison brushed off the landmark report, saying it didn't provide recommendations for Australia—'this is dealing with the global program,' he said. Morrison repeated the mantra Australia's leaders had used since Howard's time: 'Let's not forget that Australia accounts for just over 1 per cent of global emissions, so there are a lot bigger players than us out there impacting on these arrangements.'

Yet the new IPCC report clearly spelt out that the risks from dangerous climate change were likely to be significantly lower for countries like Australia if warming was kept to 1.5 degrees rather than 2 degrees. These were risks to food production, water resources, plants, animals and coral reefs. At 1.5 degrees of warming, fewer people were likely to be exposed to droughts and heatwaves and the serious health impacts that came with them.

The new IPCC report also raised profound implications for Australia's rich fossil fuel industry. To keep to 1.5 degrees of warming, the report detailed that countries would need to commit to far more ambitious cuts to greenhouse gas emissions. Global annual investment in clean energy and energy efficiency would probably need to overtake investment in fossil fuel energy by around 2025—just seven years away. Annual investment in low-carbon energy technologies and energy efficiency would also probably need to go up roughly by a factor of six by mid-century. If that happened, it would have serious impacts on Australia's coal and gas industries—its workers, the superannuation funds that invested in them, and the government budgets that relied on their royalties and taxes.

The IPCC report also red-flagged that countries like Australia would have to increase their weak Paris targets to cut emissions. The current global Paris pledges were nowhere near enough to keep warming to 1.5 degrees, or even 2 degrees.

But these stark findings made little impression on the Morrison government. Interviewed by the ABC about the special IPCC report, the new environment minister, Melissa Price, stoutly defended Australia's continued heavy use of coal-fired electricity, saying 'we make no apology for the fact that our focus at the moment is on getting electricity prices down'. Asked about the IPCC's timeline to phase out fossil fuels, particularly coal, the environment minister replied, 'I think, you know, to say that it's got to be phased out by 2050 is drawing a very long bow.'

———

Two days after the IPCC released its report, Morrison's new energy minister, Angus Taylor, arrived at Sydney's Sofitel Wentworth Hotel for the annual *Australian Financial Review* National Energy

Summit. Surveying the room full of power company executives and government officials, Taylor delivered a simple message: 'When the prime minister, Scott Morrison, gave me this portfolio, appointing me as the Minister for Energy, he gave me one very clear KPI—one KPI. That was to lower power prices while we keep the lights on,' Taylor said bluntly. 'If you look over the most successful campaigns I've seen in politics, and indeed in business over the course of my career, there is typically one common characteristic—clarity.'

Taylor certainly made himself clear that morning. Gone was any stress on cutting greenhouse gas emissions. Taylor told the executives the Morrison government was backing investment in 'fair-dinkum, affordable and reliable generation'. To many that translated as a thumbs-up for coal-fired power plants, either by propping up ageing coal plants or supporting new ones. Gas plants and pumped hydro were also okay, but there was little enthusiasm for wind and solar.

The new minister was bristling with confidence and this was not surprising. As one of his rivals once quipped, Taylor was a FIGJAM—a 'Fuck I'm good, just ask me'—guy. After securing his law and economics degrees from the University of Sydney, Taylor had gone to Oxford as a Rhodes Scholar before becoming a highly paid energy and water consultant. By the time he arrived in Canberra, Taylor's views on the energy industry were pretty much formed. He was a big critic of the Renewable Energy Target and wind farms in particular—especially after one was built next to a family property in southern New South Wales.

Morrison put Taylor in as energy minister as a way of healing the rifts in the Coalition. In the fraught battle over the National Energy Guarantee (NEG), pushed by Malcolm Turnbull and his then energy minister, Josh Frydenberg, Taylor had been an intellectual beacon for Abbott and his supporters in the MPs' pro-coal lobby, the Monash Forum.

'He's always been close to these guys,' Turnbull said when asked about Taylor. 'He's very ambitious. He thinks it's a great injustice that he's not already prime minister. Hugely ambitious guy and so I never had any doubt that Angus was involved with all of that. But he was telling Josh he supported the NEG. He told me too. But we knew, and Josh knew particularly, that he was actively briefing and advising the Monash Group and others of the same mind.'

Abbott subsequently denied that Taylor ever plotted with him against the NEG. 'I'm a great fan of Angus. So we would talk a lot and he obviously had some question marks, some big hesitations about the National Energy Guarantee. But did we plan an attack on the NEG? No. Did we coordinate speeches or observations to the media? No.'

Unlike Abbott, Taylor insisted he was not a climate sceptic. When he became minister, he deflected questions about Australia's rising emissions by saying he was very confident the government would reach its Paris target, 'in a canter'.

Soon after the *Financial Review* summit, Taylor and Morrison unveiled the government's new policy, 'A Fair Deal on Energy'. It foreshadowed 'big stick' laws to stop price-gouging by energy companies that ripped off customers. It also increased pressure on the companies to deliver reliable energy supplies. But this new policy had no price signal to encourage new investment in clean energy and no serious plan to work with the states to make the transition to a clean energy economy in the 21st century.

The prime minister did send one signal, and it was political: the climate wars were still on. He and Taylor were again turning the climate change debate into a fight with Labor over electricity prices in time for the next election. Just like his predecessors, John Howard and Tony Abbott, Morrison was reverting to the old 'No Regrets' policy. The Coalition government would take no action on climate change that would impose any significant costs on Australia's carbon-intensive economy. And 'No Regrets' worked for the Coalition—so long as the majority of voters were encouraged to worry more about power prices and potential job losses than about the looming costs of inaction on climate change.

Morrison's thinking was reflected not only in his choice of minister, but his choice of advisers. His chief of staff, John Kunkel, had been a long-time senior executive with the peak mining lobby, the Minerals Council of Australia, and a former head of government relations with mining company Rio Tinto. Morrison's principal private secretary, Yaron Finkelstein, was a long-time lobbyist with Crosby Textor and chief executive of its Australasian branch during the time the firm took on Project Caesar, the pro-coal campaign for Glencore. Both Finkelstein and Kunkel had cut their teeth as staffers in the Howard government.

Morrison boldly drew the battlelines with Labor on energy and climate change just three days after his government suffered a humiliating defeat in the by-election for Malcolm Turnbull's old seat of Wentworth in Sydney's eastern suburbs. The Liberal candidate, former ambassador Dave Sharma, was beaten by Dr Kerryn Phelps, an independent candidate who ran strongly on climate change. The defeat meant Morrison also lost his one-seat majority in parliament and was now relying on the key cross-benchers to stay in power until the next election.

But Morrison and Taylor were not about to switch course because of Wentworth. A month later they doubled down when opposition leader Bill Shorten launched Labor's 'energy future' policy. Shorten reaffirmed his promise to increase Australia's Paris target to a 45 per cent cut in emissions by 2030—significantly above the government's 26–28 per cent cut. He also fleshed out Labor's promise to make the energy network 50 per cent renewable by 2030, vowing to make Australia a renewable energy 'powerhouse'.

Taylor and Morrison dug in for battle. While Shorten announced his push for more renewables, Taylor paid a visit to the Tomago aluminium smelter in the NSW Hunter Valley, part-owned by Rio Tinto. The smelter was both a big local employer and the single largest consumer of electricity in a state heavily dependent on coal-fired power. Taylor had one simple message: Labor's targets would kill the smelter and destroy the jobs it supported.

'The truth is these are reckless policies and Labor will need to explain to every farmer, every truck driver, to every worker who works here at Tomago, to every worker who works in every other energy-intensive business in Australia, how they are going to cut their jobs,' said Taylor as he stood shoulder to shoulder with the smelter boss, Matt Howell.

It was Abbott's 2013 scare campaign all over again, but this time Labor was convinced it wouldn't work. Nearly every poll was running Labor's way. The smart money was on Bill Shorten becoming the next prime minister at the 2019 election, and transforming Australia's climate change policy.

But that didn't happen.

———

The 2019 election campaign fired up Australia's climate wars, and the coalfields of Queensland's Galilee Basin became ground zero. That was inevitable when, two days before the election was called, Morrison's environment minister, Melissa Price, approved the final groundwater plans for the Adani coalmine in the Galilee. The Adani decision impassioned climate activists around the country, just as some in the Coalition hoped it would.

'How dare this government give its stamp of approval to this appalling project,' said an outraged Tim Flannery, the head of the Climate Council. 'This project doesn't stack up economically, Australians don't want it and none of the major banks will back it. It's the exact opposite of responsible climate policy. Fortunately, there's still time to stop this insanity.'

Price's approval came after intense lobbying from her Queensland Coalition colleagues, who understood what the Liberal Party's campaign headquarters also understood: in marginal seats all over regional Queensland, where coalmines meant work, Adani could be framed as a rallying cry for jobs against Labor's climate change policies.

And that's what happened. 'We lost credibility with coal industry workers,' said Jay Weatherill, Labor's former South Australian premier, who ended up conducting the post-mortem on Labor's failed campaign with his colleague Craig Emerson. 'We were associated with the Greens' policy and the LNP clipped us with the Greens as anti-coal and anti-jobs.'

From the outset, the Coalition's attacks on Labor's climate policy were crude, effective and supercharged by the Murdoch media. Morrison aggressively slammed Labor's plans to cut greenhouse emissions as economy-wrecking—borrowing language from the Business Council's CEO, Jennifer Westacott—and job-destroying. He even argued they were an assault on the Australian way of life.

When Shorten launched Labor's plan to cut soaring transport emissions by encouraging electric cars to make up 50 per cent of new sales by 2030, Morrison accused him of wanting to 'end the weekend'. Morrison claimed that an affordable electric car could not 'tow your boat'—killing off holidays and camping trips—as if electric vehicles would become compulsory. He belted Labor's plan to raise Australia's vehicle emissions standards—some of the weakest in the world—claiming it was a $5000 carbon tax on new cars. Or, as Murdoch's Sydney tabloid the *Daily Telegraph* put it, 'Bill's $5K CAR-BON TAX'.

Shorten proved incapable of defending his climate policies

against the barrage. Scarred by Tony Abbott's 'carbon tax' campaign in 2013, Labor had dropped its policy of putting a price on carbon through an emissions trading scheme. So while Shorten promoted his ambitious targets on renewable energy and cutting emissions, he and his shadow environment minister, Mark Butler, failed to explain the cost of reaching their goals. By giving no real costings on their key policy, they allowed Morrison and the media to fill in the numbers. Headlines like 'Labor's carbon plan "to cost miners $2bn"' came thick and fast.

More importantly, Labor could not resolve its tortured position on opening up the Galilee Basin to coalmining. A year earlier, when Shorten was fighting to hold the Melbourne seat of Batman from the Greens in a by-election, he hinted Labor would try to stop the Adani coalmine. By the 2019 federal campaign, some of his Labor candidates in North Queensland were openly backing coalmining in the Galilee Basin. These candidates were aligned with the powerful mining and construction union, the CFMEU, a big donor to the Labor Party, and the AWU, the Australian Workers' Union. It made Shorten's prevarication on Adani look duplicitous. The tagline the Liberal Party had workshopped on social media for more than a year, 'Shifty Shorten', hit home. By late April, just weeks before the election, Labor was badly on the defensive in Queensland over climate change and coalmining.

Adding to Labor's problems, the former Greens leader, Bob Brown, organised a Stop Adani 'convoy' of climate activists to drive from Tasmania up to the Galilee. Hailing the convoy as a public showdown with the coal industry and its political backers, Brown and his supporters held rallies and picked up activists along the way. By the time the convoy got to North Queensland, Coalition MPs were rubbing their hands with glee.

One of those was the then resources minister, Senator Matt Canavan, a hardline climate sceptic from Queensland, whose brother was a senior executive in the coal industry. Canavan backed a 'resistance rally' to Brown's convoy when it rolled into the coal town of Clermont. The town is in the seat of Capricornia, where the local Labor candidate, a coalminer, was trying to dislodge the Coalition MP. Local residents, miners and Liberal National Party (LNP) members donned 'Start Adani' T-shirts and joined the rally against the climate activists. 'I want to thank Bob Brown for getting us all together, he's done an excellent job of just combining and uniting the whole Central Queensland community in favour of jobs,' said a pugnacious Canavan.

Pauline Hanson, the far-right One Nation leader, stood in the rally with a CFMEU placard reading, 'I support jobs and rights for coalminers'. Labor was furious. It was wedged between the Coalition and the Greens. But Brown dismissed Labor's anger. He was trying to stop the Galilee mines, rally the Greens' base and save its Senate seat in Queensland.

One familiar face in the crowd at the pro-Adani rally was Clive Palmer, the Queensland billionaire mining entrepreneur. Palmer was an avid supporter of opening up the Galilee for coalmining. He by now had state approvals for a big coalmine there and he wanted to get it off the ground. In late 2018, Palmer had re-launched his political career, turning his defunct Palmer United Party into the United Australia Party (UAP). Already Palmer had spent over $30 million promoting his lower house candidates in this election and his own bid to win a Senate seat.

'The United Australia Party is happy to support anything that we can sell on the market to the world that can create jobs for Australia,' Palmer cheerily told reporters at the pro-Adani rally.

Just days earlier, Palmer had struck a deal with the Coalition campaign heavies to swap preferences in the election. This meant both sides would list the other ahead of Labor on their how-to-vote cards. Palmer's party was polling 5 per cent, and higher in some seats, so their preferences could be critical in getting Coalition candidates over the line in a tight race, especially in Queensland.

The preference deal was a delicate issue for Morrison because of Palmer's recent turbulent history. His old North Queensland nickel refinery had been put into liquidation in 2016, leaving a string of debts and shutting down over 700 jobs. While Palmer technically was not on the board of the company, liquidators on behalf of the government were chasing him in court for up to $66 million for sacked workers' entitlements. Around the time the preference deal was struck with the Coalition, Palmer promised to settle the outstanding worker's entitlements.

In the two weeks before election day, Palmer's well-funded UAP campaign changed decisively. Rather than attacking both parties, his campaign began to dovetail with the Coalition's 'Kill Bill' strategy. As Palmer explained it, his stepped-up attacks on Shorten were partly motivated by Labor's tax policies. But he also had an antagonism towards Shorten's former union—the Australian Workers' Union, which represented many nickel refinery workers—for exacerbating his ordeal in the courts. Just as importantly, Palmer wanted the Galilee Basin open for mining.

Palmer said he made a calculated decision. It was clear his UAP was not going to win the balance of power in the Senate. At that point, recalled Palmer, 'I made a decision not to have a Labor government.' In the final week of the campaign, Palmer shifted the majority of his advertising to an anti-Labor and anti-Shorten pitch: 'I made the decision to polarise the electorate and sacrifice our own position.'

Palmer wrote the UAP ads himself, taglined 'Tell Shifty He's Dreaming'. With his deep pockets Palmer had already bought up space in the crucial TV markets and on digital platforms, making it too costly or too late for Labor to compete. On election night the majority of Palmer's preferences went to the Coalition, but that was a secondary issue. The anti-Shorten advertising blitz in the last week of the campaign inflicted real damage, especially in Queensland, where Labor was left holding only six out of 30 seats.

'I don't regret it, it was big stakes,' Palmer said later. 'We spent more money than a Coca-Cola campaign or a Foster's.' Palmer's companies donated over $83 million to the UAP, much of it spent on advertising. 'But what price do you put on saving Australian democracy?' the billionaire asked.

Labor's post-mortem analysis made it clear it did not lose the election just because of its climate change policy or because of Palmer. But the battle over climate change in the campaign did deepen divisions in the country. Labor's analysis revealed a disturbing voting pattern. Low-income voters in regional and rural communities continued to abandon Labor. The party had failed to make the case that the cost of inaction on climate change affected regional and rural Australians as much, if not more, than inner-city voters. Instead, in the coal towns of Queensland, many blue-collar and low-income voters just felt alienated by Labor.

By contrast, in the cities, better educated voters shifted to Labor and the independents, many driven by concerns over climate change. That certainly happened in Sydney's northern beaches. In a historic defeat for the Coalition sceptics, Tony Abbott, their long-time hero, lost to independent Zali Steggall, who campaigned strongly on climate change.

Despite Abbott's thumping, the government's return was a stunning victory for Morrison and the Coalition. The election was seen by many as an endorsement of its 'No Regrets' climate policy—no action that would harm the greenhouse-heavy industries and the jobs they created. In reality, Morrison had scraped home by just a slim majority but, in the eyes of his federal colleagues, he was a miracle maker. And on the night of his victory he spoke like one, repeating an evangelical exhortation he had used on the campaign trail.

'I said that I was going to burn for you and I am—every single day,' he promised.

Morrison had no idea that within months it would be Australia on fire, not him.

———

By September 2019 it was clear the bushfire season was going to be bad. When the Australian Bureau of Meteorology warns the country about the danger of bushfires, it uses the Forest Fire Danger Index—the FFDI. The index uses a combination of temperature, humidity, wind speed and drought to calculate bushfire danger 'values'. That spring, the FFDI was off the charts. Almost 60 per cent of Australia had the highest FFDI values on record. In some parts of New South Wales that spring, fire authorities already rated bushfire conditions as 'catastrophic'.

By early September the fires were raging in south-east Queensland and northern New South Wales, where the long drought had left the country tinder-dry. Queensland Fire and Emergency Services inspector Andrew Sturgess warned that his state had never seen conditions so bad so early. 'We've never seen fire danger indices, fire danger ratings, at this time of the year, as we're seeing now,' he told reporters. 'Never seen this before in recorded history.'

With the fires still burning, Scott Morrison headed to the Queensland fire zone offering help. 'They can expect the same support that we were able to provide so quickly and swiftly to the people of Queensland in the floods. We'll also be there for those affected by fires,' he promised. For the prime minister, the latest bushfires were one more natural disaster, like many Australia had weathered before. But for others, this fire season was unprecedented. They were determined that the role of climate change would no longer be ignored.

Earlier that year, 23 of Australia's former fire and emergency leaders from across the country had banded together in a group called the Emergency Leaders for Climate Action. Between them, they had 600 years' experience in fighting fires. In April 2019 they wrote a public statement calling on the prime minister to meet a delegation from the group. They warned that Australia was unprepared for worsening extreme weather. Climate change, they said, was increasing the frequency and intensity of these events, in particular catastrophic bushfire weather. Bushfire seasons were lasting longer and dangerous fire days were getting worse. They wanted the meeting so they could explain 'how climate change risks are rapidly escalating', and they wanted a parliamentary inquiry.

Morrison didn't want to meet them. Greg Mullins, a former commissioner for Fire and Rescue NSW and a volunteer firefighter, was spokesperson for the group. He was a prominent member of the Climate Council headed by Tim Flannery. As Mullins and his colleagues saw it, the government needed to do two things: put up more money and resources to deal with disasters like bushfires, which were going to be supercharged by climate change; and step up efforts to reduce greenhouse emissions. Adaptation and mitigation—these two responses to climate change needed to be acted on together.

Mullins and the former fire chiefs never got their meeting with the prime minister. Mullins believed Morrison's attitude was, 'We will not negotiate with climate change activists.' Instead Morrison wrote to Mullins that July, referring him to Angus Taylor. Since the election, Morrison had effectively put Taylor in charge of Australia's climate change policy, making him the

Minister for Energy and Emissions Reduction. By October, the former fire chiefs still hadn't met Taylor or David Littleproud, the federal emergency management minister.

On 6 November, a frustrated Mullins told the ABC, 'We're coming into what I think is the most dangerous build-up to a fire season I've seen since 1994, when New South Wales was devastated,' he said. 'And there's not even platitudes—there's just closed doors and closed minds as far as I am concerned.'

Days later, three people died in intense bushfires in northern New South Wales, including a grandmother of six trying to protect her home. Fires were now burning from Queensland through New South Wales towards the Victorian border. Four hours north of Sydney, in the coastal resort town of Port Macquarie, ash from the fires washed up on the beaches and the town's air quality registered as the most hazardous in the world. Schools were shut down and people were advised to stay indoors. Fires tore through nearby bushland, devastating the area's famous koala populations, and hundreds of the animals were feared dead.

That same week, the NSW Rural Fire Service issued catastrophic fire warnings for Sydney and the Hunter Valley. Resources were stretched as volunteer firefighters battled blazes, often started by dry lightning, that quickly escalated out of control. Bushfire smoke blanketed Sydney. Hundreds of thousands of people were breathing air rated from poor to hazardous and worse.

But the prime minister was undaunted. Asked why he hadn't met the former fire chiefs, he insisted the federal and state response to the bushfires was well in hand and indeed was 'outstanding'. Asked if Australia should do more to cut its greenhouse emissions, Morrison dismissed the idea: 'the suggestion that any way, shape or form with Australia accountable for 1.3 per cent of the world's emissions, that the individual actions of Australia are impacting directly on specific fire events, whether it's here or anywhere else in the world, that doesn't bear up to credible scientific evidence either,' he said.

The prime minister was being disingenuous. None of the former fire chiefs had suggested a direct link between Australia's emissions and a specific fire. What they had said was that Australia needed to do much more to cut its emissions as part of the effort to keep global warming from rising above 1.5 degrees Celsius. But on that point Morrison was adamant. Australia, he said, was doing 'our bit' on climate, indeed overachieving on its commitments. He was bullish that Australia would meet its 2030 targets. 'I'm up for taking action on it, not just jabbering on about it,' he said.

In early December, ministers Littleproud and Taylor finally met the former fire chiefs; but only to reassure them, again, that everything was in hand. Taylor also assured Mullins that Australia would meet its 2030 Paris target to cut emissions. The former fire chief wasn't impressed and recalled telling him: 'Your lack of policy is going to doom future generations.'

Taylor had every reason to be confident the government would meet its 2030 target. Not only was it a weak target, but the government had again decided to use dubious carbon credits to meet it. This was the position Taylor was taking to the latest UN climate summit that had just opened in Madrid.

———

Angus Taylor didn't arrive in the Spanish capital in time for the opening speech by the UN secretary-general, which was probably just as well. António Guterres's message to the UN's 25th Conference of the Parties (the COP) was not one Taylor or his prime minister wanted to hear.

By the end of the coming decade, in just ten years, the world would be on one of two paths in its response to climate change,

Guterres told the delegates. One was the path of surrender, where people had sleepwalked past the point of no return, jeopardising the health and safety of everyone on the planet. The other path was one of hope. 'Do we really want to be remembered as the generation that buried its head in the sand, that fiddled while the planet burned?' he asked them. The path of hope, Guterres said, was a path 'where more fossil fuels remain where they should be—in the ground—and where we are on the way to carbon neutrality by 2050'. This was the only way to limit the global temperature rise to 1.5 degrees Celsius by the end of this century.

Guterres pointed to the latest data from the World Meteorological Organization, showing that greenhouse gases in the atmosphere had reached a new record high. The signs the world was being impacted by climate change were unmissable, he warned. The last five years were the hottest on record, and the consequences were being seen in more extreme weather events and disasters, from hurricanes, droughts, floods and wildfires. Greenland's ice cap was melting faster than predicted; permafrost in the Arctic was thawing 70 years ahead of projections; Antarctica was melting three times as fast as a decade ago; and ocean levels were rising quicker than expected.

Yet, said Guterres, coal-fired power plants continued to be planned and built in large numbers. 'Either we stop this addiction to coal or all our efforts to tackle climate change will be doomed.' On current trends, the world was looking at global warming of between 3.4 and 3.9 degrees Celsius by the end of the century. 'The impact on all life on the planet—including ours—will be catastrophic,' the secretary-general concluded. 'My call to you all today is to increase your ambition and urgency.'

For Australia, the world's biggest coal exporter and, in 2019, the world's biggest liquefied natural gas (LNG) exporter, these were confronting words. But when Taylor took the podium, he didn't

speak of new ambition or new urgency. Nor did he mention the bushfires raging across his country or the smoke that had taken Sydney's
air quality that day ten times above the hazardous threshold.

Instead, Taylor offered characteristic pragmatism: Australia
would look for a solution in technology. 'We can only reduce
emissions as fast as the deployment of commercially viable
technologies allow,' he told the delegates. He repeated the government's mantra on its Paris target to cut greenhouse emissions. 'We
are already on track to meet and beat the target we have set for
2030,' he told the delegates, 'just as we are meeting and beating
our Kyoto targets.'

But many delegates knew those words meant Australia
planned to meet its weak Paris target by using so-called 'carry-
over' carbon credits. These credits had long been condemned
as 'hot air' credits by EU delegates and environmentalists. They
were carbon credits created when developed countries beat their
emissions targets set under the two-decades-old Kyoto Protocol
climate agreement. Most developed countries simply let these
credits lapse. But not Australia.

Australia was doing what it had always done: set weak targets
to cut emissions, beat those targets, and then demand to use the
'over-achieved' credits to help meet the next weak target. Taylor
put it more positively to the media: 'We think Australia should be
given credit where it's due. As always, we will seek to do more
where we can,' he told the *Australian Financial Review*.

This time Australia was called out. The main job of this UN
conference was to work out the rules for a new, credible, global
carbon market for emissions trading under the Paris Agreement.
China, Brazil and some other developing countries were already
trying to undermine the rules; they were also pushing to use dubious
credits from the Kyoto period. In this atmosphere, Australia's efforts
to use its 'hot air' credits did not go down well with many delegates.

One of those following the diplomatic brawl in Madrid was Frank Jotzo, from the ANU's Centre for Climate and Energy Policy. Australia's intransigence over carry-over credits, he believed, was damaging its international credibility: 'All the time it is costing Australia diplomatically, because it is just a deeply unpopular thing.'

The Madrid climate summit made little headway. Looming over everything was President Trump's formal notice that the US would withdraw from the Paris Agreement in 2020. Without the White House pushing the big players like China, India and Brazil, the pleas from the secretary-general for more action on climate change made little impact. Governments rarely acted on climate change unless they were under real pressure to do so. That was certainly true in Australia.

———

In mid-December 2019, Scott Morrison and his family boarded a flight to Hawaii for a pre-Christmas holiday. The prime minister made no formal announcement he was leaving the country, hoping to slip away unnoticed as the bushfire crisis raged on. The day he left, 2000 volunteer firefighters were battling over a hundred blazes across his home state of New South Wales. Fires were also burning in Queensland and Western Australia. Ominously, the Bureau of Meteorology was predicting a brutal new heatwave that would smash December temperature records across the country.

The deaths of two NSW volunteer firefighters forced Morrison's hasty return and brought an uncharacteristic apology. The country was in uproar at the prime minister abandoning the country in a crisis. Almost four months into the devastating bushfires, Morrison was suddenly on the back foot, and not just over his handling of the disaster. The politics of climate change in Australia were once again shifting. As the summer went on and the crisis deepened, the shift became more profound.

After a brief Christmas reprieve, fire conditions in New South Wales and Victoria again turned deadly. Victorian fire authorities ordered the evacuation of a vast area of East Gippsland up to the New South Wales border. Thousands of holidaymakers had arrived in the small towns and villages nestled in the bushland on the Gippsland coast and on the NSW south coast. Firefighters in both states braced for the worst.

At 6 am on New Year's Eve, Greg Mullins joined a group of volunteer firefighters heading to Bateman's Bay on the NSW south coast. On the drive down they passed long lines of cars carrying people fleeing north. By the time they arrived on the outskirts of town at around ten that morning, Mullins remembered, the fire had leapt the main highway that cuts through the town. Some of the houses were already burning. 'If the roof was on fire, you'd move on to the next house to try to save it,' he said of that terrible day. Up and down the coast, cafes and gift shops, homes and farms went up in flames. In the village of Cobargo, a bit over an hour south of Batemans Bay, historic main-street buildings were burnt to the ground. A father and son died trying to protect their farmhouse.

Across the Victorian border in the holiday town of Mallacoota on the Gippsland coast, thousands of locals and holidaymakers were forced to flee to the beach when the sky turned red, then black, and embers rained down on the houses. Many took to small boats, seeking safety in the water. The image of Finn Burns—an eleven-year-old Mallacoota boy in his breathing mask, bathed in an eerie orange light and steering his family's boat to safety—flashed around the world and became the face of Australia's bushfire crisis. The Victorian premier, Daniel Andrews, called on the prime minister to get the navy to evacuate the town.

By 1 January, the fires across the country had burned an area almost twice the size of Belgium. Ultimately, the 'Black Summer' fires, as they became known, would burn over 10 million hectares, or 1.6 per cent of Australia. The death toll was put at 33 but it was later estimated that another 445 people died because of the smoke from the fires. Over 3000 homes and 7000 structures were burnt, along with 80,000 head of livestock. The loss of native animals, judged to be in the many millions, was described by an expert panel on threatened species as 'an ecological disaster'.

What Mullins and the former fire chiefs had predicted for months had come to pass. Bushfires in Australia are normal, but this season was not normal. The record-breaking heat, and the record lack of rain, underscored what climate scientists had been saying for years: climate change was expected to make Australia's bushfire season longer and more intense. For many Australians, climate change was no longer a distant threat on the horizon, it was at their front doorstep.

In the days after the New Year's Eve fires, anger exploded over the government's tardy response. Suddenly Morrison and his ministers leapt into action. Three thousand defence force reservists were called up; ships, planes, choppers, water bombers were all on offer. A $2 billion national bushfire recovery agency and a federal inquiry were on the table. Climate adaptation was the new urgent priority for senior cabinet ministers, including the prime minister.

But when it came to making more ambitious cuts to Australia's greenhouse gas emissions—climate mitigation—Morrison still drew a line. In a round of interviews the week after the fires, the prime minister made it clear there would be no big increase in the target to cut Australia's emissions. Morrison insisted he would not jeopardise Australia's carbon-intensive industry by ambitious targets he described as 'reckless'. 'What our commitment is to do is to reduce emissions and reduce it in the way that I've said. That is, to ensure that we protect Australians from reckless targets, from reckless policies that can destroy their livelihoods and their incomes, and the future of their towns and of their regions, that force up their electricity prices and then force up their costs of living,' said Morrison.

His government, he promised, would 'stay true to the policy I took to the last election'. No carbon tax, no price on carbon. No change.

For economist Frank Jotzo, who had followed the fraught climate and energy battles since the Howard years, the government was not for turning. 'They've made that quite clear, that their prime objective is to protect Australian fossil fuel production and export industry interests,' he said.

There was one big change. In the weeks after the fires, Morrison, who couldn't talk about climate change on his first outing as prime minister, now found his voice. He and his ministers scrambled to distance themselves from high-profile backbench sceptics like Craig Kelly and their former leader, Tony Abbott. The government was now promoting 'climate action'.

The loud cheerleaders of the carbon club—the Murdoch-media commentators, the radio shock jocks and the IPA's experts who had championed the sceptic cause for decades—were suddenly kicked to the kerb. In 2020, arguing over the science was seen as a political dead end.

On closer scrutiny, Morrison's strategy was 'back to the future'. It mirrored the desperate backflip by John Howard over a decade earlier when his prime ministership was under pressure from Kevin Rudd: accept the science, stop arguing about it, but back Australia's carbon intensive industries. Australia's competitive advantage as the world's largest coal exporter and the world's largest LNG exporter could not be challenged.

Yet by discarding the powerful voice of the climate sceptics who for years had questioned the science and dismissed the IPCC as alarmist, Morrison and his senior ministers created a political dilemma. If the science was right and dangerous climate change was a major problem, those demanding the government do more to protect them and their children from its impacts had the moral imperative on their side. If the government was now accepting that climate change would cost Australians billions of dollars in adaptation costs and claim untold lives and livelihoods, then championing the fossil fuel industry, refusing to make more ambitious cuts to emissions, and locking Australia into its dubious position as the biggest per capita emitter in the developed world, looked like protecting vested interests over the national interest.

Morrison tried to grapple with this dilemma when he addressed the National Press Club in Canberra the week before federal parliament opened for the first time since the deadly bushfire season. Mustering his marketing skills, he tried to mount a credible argument for Australia's weak targets to cut its emissions. In the end, he fell back on an old defence, one used so often during the Howard years: Australia's share of global greenhouse emissions was only 1.3 per cent. It was up to others to do more.

A few weeks later even the main business lobby, the Business Council of Australia, under pressure from its members, called for a price on carbon. It also called for Australia to transition to net zero emissions by 2050. Some 70 countries had already made a similar commitment. But Morrison and Taylor refused to budge. In a political fix from the playbook of his old predecessor John Howard a decade and a half earlier, Morrison promised a new technology solution—a 'technology roadmap' to meet the challenge of climate change. He would offer this as Australia's signature contribution to the next UN climate conference.

But Australia would not be adding its voice to the global push for urgency and ambition on climate change. Instead, it seemed, the country was prepared to be a laggard and continue to profit from the fossil fuel industry as long as possible. From the prime minister of a nation so vulnerable to dangerous climate change, this did not sound like leadership. It sounded like the government was running up the white flag.

On the eve of the new decade, the UN secretary-general warned world leaders there were two paths ahead in the next ten years as the planet faced the threat from dangerous climate change: one of hope or one of surrender. The Australian government, under Scott Morrison, decided it would take a gamble on the future and stick with the path it knew best.

EPILOGUE

THE ROAD AHEAD

AS AUSTRALIA'S BLACK Summer finally drew to a close in 2020, the coral scientist Professor Terry Hughes once again crisscrossed the Great Barrier Reef in a light plane. Seawater temperatures on parts of the Reef had reached their highest level ever recorded that February. Hughes and his colleagues feared another mass coral bleaching was underway. What he saw as he flew over the vast natural wonder he called 'an utter tragedy'. The Reef had been hit with its third mass bleaching in just five years. Around a quarter of all the coral reefs he surveyed, from the far north down to the south, were severely affected.

Around the same time, the royal commission investigating the devastating summer bushfires was visiting burnt-out communities across Australia, from Kangaroo Island in South Australia to Marcoola on Queensland's Sunshine Coast, recording harrowing accounts of traumatised residents who had suffered one of the longest, most intense fire seasons ever recorded.

But neither bushfires nor bleachings made headlines that month. Climate change had suddenly disappeared from the news. Australia was reeling in the face of the COVID-19 pandemic. The unprecedented health crisis was engulfing the world and threatening the biggest economic downturn since the Great Depression, upending the lives and livelihoods of people everywhere.

Despite some early missteps, Australia's health response to the pandemic was remarkable. Political leaders listened to their medical science experts and acted on their advice. Prime Minister Scott Morrison, his key cabinet ministers and state premiers stood shoulder to shoulder with the experts as they wrestled with life-and-death decisions on nationwide lockdowns, intensive care bed numbers and supplies of personal protective equipment.

The response was in stark contrast to the mistrust faced by climate scientists from so many Australian politicians, who had dismissed or downplayed their advice for decades. For some who had spent years working for an effective response to climate change, it sparked a flicker of hope that their warnings would now also be treated with the urgency they deserved.

Those warnings were clear. Climate change was no black swan event, unpredictable or unforeseen. To avoid accelerating dangerous climate change the world needed to keep the global temperature rise to around 1.5 degrees Celsius from pre-industrial levels or at least well below 2 degrees. That advice meant global emissions needed to be cut by some 45 per cent within ten years—by 2030—and reach net zero by 2050 or even earlier. For a rich, developed country like Australia that meant rapid and transformative change. But the flicker of hope that the Morrison government would seriously rethink its climate policy after COVID-19 was quickly extinguished.

The economic shock caused by the pandemic had prompted governments around the world to raise trillions of dollars for stimulus plans to jump-start their recoveries. The European Union, the UN Secretary-General and the International Energy Agency all called for these plans to support a speedier transition to a clean energy planet. One eminent group of economists, including the

UK's Sir Nicholas Stern, put it graphically: the world should not leap 'from the COVID frying pan into the climate fire'.

But the Morrison government's plans did not seriously canvas this thinking. Indeed, some of its handpicked advisers on the recovery pushed a dramatic expansion of Australia's domestic gas industry.

The chair of the National COVID-19 Coordination Commission, Nev Power, publicly advocated a gas-led recovery. Power was a long-time iron ore executive who also sat on the board of a West Australian gas exploration company, Strike Energy—a position he stood aside from after the possibility of a conflict of interest was raised.

The commission's special adviser on manufacturing, Andrew Liveris, also supported a big gas expansion. The former global head of Dow Chemicals, he sat on the board of the world's largest oil and gas company, Saudi Aramco, and was Deputy Chair of Worley, one of the world's biggest energy engineering companies, whose business included gas pipelines. The first draft report from Liveris's panel was a sweeping vision for a gas-powered bonanza that called for slashing environmental 'green tape', unlocking bans on onshore gas drilling and big investment in new gas pipelines to promote jobs and growth. The gas vision was also backed by the Labor Party's key trade union affiliate, the Australian Workers Union.

This vision was not shared by all on the advisory commission but the big gas agenda was encouraged by the top levels of the Morrison government. This was clear when Angus Taylor, the Energy and Emissions Reduction Minister, released a discussion paper in May on his long-promised 'Technology Investment Roadmap' as part of Australia's response to climate change. Domestic gas use and the Liquefied Natural Gas (LNG) export business figured prominently.

'As the world's largest exporter of LNG, Australia will continue to capitalise on this important low emissions export opportunity,' the paper pronounced. 'Some of our key trading partners, including Japan and South Korea, have indicated that LNG will play an important role in decarbonising their electricity systems. LNG represents a continuing export opportunity for Australia.'

While gas is cleaner than coal, it is a heavily polluting fossil fuel. The International Energy Agency sees gas playing a role as a transition fuel to push out coal-fired power but its use would need to peak in the late 2020s, less than a decade away. By then, gas use would have to steadily decline if global emissions were going to seriously fall—unless gas producers and users found an economic way to neutralise their emissions.

The government's answer was clean gas. It promoted Carbon Capture and Storage (CCS) technology to reduce carbon emissions from gas. Well over a decade earlier, then Prime Minister John Howard had pushed CCS to make 'clean coal', where carbon dioxide emissions from coal-fired power stations would be stored deep underground. But, despite spending hundreds of millions of taxpayer dollars on it, clean coal was still not financially feasible. The Morrison government's first Technology Investment Roadmap discussion paper pretty much abandoned clean coal. Instead Taylor touted CCS as a lifeline to clean up the gas industry and manufacturers who depend on gas.

But as the climate crisis escalates, the Morrison government and the fossil fuel companies are in a race against time. Local gas companies like Woodside and Santos and the global giants like Chevron, Shell and ExxonMobil are under mounting pressure from shareholder activists over their soaring emissions from gas production and processing, and the emissions of their customers. Financial markets are growing increasingly wary of big fossil fuel investments. The risk is that giant gas projects, like coal-fired power stations, will become stranded assets in a carbon-constrained future.

Tellingly, when Taylor pitched the Roadmap paper to the media he made scant reference to renewable energy. Yet the renewable energy industry had outperformed Australian government expectations for years, thanks to the Renewable Energy Target—a policy that Canberra is winding up. By 2019, renewable energy made up well over 20 per cent of the country's electricity generation, helping to push down energy prices. Australians had embraced rooftop solar panels in record numbers and large-scale solar and wind generation had forged ahead. Renewable energy costs were falling rapidly.

The vexing problems around how to make the electricity grid work with renewables were slowly being overcome. While gas generation would still be needed to back up wind and solar energy in the short to medium term as coal generators shut down, it was possible battery storage, pumped-hydro plants and advances to the grid could push out most gas by the end of the decade.

The government's own Roadmap paper recognised that if the cost of storing renewable energy could be significantly reduced, it could be used in Australia's energy-intensive manufacturing industries instead of coal and gas. Renewable electricity, and clean hydrogen produced from renewable energy, could also slash emissions from transport. Indeed, Taylor had a few weeks earlier launched a $300 million fund to promote the government's National Hydrogen Strategy.

Few doubt that new technology is critical to solving the climate crisis, but picking winners is a tricky business. Investors are unlikely to risk big money in a policy vacuum. But after more than two decades of bitter climate wars, the Morrison government was not prepared to re-open the climate and energy policy debate. It steadfastly refused to put a price on carbon emissions or set ambitious targets to cut emissions. As Taylor put it, 'this is about technology not taxes. It means reducing emissions, not reducing jobs and the economy.'

Yet this old rhetoric played out against a new reality. The COVID-19 pandemic gave Australians a disturbing glimpse into a future when fossil fuel companies are put under real stress.

The early global lockdowns in China and Europe temporarily slashed the world's demand for energy. Prices for gas, coal and oil plummeted, resulting in the most profound shock to global energy markets since World War II. As the world's biggest coal and LNG exporter, Australia was exposed as energy companies began cutting jobs and investment plans.

The energy shock and the economic impact of the pandemic saw global emissions predicted to fall in 2020 by more than in any other year on record. And while some cheered this on, climate scientists cautioned that such dramatic falls in emissions would have to be repeated every year until 2030 to keep global temperature rises to well below 2 degrees Celsius.

This underscored Australia's enormous challenge—to cut its greenhouse gas emissions without killing jobs and destroying livelihoods. For one of the world's most carbon-intensive countries, the scale of the challenge reinforced calls from growing numbers of business leaders, financial experts, farmers and scientists that climate policy needs to be at the centre of government decision-making, not left on the sidelines.

Australia's vulnerability to dangerous climate change has been laid out in untold numbers of government and scientific reports: more intense bushfires, longer and more devastating droughts, more severe storms, sea-level rise that threatens coastal cities and towns, heatwaves that kill the elderly and vulnerable, the destruction of natural ecosystems like the Great Barrier Reef, and the arrival of waves of climate refugees from low-lying cities and towns in Asia and the Pacific.

For over two decades, Australia's politicians have been fighting the climate wars fuelled by the carbon club. But despite the political carnage, the science of climate change has not been defeated. The carbon club is breaking up as the climate crisis becomes more urgent. The next decade will determine whether the country and its leaders can rise to the challenge that lies ahead.

ACKNOWLEDGEMENTS

THIS BOOK WOULD not have been possible without Matthew Moore, a brilliant journalist and luckily my husband, because the support, advice and love he gave me during this project can never be repaid.

Two generous and talented colleagues, David Marr and Deborah Snow, gave me insightful feedback, friendship and encouragement.

Big thanks must go to Richard Walsh, who picked up this book when it needed a backer. Not only was he a delight to work with, but his sharp eye and political brain improved the manuscript immeasurably. To the talented team at Allen & Unwin, Tom Gilliatt, Angela Handley, Nicola Young, Julia Cain, Phil Campbell, thank you so much for your hard work, deft touch, creativity and patience.

To the many, many people who agreed to be interviewed both in Australia and the US, on and off the record, my sincere thanks for your insights from all perspectives. You made this a much better book.

This is very much a journalist's account and my own very particular take on these events. But I would like to mention three authors who have impressed me over the years: Guy Pearse for his groundbreaking *High and Dry*; Paul Kelly, whose political histories are classics, in particular *The End of Certainty* and *Triumph and Demise*; and the late Philip Chubb for his richly researched *Power Failure*.

A shout-out to the reporters covering this area who kept me informed: *The Guardian's* generous and indefatigable Graham

Readfearn, the *Australian Financial Review's* Angela Macdonald-Smith, and the *Sydney Morning Herald's* Peter Hannam.

To my former colleagues at the *Sydney Morning Herald* and ABC's *Four Corners,* who worked with me on so many of the dramas that appear in this book, it was a pleasure. Special thanks to Sarah Ferguson, Trish Drum and Anne Davies, who helped with unpicking some old history.

Thanks to Kert Davies at the US Climate Investigations Center, who first guided me through a maze of FOIA documents in 2005 and again in 2019.

To my good friends Jenni Hewett, Pam Williams and Judith Hoare, apologies for too many dinners where we argued about climate change. Thanks to Jenni, they were usually resolved the next morning over coffee. You always made me think twice.

To Anna and Sybil, thanks for trying to keep me healthy through this, and to the marvellous bridge-cum-gossip club, Kate, Deb and Penny, thanks for holding my place.

Special thanks to the loyal Wilkinson tribe and their partners for your unquestioning support, and to Ross and Elaine for your on-the-ground reports.

And for those who matter so much—the next generation—the wonderfully resilient Lewis, Leon and Adele, and the cheerfully resourceful Alex and Red, you make me ever optimistic for our future.

CHAPTER NOTES

Chapter 1: Hearts and minds

Page 1: 'Once inside, he mingled discreetly . . .' The protests that occurred at the Countdown to Kyoto conference were described in the author's interviews with former Greenpeace activists Simon McRae, Nic Clyde, Teri Calder and Toby Hutcheon in May 2019. Also from the author's interviews with attendees Myron Ebell, 25 March 2019, and Dr Patrick Michaels, 27 March 2019.

Page 2: 'Wallop landed with a who's who . . .' The speaker list for the Countdown to Kyoto conference in August 1997 and quotes from speeches are from the conference papers held in the National Library of Australia, Nq 363.73874 C855.

Page 2: '. . . Hagel, with Democrat senator Robert Byrd . . .' The Byrd–Hagel Resolution was a non-binding 'sense of the Senate' resolution stating that the US should reject any climate agreement that did not also impose obligations on developing countries like China and India as well as developed countries like the US or result 'in serious harm to the economy'. It passed unanimously, 95–0 votes, in July 1997. See Eric Pooley, *The Climate War: True believers, power brokers, and the fight to save the earth*, New York: Hyperion, 2010, pp. 90–1.

Page 2: '. . . recalled the American climate sceptic Myron Ebell . . .' Ebell's quotes in this chapter are all based on the author's interview, 25 March 2019.

Page 5: '. . . Hill lectured them.' Quotes from Robert Hill's speech are from the conference papers held in the National Library of Australia, Nq 363.73874 C855.

Page 5: '. . . a front-page story in the *New York Times*'. The headline story on James Hansen's testimony was reprinted in Nathaniel Rich, 'Losing earth: The decade we almost stopped climate change', *New York Times Magazine*, 1 August 2018.

Page 6: 'Hansen's testimony marked a turning point . . .' Quotes are from James Hansen's testimony from the US Senate Committee on Energy and Natural Resources, 23 June 1988.

Page 6: 'concrete action to protect the planet', George H.W. Bush, quoted in Elizabeth Kolbert, *Field Notes from a Catastrophe: Man, nature and climate change*, New York: Bloomsbury Publishing, 2006, p. 151.

Page 7: 'Exxon, Texaco Oil, Peabody Coal . . .' For the Global Climate Coalition membership and dissolution, see Ross Gelbspan, *Boiling Point: How politicians, Big Oil and Coal, journalists, and activists are fueling the climate crisis—and what we can do to avert disaster*, New

York: Basic Books, 2005, p. 413. Original documents on the forma-
tion and strategy of the Global Climate Coalition are housed on the
website of the Climate Investigations Center, climateinvestigations.org/
global-climate-coalition-documents-index.

Page 7: '. . . a multimillion-dollar advertising campaign . . .' For the pre-
Kyoto advertising campaign, see Pooley, *The Climate War*, p. 41.

Page 7: 'This attack was echoed loudly . . .' The quote from Malcolm Wallop
is from his speech at the Countdown to Kyoto conference.

Page 8: 'Australia's environment minister put the problem . . .' The Robert
Hill quote is from the author's interview, 20 December 2018.

Page 8: '. . . Roger Beale, then head of the Environment Department . . .' The
Roger Beale quote is from the author's interview, 20 February 2019.

Page 8: 'The scientific advice from the IPCC . . .' For the IPCC's 1997
predictions for Australia, see R.T. Watson, M.C. Zinyowera and R.H.
Moss (eds), *The Regional Impacts of Climate Change: An assessment of
vulnerability*, Cambridge: Cambridge University Press, 1997, Chapter 4
'Australasia', archive.ipcc.ch/ipccreports/sres/regional/index.php?idp=58.

Page 9: 'For twenty years, from 1990 . . .' Australia's deforestation rate
between 1990 and 2009 is cited in Andrew Macintosh, ANU Centre for
Climate Law and Policy, 'The Australia clause and REDD: A cautionary
tale', *Climate Change*, 2012, vol. 112, no. 2, pp. 169–88, doi:10.1007/
s10584-011-0210-xs.

Page 9: 'Howard's top advisers were warning . . .' Cabinet Minute, 'Climate
change: Approach to international negotiations', 4 June 1996, Howard
Government Cabinet Papers, 1996–97. The projection for Australia's
34 per cent increase in emissions from the energy sector is in Attachment
C of the Cabinet Minute, p. 16.

Page 9: 'Inside the Howard cabinet an intense debate . . .' The cabinet debate
on the Kyoto strategy is set out in the declassified Howard Government
Cabinet Papers, 1996–97. The quote on the trade impact for Australia
is from the 4 June 1996 Cabinet Minute, 'Climate change: Approach to
international negotiations', p. 35.

Page 10: 'By Morgan's own calculation . . .' The estimated cost of Australia's
Kyoto agreement to Western Mining is from Hugh Morgan's address to the
Lavoisier Group's opening conference in May 2000, held in the Lavoisier
Group's archived speeches, www.lavoisier.com.au/articles/climate-policy/
science-and-policy/morgan2000-c1.php.

Page 10: 'He had the respect of the prime minister . . .' Hugh Morgan's relationship with John Howard is based on confidential author interviews
with former cabinet colleagues and is also referenced in 'Different shades
of Hugh', *Sydney Morning Herald*, 17 August 2002, www.smh.com.au/
business/different-shades-of-hugh-20020817-gdfjvs.html.

Page 10: 'In an infamous speech . . .' The quotes from Hugh Morgan at
the Mining Industry Council in May 1984 are from the original speech
supplied to the author. For an analysis of the speech, see Andrew Markus,

Race: John Howard and the remaking of Australia, Sydney: Allen & Unwin, 2001, pp. 60–1.

Page 10: 'Morgan had claimed it put at risk . . .' For Hugh Morgan's quote on Mabo, see Markus, *Race*, p. 72.

Page 11: 'Morgan backed Evans' crusade . . .' For Hugh Morgan's quote on labour unions and former prime minister Bob Hawke's quote, see Dominic Kelly, *Political Troglodytes and Economic Lunatics: The hard right in Australia*, Melbourne: La Trobe University Press, 2019, pp. 72, 6.

Page 11: 'Morgan's fellow directors on Cormack . . .' Details on Morgan and Charles Goode's role in the Cormack Foundation are from Hugh Morgan's witness statement, 20 February 2018, in Federal Court Case VID1270/2017 *Alston v Cormack Foundation* and the judgment in that case by Beach J, 14 June 2018.

Page 11: 'One of its recipients was the Institute of Public Affairs . . .' For Cormack's funding of the IPA, see Morgan's witness statement cited above.

Page 12: 'I saw no sign of Howard saying . . .' Roger Beale's quote is from the author's interview, 20 February 2019.

Page 12: 'We were faced with big choices . . .' Michael Thawley's quote is from his comment to the author, 8 February 2019.

Page 12: '. . . Liberal and Labor governments had followed one rigid rule . . .' The 'No Regrets' policy is referenced in the Cabinet Papers 1996–97, Cabinet Minute, 'Memorandum climate change: Approach to international negotiations—IDC Report', 29 July 1997, pp. 1–2. It is also referenced in Cabinet Paper, 'An analysis of the cost of greenhouse reductions policies', by the Treasury, 12 September 1997.

Page 13: 'New advice to cabinet . . .' The 40–50 per cent figure is referenced in the Cabinet Papers Cabinet Minute on Climate Change Overview, 12 September 1997, p. 3.

Page 13: 'There would be some short-term political pain . . .' The quote on the impact of Australia not signing Kyoto is from Cabinet Minute, 'Climate change: Approach to international negotiations: Further submission', 2 June 1997, p. 9, in Cabinet Paper, 'Climate change: Approach to international negotiations'.

Page 13: 'Thawley by nature was a hawk . . .' Beale's quotes are from the author's interview, 20 February 2019.

Page 14: '. . . the largest and most far-reaching package . . .' Prime minister's statement to parliament, *Hansard*, House of Representatives, 20 November 1997.

Page 14: 'We had very tense negotiations in cabinet . . .' The Beale quote is from the author's interview, 20 February 2019.

Page 15: 'The so-called "first commitment period" . . .' For the first and second Kyoto commitment targets, see 'What is the Kyoto Protocol?', UN Climate Change, unfccc.int/kyoto_protocol.

Page 15: 'It meant Australia could claim credits . . .' The argument that Australia was claiming windfall credits from land clearing is set out in Macintosh, 'The Australia clause and REDD'.

Page 16: 'I think I was really the most convinced person . . .' The quote from Howard Bamsey is from the author's interview, 27 November 2018.

Page 16: 'On the plane over, Beale argued with Hill . . .' The debate over the target between Beale and Hill is from Guy Pearse, *High and Dry: John Howard, climate change and the selling of Australia's future*, Melbourne: Penguin Books, 2007, p. 276.

Page 17: 'He and President Clinton had been . . .' Al Gore's quotes on 'burning up the phone lines' and 'the meaningful participation of developing countries' are from his speech to the Kyoto summit, clintonwhitehouse2. archives.gov/WH/EOP/OVP/speeches/kyotofin.html.

Page 17: 'Gore's gone rogue'. The quote about vice-president Al Gore is from a confidential interview with a senior Howard government official.

Page 17: 'They began to work out what had happened . . .' The Beale quote on the Australia Clause is from the author's interview, 20 February 2019.

Page 18: 'The judgement at the end was that Australia . . .' The Bill Hare quotes on the Australia Clause are from the author's interview, 22 February 2019.

Page 18: 'A great result, a splendid result . . .' Howard's quote is from his press conference, 11 December 1997.

Page 19: 'The Cooler Heads Coalition very quickly decided . . .' Myron Ebell quotes on the Cooler Heads Coalition and Bill O'Keefe are from the author's interview, 25 March 2019.

Page 20: 'A lengthy memo was circulated . . .' The API strategy memo is available in full on the Climate Investigations Center website, climate investigations.org/global-climate-coalition-documents.

Page 21: 'Ray decided that the science . . .' The Ebell quotes on climate science and Evans setting up the Lavoisier Group are from the author's interview, 25 March 2019.

Page 21: '. . . the economic dislocation which must follow . . .' The Ray Evans quote on Kyoto's impact is cited in Kelly, *Political Troglodytes and Economic Lunatics*, p. 168.

Page 22: 'In this regard I applaud the objectives . . .' Morgan's quote on Robert Hill is from his speech at the opening Lavoisier Group conference, www.lavoisier.com.au/articles/climate-policy/science-and-policy/ morgan2000-c1.php.

Chapter 2: Old allies

Page 23: 'He put on a good party . . .' Details on the prime minister's Washington visit in September 2001, including his quote on the ambassador, are from John Howard, *Lazarus Rising: A personal and political autobiography*, Sydney: HarperCollins, 2010, p. 376. Details are also drawn from Leigh Sales, *Detainee 002: The case of David Hicks*, Melbourne: Melbourne University Press, 2007, pp. 2–3.

Page 24: 'Arthur "Randy" Randol, faxed a memo . . .' The fax from ExxonMobil's Randy Randol was originally released under the US

Freedom of Information Act (FOIA) and is available from the Climate Investigations Center at www.climatefiles.com/exxonmobil/2001-exxonmobil-randol-white-house-ipcc.

Page 25: 'I would say that it behoves us . . .' Dr Richard Lindzen's testimony was given to the Australian Parliament's Joint Standing Committee on Treaties on 3 November 2000.

Page 25: 'Other climate scientists had attacked Lindzen . . .' Dr Lindzen had rejected assertions that he denied links between smoking and lung cancer but said 'there was a reasonable case for the role of cigarette smoking in lung cancer, but that the case was not so strong that one should rule that any questions were out of order. I think that the precedent of establishing a complex statistical finding as dogma is a bad one. Among other things, it has led to the much, much weaker case against second hand smoke also being treated as dogma.' See www.climateconversation.org.nz/2011/05.

Page 26: '. . . the draft report estimated . . .' The details of the IPCC's Third Assessment Report are available on the IPCC's archive website at archive.ipcc.ch/publications_and_data/publications_and_data_reports.shtml.

Page 26: '. . . Lee Raymond was one of the most powerful . . .' For further details on Lee Raymond's influence and strategy as ExxonMobil's chairman and chief executive, see Steve Coll, *Private Empire: ExxonMobil and American Power*, London: Penguin Books, 2013.

Page 26: 'It explored for oil and gas across six continents . . .' For Exxon-Mobil's history in Australia, see ExxonMobil Australia submission to the Senate Economics Legislation Committee Inquiry into the *Taxation Laws Amendment (2011 Measures No. 8) Bill 2011*, 11 November 2011.

Page 26: 'Exxon's attitude is that they're the big boys . . .' Senator Chuck Schumer's quote is from Nelson Schwartz, 'The biggest company in America . . . is also a big target', *Fortune Magazine*, 17 April 2006, fortune.com/2006/04/17/exxon-mobil-rex-tillerson.

Page 27: '. . . Lee Raymond had delivered a landmark speech . . .' Lee Raymond's 1997 speech to the World Petroleum Conference is available from the Climate Investigations Center at www.climatefiles.com/exxonmobil.

Page 27: 'We did get substantial funding from ExxonMobil . . .' The quote from Myron Ebell is from the author's interview, 25 March 2019.

Page 27: 'Dick Cheney had known Lee Raymond . . .' For details on the relationship between Lee Raymond and Dick Cheney, see Coll, *Private Empire*.

Page 28: '. . . after his new head of the Environmental Protection Agency . . .' Details on EPA chief Whitman's defeat over power plant emissions are taken from her memoir, Christine Todd Whitman, *It's My Party Too*, New York: Penguin Books, 2006, p. 173.

Page 28: '. . . right-wing Republicans, led by Senator Chuck Hagel . . .' Senator Hagel's objections to Whitman's actions are stated in a letter, signed by him and three other senators, released by the US Senate on 6 March 2001.

Page 28: 'Significantly, Bush explained . . .' The quotes from Bush come from 'Text of a letter from the President to Senators Hagel, Helms, Craig, and Roberts', released by the White House Press office, 13 March 2001, georgew bush-whitehouse.archives.gov/news/releases/2001/03/20010314.html.

Page 28: '. . . he gave a big boost to the climate sceptics . . .' For the global reaction to Bush's statement on Kyoto, see 'Bush kills global warming treaty', *The Guardian*, UK edition, 29 March 2001, www.theguardian. com/environment/2001/mar/29/globalwarming.usnews.

Page 28: 'ExxonMobil's scientists and executives drew up a sweeping proposal . . .' ExxonMobil's recommendations on climate science research, 15 June 2001, were attached to a cover letter from its head of Science Strategy, Brian Flannery, and sent to the White House assistant to the president for science and technology, Dr John Marburger, on 18 March 2002. The documents and a cover fax from ExxonMobil's Randy Randol were released under the US FOI Act and are available from the Climate Investigations Center at www.climatefiles.com/exxonmobil.

Page 29: '. . . with the support of Japan and Australia . . .' On Australia at Bonn, see ABC Science report, 'Breakthrough on climate negotiations', 25 July 2001, www.abc.net.au/science/articles/2001/07/25/334707.htm.

Page 29: 'By the end of the Bonn talks . . .' 'The quotes from UK prime minister Tony Blair and his environment secretary, Michael Meacher, are from *The Telegraph*, UK, 23 July 2001, www.telegraph.co.uk/news/1334985/Deal-saves-Kyoto-agreement.html.

Page 29: 'The *New York Times* scathingly . . .', see 'Clueless on global warming', *New York Times*, 21 July 2001, www.nytimes.com/2001/07/19/opinion/clueless-on-global-warming.html.

Page 30: 'Flanked by his senior advisers . . .' Details on the trade delegation came from former senior government officials and from Michelle Grattan, 'Dreadful, just appalling, awful: Sombre Howard horrified by senseless carnage', *The Age*, 12 September 2001. Details of the Australia–US Free Trade Agreement are from the Department of Foreign Affairs and Trade annual report for 2001.

Page 30: 'The President and I have a great similarity . . .' John Howard's quote on Bush is from the transcript of their joint press conference, White House Press Office, 10 September 2001, georgewbush-whitehouse.archives.gov/news/releases/2001/09/text/20010910-4.html.

Page 31: '. . . requiring a global approach . . .' The quote on climate change is from 'Joint statement between the United States of America and Australia', White House Press Office, 10 September 2001, georgewbush-whitehouse. archives.gov/news/releases/2001/09/text/20010910-8.html.

Page 31: 'We're both of the view that Kyoto . . .' John Howard's quote on Kyoto is from his Doorstop Interview at the White House, 10 September 2001, pmtranscripts.pmc.gov.au/release/transcript-11767.

Page 31: 'I got a call from a limo . . .' The quote from Roger Beale is from the author's interview, 20 February 2019. Max Moore-Wilton did not recall the exchange.

Page 32: 'You watch that Hill.' The quote from Robert Hill is from the author's interview, 20 December 2018.

Page 32: 'He was running a Republican line.' The quote from Roger Beale is from the author's interview, 20 February 2019.

Page 32: 'Thawley was deep in conversation . . .' Details of the meeting between Hugh Morgan, David O'Reilly and Michael Thawley were confirmed by confidential sources.

Page 33: 'The prime minister and his party were soon . . .' Details on Howard's move to the embassy bunker after the attacks on the Twin Towers are from Howard, *Lazarus Rising*, pp. 379–80, and Sales, *Detainee 002*, p. 3.

Page 34: '. . . resolute solidarity with the American people . . .' The quote from John Howard's letter is from https://pmtranscripts.pmc.gov.au/release/transcript-11783.

Page 34: '. . . the only other leader in the developed world . . .' While Australia was ultimately the only other developed country not to ratify Kyoto with the US, Canada under Conservative PM Stephen Harper formally withdrew from the Kyoto Protocol in 2011 when Harper announced Canada would not meet its 2012 targets. Canada would have faced $14 billion in penalties under the protocol.

Page 34: 'Australians brushed off the news . . .' Australia's per capita emissions data is from the Australian Senate report, 'The heat is on: Australia's greenhouse future', November 2000, Commonwealth of Australia, p. 107, www.aph.gov.au/Parliamentary_Business/Committees/Senate/Environment_and_Communications/Completed inquiries/1999-02/gobalwarm/report/contents.

Page 34: 'They barely noticed a report . . .' The CSIRO figures are from, 'Climate change projections for Australia', CSIRO, 2001.

Page 35: 'We weren't absolutely sure what was happening . . .' The quote from Roger Beale is from the author's interview, 20 February 2019.

Page 35: 'I don't see myself in a category . . .' David Kemp's quotes throughout this chapter are from his comments to the author, 1 June 2019.

Page 35: 'a really decent guy'. This quote is from Howard Bamsey, from the author's interview, 27 November 2018.

Page 37: 'They were a bunch of loonies . . .' Roger Beale's quotes are from the author's interview, 20 February 2019.

Page 37: 'He still sat on the board . . .' Hugh Morgan's position with the Cormack Foundation and his shareholding is from Hugh Morgan's witness statement, 20 February 2018, in Federal Court Case VID1270/2017 *Alston v Cormack Foundation* and the judgment in that case by Beach J, 14 June 2018.

Page 37: 'Cormack had donated . . .' The donation figures for the Cormack Foundation are from the Australian electoral commission returns, as are those for the Western Mining Corporation (WMC) donations.

Page 38: 'I think at that stage the US was the biggest emitter . . .' Nick Minchin's quotes are from the author's interview, 28 February 2019.

Page 39: 'Unfortunately, America's greenhouse gas emissions . . .' US emissions are from the year 2001 and are quoted in Whitman, *It's My Party Too*, p. 178.

Page 39: 'Bush was promising to fund new technology . . .' President Bush's plan on emissions intensity reductions for the US was announced by the White House Council on Environmental Quality as, 'An ambitious national goal to reduce emissions intensity', 14 February 2002, see georgewbush-whitehouse.archives.gov/ceq/clean-energy.html.

Page 39: 'We met with a deputy head of Exxon . . .' Beale's quote on the meeting with ExxonMobil is from the author's interview, 20 February 2019.

Page 40: 'Instead they offered Australia . . .' The US–Australia Climate Action Partnership was set out in detail by the US State Department in, 'The U.S.–Australia Climate Action Partnership moves forward', 9 July 2002, 2001-2009.state.gov/r/pa/prs/ps/2002/11744.htm.

Page 40: 'In May, the new Japanese prime minister . . .' The call for Australia to stay in Kyoto is from the transcript of the joint press statement by John Howard and Japanese prime minister Koizumi, 1 May 2002, https://parlinfo.aph.gov.au/parlInfo/search/display/display.w3p;query=Id%3A%22media%2Fpressrel%2FHQF66%22.

Page 41: 'A month later, on 4 June . . .' The call for all developed nations again to ratify Kyoto is from, 'Japan officially ratifies Kyoto climate protocol', *Japan Times*, 5 June 2002, www.japantimes.co.jp/news/2002/06/05/national/japan-officially-ratifies-kyoto-climate-protocol.

Page 41: 'It is not in Australia's interests . . .' John Howard's comments on not ratifying the Kyoto Protocol are from *Hansard*, House of Representatives, 5 June 2002.

Page 41: 'Neither of us knew a thing . . .' Howard Bamsey's quotes are from the author's interview, 27 November 2018.

Page 42: 'It was a "captain's call" . . .' David Kemp's quote is from comments to the author, 1 June 2019.

Page 42: 'Where Howard switched, I think . . .' Hill's quote on Howard is from the author's interview, 20 December 2018.

Page 42: '. . . the problem with this debate . . .' Howard's quote is from his speech for the UK 2013 Annual Global Warming Policy Foundation lecture, 'One religion is enough', 5 November 2013, www.thegwpf.org/content/uploads/2013/12/Howard-2013-Annual-GWPF-Lecture.pdf.

Page 42: 'There was deep shock that Australia . . .' The Bill Hare quote is from the author's interview, 22 February 2019.

Page 42: 'Hugh Morgan was one of the first . . .' Hugh Morgan's opinion piece, 'Carbon blackmail doesn't lead to greener future', was published in an abridged form in *The Australian*, 10 June 2002; the original is on the Lavoisier Group's website, www.lavoisier.com.au/articles/climate-policy/politics/morgan2002-3.php.

Chapter 3: Division in the ranks

Page 44: 'Our strategy is to move ahead . . .' David Kemp's speech, 'Moving forward on climate change: alternative perspectives', was delivered at Chatham House, London, on 15 July 2002, www.chathamhouse.org/sites/default/files/public/Research/Energy, Environment and Development/david_kemp_climate_change_speech_15_july_02.pdf.

Page 45: 'Climate change was a policy issue . . .' Kemp's quotes on climate change policy and socialism are in response to the author's questions, 1 June 2019.

Page 45: 'I said to David that we had to . . .' Roger Beale's quote is from the author's interview, 20 February 2019.

Page 46: 'Put simply, they work like this . . .' A description of the workings of emissions trading schemes can be found in the *Prime Ministerial Task Group on Emissions Trading* [Shergold] *Report*, Canberra: Commonwealth of Australia, 2007, archived at webarchive.nla.gov.au/awa/20070604000621/http://pandora.nla.gov.au/pan/72614/20070601-0000/www.pmc.gov.au/publications/emissions/index.html.

Page 46: 'one of the most influential businessmen . . .' The Kemp quote on Morgan is from his responses to the author, 1 June 2019.

Page 47: 'This is the Beale plan . . .' Howard Bamsey's quote is from the author's interview, 27 November 2019.

Page 47: 'The climate-sceptics headquarters, the Lavoisier Group . . .' The open letter from the Lavoisier Group to Prime Minister Howard, written on 28 May 2003, was signed by the then Lavoisier president, Peter Walsh, a former Labor senator and resources minister. Ray Evans was still on the board; Morgan was still a supporter of the group. See www.lavoisier.com.au/articles/climate-policy/science-and-policy/openletter.pdf.

Page 47: '. . . including Malcolm Broomhead . . .' Malcolm Broomhead's role in the opposition to the emissions trading scheme is cited by Clive Hamilton in 'The dirty politics of climate change', Climate Change and Business Conference, 20 February 2006, clivehamilton.com/the-dirty-politics-of-climate-change.

Page 47: 'Some of the big miners hired . . .' Andrew Robb's role in lobbying against the emissions trading scheme was confirmed by Beale in the author's interview, 20 February 2019. Robb also is reported lobbying for the mining companies BHP Billiton and Rio Tinto, against the ratification of Kyoto and a carbon tax in Angus Grigg, 'The player', *Australian Financial Review Magazine*, 29 September 2006.

Page 48: 'a cause celebre for the Left . . .' Robb's comments on climate change are from Grigg, 'The player'.

Page 48: 'Weighing in from Washington . . .' Michael Thawley's opposition to the Beale plan is from an author interview with a confidential source.

Page 48: 'But despite the opposition from Morgan . . .' The description of the split in the BCA comes from confidential correspondence made available to the author and also from the author's interview with Greg Bourne,

16 July 2019. See also Sophie Black, 'While the world burns, business leaders fiddle', Crikey, 30 October 2006, www.crikey.com.au/2006/10/30/while-the-world-burns-business-leaders-fiddle, for the BCA opposition to Kyoto from Alcoa, Rio and ExxonMobil.

Page 49: 'Morgan was my nemesis . . .' The Greg Bourne quote comes from the author's interview, 16 July 2019.

Page 50: 'The Liberal Party's own cash cow . . .' The Cormack Foundation's investments in BHP and Rio Tinto are set out in its Australian Electoral Commission (AEC) associated entity annual return 'Receipts from investments' for 2002–03, available on the AEC's Transparency Register website, transparency.aec.gov.au/AnnualAssociatedEntity/ReturnDetail?returnId=5668.

Page 50: 'Howard's nephew Lyall . . .' Lyall Howard's role with Rio Tinto was reported by the ABC's *PM* program's Andrew Fowler, 7 September 2004.

Page 50: 'Kemp's former chief of staff . . .' John Roskam's role with Rio Tinto was reported in Lucinda Schmidt, 'Profile: John Roskam', *Sydney Morning Herald*, 3 February 2012, www.smh.com.au/money/investing/profile-john-roskam-20120201-1qsod.html.

Page 50: '. . . Institute of Public Affairs, which in turn . . .' Cormack Foundation's donations to the IPA from Hugh Morgan's witness statement, 20 February 2018, in Federal Court Case VID1270/2017 *Alston v Cormack Foundation*.

Page 50: 'They all shared what Pearse called . . .' The quote on Howard's 'quarry vision' is from Guy Pearse, *High and Dry: John Howard, climate change and the selling of Australia's future*, Melbourne: Penguin Books, 2007, p. 137.

Page 51: 'What happened was, at a certain point . . .' The Howard Bamsey quote is from the author's interview, 27 November 2018.

Page 51: 'A minority of big, dirty polluters . . .' Don Henry's quote was reported in, 'Government appears to back away from domestic carbon trading', *Wind Power Monthly*, 1 October 2003, www.windpowermonthly.com/article/960002/government-appears-back-away-domestic-carbon-trading.

Page 51: 'I think it was probably just . . .' The quote from Greg Bourne comes from the author's interview, 16 July 2019.

Page 51: 'I've seen in his eyes a glint . . .' Andrew Bolt, 'Lost in green daze', *Herald Sun*, 13 October 2003.

Page 52: 'Greenhouse policy is in a state of chaos . . .' Email from Barry Jones supplied to the author by a confidential source.

Page 52: 'In a touching farewell message . . .' John Howard's farewell message for Roger Beale is archived on the PM Transcripts website, web.archive.org/web/20140119030850/http://pmtranscripts.dpmc.gov.au/browse.php?did=21085.

Page 53: 'It would eventually morph . . .' The origins of the Asia-Pacific Partnership can be found in 'The U.S.–Australia Climate Action Partnership moves forward', US State Department, 9 July 2002, 2001–2009.state.gov/r/

pa/prs/ps/2002/11744.htm. The purpose of the partnership was described by Foreign Minister Alexander Downer in his press release, 11 August 2005, webarchive.nla.gov.au/wayback/20190808195945/https://foreign minister.gov.au/releases/2005/js_cdc.html.

Page 53: 'The fairness and effectiveness . . .' The quote from John Howard comes from his speech delivered at the announcement of the Asia-Pacific Partnership, BBC News, 28 July 2005, news.bbc.co.uk/2/hi/science/nature/4723305.stm.

Page 53: 'To its critics, like US Republican . . .' Senator John McCain's criticism was quoted by Shadow Environment Minister Anthony Albanese in a speech to parliament on the Asia-Pacific Partnership, *Hansard*, House of Representatives, 11 August 2005.

Page 53: 'There is no doubt in my mind . . .' Peter Shergold's quote on John Howard's interest in CCS is from the author's interview, 24 April 2019.

Page 54: 'I'm very careful to separate the jobs . . .' The debate over Dr Batterham's dual roles as chief scientist and as chief technologist at Rio Tinto is explained in the report produced for Senator Bob Brown by Margaret Blakers, 'The chief scientist, global warming and potential conflicts of interest', October 2003, and covered in the *Canberra Times*, 7 October 2003 and the ABC's *PM* program, 9 December 2003.

Page 54: 'So people like myself, who actually drill . . .' Greg Bourne's criticism of clean coal is from the author's interview, 16 July 2019.

Page 55: 'The Low Emissions Technology Advisory Group . . .' The membership of the group comes from the media release by Ian Macfarlane, industry minister, 20 February 2004.

Page 55: 'A leaked confidential record of a private meeting . . .' The full notes of the meeting with Howard on 6 May 2004 were provided by a confidential source. The details of the meeting were originally reported by Andrew Fowler on the ABC's *PM* program, 7 September 2004.

Page 56: 'The speech stared down Howard's . . .' The quotes from 'Securing Australian's energy future' are from Prime Minister Howard's speech to the National Press Club, 15 June 2004, pmtranscripts.pmc.gov.au/release/transcript-21569.

Page 57: 'Instead of providing what the Business Council . . .' The Ian Lowe quote is from the Senate Environment Committee's report, 'Lurching forward, looking back—budgetary and environmental implications of the government's energy white paper', May 2005, www.aph.gov.au/Parliamentary_Business/Committees/Senate/Environment_and_Communications/Completed_inquiries/2004-07/energywhitepaper/report/index.

Page 57: 'Australians yet again were vying . . .' The stationary energy figure is from Australia's Fourth National Communication on Climate Change to the UNFCCC, Australian Greenhouse Office, Department of Environment and Heritage, 2005, unfccc.int/resource/docs/natc/ausnc4.pdf.

Page 57: 'The PM's Office and the treasurer . . .' The calls for Robb to run for Goldstein are from Grigg, 'The player'.

Chapter 4: Scorched earth

Page 59: 'I think what you've got there . . .' The interview with farmer Trevor Smith is from the ABC's *PM* program, 'Howard tours drought-ravaged NSW', 26 October 2006, www.abc.net.au/radio/programs/pm/howard-tours-drought-ravaged-nsw/1295936.

Page 59: 'Much of the nation's farmland was in drought . . .' Descriptions of the Millennium Drought come from Albert I.J.M. van Dijk, Hylke E. Beck, Russell S. Crosbie et al., 'The Millennium Drought in southeast Australia (2001–2009): Natural and human causes and implications for water resources, ecosystems, economy, and society', *Water Resources Research*, 2013, vol. 49, no. 2, pp. 1040–57, doi:10.1002/wrcr.20123.

Page 60: 'We cannot afford to lose our farm sector . . .' Howard's quote is from his 'Address to the Rural Forum "Pine Park", Forbes', 26 October 2006, pmtranscripts.pmc.gov.au/release/transcript-22541.

Page 60: '. . . years later researchers from the CSIRO . . .' The role of climate change in the Millennium Drought is discussed in Wenju Cai, Ariaan Purich, Tim Cowan et al., 'Did climate change-induced rainfall trends contribute to the Australian Millennium Drought?', *Journal of Climate*, 2014, vol. 27, no. 9, pp. 3145–68, doi:10.1175/JCLI-D-13-00322.1.

Page 60: 'What we're seeing with this drought . . .' Mike Rann's quote is from 'Drought declared "worst in millennium"', *Sydney Morning Herald*, 7 November 2006, www.smh.com.au/national/drought-declared-worst-in-millennium-20061107-gdorys.html.

Page 60: 'In 2006 almost 68 per cent of Australians . . .' The polling on climate change is from the Lowy Institute, www.lowyinstitute.org/publications/media-release-2019-lowy-institute-poll-Australian-attitudes-climate-change.

Page 60: 'He still doubted the scientific warnings . . .' John Howard's views on climate science are from his political memoir *Lazarus Rising*, p. 553.

Page 61: 'What makes me suspicious . . .' John Howard's quote is from *Lazarus Rising*, p. 553.

Page 61: '*Time* magazine ran a seven-page spread . . .' 'The tipping point', *Time* magazine, 3 April 2006.

Page 62: 'Even the former climate sceptic . . .' Senator Chuck Hagel was shifting his position on climate change by 2005. See Amanda Little, 'New Republican leaders emerging in battle against climate change', *Grist*, 5 February 2005, grist.org/article/little-repubclimate.

Page 62: 'Lee Raymond was succeeded . . .' The quote from Myron Ebell is from the author's interview, 25 March 2019. ExxonMobil's new chief executive, Rex Tillerson, appeared to soften his predecessor's stance but not fundamentally change ExxonMobil's opposition to climate change legislation. See John Schwartz, 'Tillerson led Exxon's shift on climate change; some say it was all P.R.', *New York Times*, 28 December 2016, www.nytimes.com/2016/12/28/business/energy-environment/rex-tillerson-secretary-of-state-exxon.html.

Page 63: 'We believe that climate change is a major . . .' 'The business case for early action' report was released by the Australian Business Roundtable on Climate Change in April 2006.

Page 63: 'In a provocative move, they released their own report . . .' The discussion paper prepared by the states' National Emissions Trading Taskforce was released in August 2006, apo.org.au/node/1372.

Page 63: 'Walker and Hugh Morgan had registered . . .' On Ron Walker and Hugh Morgan's nuclear plans, see Katharine Murphy, 'Walker told me about nuclear plans, says PM', *The Age*, 28 February 2007, www. theage.com.au/national/walker-told-me-about-nuclear-plans-says-pm-20070228-ge4bhm.html.

Page 64: '. . . while Howard dismissed the film . . .' John Howard's comments on *An Inconvenient Truth* are from his memoir, *Lazarus Rising*, p. 550.

Page 64: 'He tied Howard to Bush, branding the two leaders . . .' Al Gore's comment about Howard and Bush being 'Bonnie and Clyde' reported in Ben Doherty and Marian Wilkinson, 'Paying dearly to hear Gore's climate story', *The Age*, 20 September 2007, www.theage.com.au/national/paying-dearly-to-hear-gores-climate-story-20070920-ge5v6t.html.

Page 64: 'Look at what's at risk here in Australia . . .' Al Gore's interview was on ABC's *7.30 Report*, 11 September 2006, www.abc.net.au/7.30/aust-accused-of-dismissing-global-warming-threat/2680472.

Page 65: 'It is an inconvenient truth that the Howard government . . .' Anthony Albanese's quote is from 'A Matter of Public Importance' speech, *Hansard*, House of Representatives, 13 September 2006, see anthonyalbanese. com.au/matter-of-public-importance-climate-change-2.

Page 65: 'The tipping point we reached . . .' Arthur Sinodinos's quote is from the author's interview, 31 May 2019.

Page 66: 'the greatest market failure the world has ever seen.' *Economics of Climate Change: The Stern Review* was released for the government of the United Kingdom on 30 October 2006. See also 'Ten years after the Stern Review on the Economics of Climate Change: Looking back, looking forward', Nicholas Stern, Jubilee Development Lecture, University of Newcastle (UK), 2 February 2017, www.ncl.ac.uk/events/public-lectures/archive/item/2017 tenyearsafterthesternreviewontheeconomicsofclimatechange.html.

Page 66: 'Howard would later dismiss Stern's analysis . . .' Howard's analysis of the Stern review is from *Lazarus Rising*, p. 550.

Page 66: 'My argument was always that I don't want to . . .' Shergold's quotes are from an author interview, 24 April 2019.

Page 67: 'There's one thing I am frozen in time about . . .' Howard's comments are from the prime minister's media conference, 10 December 2006, pmtranscripts.pmc.gov.au/release/transcript-22569.

Page 67: 'But Howard hand-picked every member of the business advisory group . . .' The membership of the Prime Ministerial Task Group on Emissions Trading is from the prime minister's press release, 10 December 2006, pmtranscripts.pmc.gov.au/release/transcript-22624.

Page 68: 'This is the Prime Minister's coal industry sop . . .' For Senator Bob Brown's response, see 'PM's carbon trading task force stacked, says Brown', *ABC News*, 10 December 2006, www.abc.net.au/news/2006-12-10/pms-carbon-trading-task-force-stacked-says-brown/2150220.

Page 68: 'The jury is in and the science is clear.' For Shadow Environment Minister Peter Garrett's comments to parliament, see 'Climate of hostility: Turnbull v Garrett', *Sydney Morning Herald*, 8 February 2007, www.smh.com.au/national/climate-of-hostility-turnbull-v-garrett-20070208-gdpfam.html.

Page 69: 'Malcolm was all the way with all the green agenda.' Nick Minchin's comments on Malcolm Turnbull's appointment are from his oral history recording with Susan Marsden, 19 October 2010, J.D. Somerville Oral History Collection, State Library of South Australia, OH 955.

Page 70: 'He was far more into the issues substantively . . .' Howard Bamsey's comments on Turnbull are from the author's interview, 27 November 2018.

Page 70: The IPCC 'Summary for Policymakers', *Fourth Assessment Report*, 2007, is available at www.ipcc.ch/site/assets/uploads/2018/02/ar4-wg1-spm-1.pdf.

Page 70: 'I can't put an exact time on it . . .' John Howard's interview with Tony Jones was aired on *Lateline*, ABC, 5 February 2007. For the date of the interview, see Laura Tingle in the *Australian Financial Review*, www.afr.com/policy/energy-and-climate/bear-in-mind-threats-of-climate-change-20070209-jeuuf.

Page 71: 'The next day, Labor's opposition leader . . .' The Rudd–Howard exchange is from *Hansard*, House of Representatives, 6 February 2006.

Page 71: 'Four months before the election . . .' See 'Howard announces the emissions trading scheme', *ABC News*, 17 July 2007, www.abc.net.au/news/2007-07-17/howard-announces-emissions-trading-system/2505080.

Page 71: 'The Shergold report had cushioned the scheme . . .' For the Shergold report recommendations on permits, see *Prime Ministerial Task Group on Emissions Trading Report*, p. 116.

Page 72: '. . . "a crock of shit" and "a picnic for the bankers and financiers" . . .' For Nick Minchin's comments on the emissions trading scheme, see his account to Susan Marsden, 19 October 2010, State Library of South Australia.

Page 73: 'In a provocative move, Howard promoted . . .' For Howard's quote on the APEC summit as a climate change meeting, see Marian Wilkinson, 'G8 Moves to put APEC in political shade', *Sydney Morning Herald*, 12 June 2007, www.smh.com.au/national/g8-moves-to-put-apec-in-political-shade-20070612-gdqd2f.html.

Page 73: 'Hu bluntly reminded Howard . . .' For President Hu Jintao's quote on the Kyoto Protocol and the Sydney Declaration, see Marian Wilkinson, 'Kyoto is the only way, Hu tells Howard', *Sydney Morning Herald*, 10 September 2007, www.smh.com.au/national/kyoto-is-the-only-way-hu-tells-howard-20070910-gdr2k2.html.

Page 73: 'Howard was a lame duck . . .' For a full account of Howard's leadership crisis on the eve of APEC, see Paul Kelly, 'The defeat', *Weekend Australian*, 15 December 2007.

Page 74: 'In the dying days of the government . . .' Malcolm Turnbull's attempt to get cabinet to look again at ratifying Kyoto was reported by the ABC, 'PM silent on Kyoto leak claims', 28 October 2007, www.abc.net.au/news/2007-10-28/pm-silent-on-kyoto-leak-claims/2582442.

Page 74: 'In a memorable clash with Garrett . . .' Malcolm Turnbull's quote is from *Hansard*, House of Representatives, 21 June 2007.

Page 74: 'Work Choices, healthcare and education . . .' The role of climate change in the election is discussed in Christopher Rootes, 'The first climate change election? The Australian General Election of 24 November 2007', *Environmental Politics*, 2008, vol. 17, no. 3, pp. 473–80, doi:10.1080/09644010802065815.

Chapter 5: The state of the Reef

Page 75: 'Climate change is now recognised . . .' The quote on the Great Barrier Reef and climate change is from the *Great Barrier Reef Climate Change Action Plan 2007–2012*, Great Barrier Reef Marine Park Authority (GBRMPA) and Australian Greenhouse Office, Commonwealth of Australia, 2007, p. 3, elibrary.gbrmpa.gov.au/jspui/bitstream/11017/198/1/Great-Barrier-Reef-Climate-Change-Action-Plan-2007-2012.pdf.

Page 76: 'Mass bleachings of corals in 1998 and 2002 . . .' For details of the Reef bleachings, seabird chick deaths and turtle hatchings, see *Great Barrier Reef Climate Change Action Plan 2007–2012*, p. 4.

Page 76: 'A sharp-eyed reader could see . . .' Footnote referencing the IPCC's 'Summary for policymakers', *Great Barrier Reef Climate Change Action Plan 2007–2012*, p. 4.

Page 76: 'That was when I think I saw . . .' Ove Hoegh-Guldberg discussed his first visit to the Reef with his grandfather on *Australian Story*, ABC TV, 9 February 2009, www.abc.net.au/austory/the-heat-of-the-moment/9173048.

Page 77: 'Stretching along the Queensland coast . . .' For a description of the Reef, see *Great Barrier Reef Outlook Report 2009*, GBRMPA, Commonwealth of Australia, July 2009, pp. 1–5, elibrary.gbrmpa.gov.au/jspui/handle/11017/199.

Page 77: 'It supports an astonishing variety of life . . .' For the Reef's biodiversity, see *Great Barrier Reef Climate Change Action Plan 2007–2012*, Chapter 2, pp. 8–33.

Page 77: 'Australians fought one of the biggest environmental campaigns . . .' On the campaign to save the Reef from oil drilling, see *Great Barrier Reef Outlook Report 2009*, p. 3. And the reasons for its World Heritage Listing, p. 4.

Page 77: 'As a young student in 1987 . . .' Dr Ove Hoegh-Guldberg's PhD findings were published with co-author G. Jason Smith as 'The effect of sudden changes in temperature, light and salinity on the population density and export of zooxanthellae from the reef corals *Stylophora pistillata* Esper and *Seriatopora hystrix* Dana', *Journal of Experimental Marine Biology and Ecology*, 1989, vol. 129, no. 3, pp. 279–303, doi:10.1016/0022-0981(89)90109-3.

Page 77: 'More and more people were seeing . . .' Hoegh-Guldberg's comments on the bleaching causes come from the author's interview, 11 March 2019.

Page 78: 'Most healthy corals get their colour . . .' For the explanation of coral bleachings, see *Great Barrier Reef Outlook Report 2009*, p. 92. For lethal bleaching recovery times, see p. 149.

Page 78: 'The only change that consistently . . .' Hoegh-Guldberg's comment is from the author's interview, 11 March 2019.

Page 79: 'I expected the answer would be hundreds . . .' Hoegh-Guldberg's quote on bleaching and research findings on climate change is from the author's interview, 11 March 2019.

Page 79: 'Hoegh-Guldberg put his career on the line . . .' Ove Hoegh-Guldberg, 'Climate change, coral bleaching and the future of the world's coral reefs', *Marine Freshwater Research*, 1999, vol. 50, no. 8, pp. 839–66, doi:10.1071/MF99078.

Page 80: 'In 1999, when Ove wrote his report . . .' Dr David Wachenfeld's comments on Hoegh-Guldberg's 1999 paper are from an author interview, 13 March 2019.

Page 80: 'I felt very out on a limb.' Hoegh-Guldberg on the reaction of his colleagues in 1999, and from Robert Hill, are from *Australian Story*, 9 February 2009.

Page 80: 'When he introduced himself there to Robert Hill . . .' Hoegh-Guldberg's description of his later conversation with Hill is from the author's interview, 11 March 2019.

Page 81: 'I like a good brouhaha . . .' Hoegh-Guldberg's 'brouhaha' comment is from the author's interview, 11 March 2019. The reaction by sceptics to Hoegh-Guldberg is collected in his blog, Climate Shifts, www.climate shifts.org.

Page 81: 'That made it undeniable that 1998 . . .' Comment from Dave Wachenfeld on 2002 bleaching is from the author's interview, 13 March 2019.

Page 81: 'Despite cries of protest . . .' On the park rezoning plan, see *Report on the Great Barrier Reef Zoning Plan, 2003*, GBMPRA, November 2005, elibrary. gbrmpa.gov.au/jspui/bitstream/11017/407/1/Report-on-the-Great-Barrier-Reef-Marine-Park-zoning-plan.pdf.

Page 82: 'It means the Reef is going to be . . .' The quote from David Kemp is from the ABC's *PM* program, 3 December 2003, www.abc.net.au/pm/content/2003/s1002876.htm.

Page 82: 'It's a wonderful story.' The quote from Hoegh-Guldberg on the zoning plan is from the author's interview, 11 March 2019.

Page 82: 'I remember someone telling me . . .' The quote from Hoegh-Guldberg on Gladstone Port is from the author's interview, 11 March 2019.

Page 83: 'Mentioning that we needed to reduce coal . . .' Hoegh-Guldberg on coal and coral, author's interview, 11 March 2019.

Page 84: 'When you are asked to put in a deposition . . .' Hoegh-Guldberg on appearing in court case, author's interview, 11 March 2019.

Page 84: 'The case was brought . . .' Details of the court case are in *Queensland Conservation Council Inc v Xstrata Coal Qld Pty Ltd and Ors* [2007] Queensland Court of Appeal 338.

Page 85: 'Climate change has grown from insignificance . . .' The quote is from Ove Hoegh-Guldberg's report for the Queensland Conservation Council case against Xstrata, 'Likely ecological impact of global warming and climate change on the Great Barrier Reef by 2050 and beyond', 19 January 2007, envlaw.com.au/wp-content/uploads/newlands6.pdf.

Page 85: '. . . he cited a paper co-authored . . .' The paper cited in the Queensland Land Tribunal criticism of the Stern review was R.M. Carter, C.R. de Freitas, I.D. Goklany et al., 'The Stern review: A dual critique— the science', *World Economics*, 2006, vol. 7, no. 4, pp. 167–98, www. world-economics-journal.com/Journal/Papers/The Stern Review A Dual Critique.details?ID=261.

Page 86: 'Even under the best case scenarios . . .' The report by Ove Hoegh-Guldberg for WWF and the Queensland Tourism Industry Council, *The Implications of Climate Change for Australia's Great Barrier Reef*, was published by WWF in 2004, www.cakex.org/documents/ implications-climate-change-Australias-great-barrier-reef.

Page 86: 'Global warming is obviously a concern . . .' The article with the Carter/Ridd criticism of Hoegh-Guldberg's findings is from Emma Young, 'Great Barrier Reef to be decimated by 2050', *New Scientist*, 23 February 2004, www.newscientist.com/article/dn4707-great-barrier-reef-to-be-decimated-by-2050.

Page 87: 'The IPA made Carter one of their "emeritus fellows" . . .' For details of Professor Bob Carter's role with the IPA, see the IPA's obituary for him by John Roskam, 'In memoriam Professor Robert M. Carter', 1 April 2016, ipa.org.au/ipa-review-articles/in-memoriam-professor-robert-m-carter.

Page 87: '. . . while Ridd joined the board . . .' On Peter Ridd's role with the IPA, see the articles concerning him on the IPA website, ipa.org.au/tag/ peter-ridd. For Carter and Ridd's role with the Australian Environment Foundation, see www.australianenvironment.org.

Page 87: 'Imagine a well-provendered and equipped . . .' Professor Carter's essay, 'It's good sense to avoid consensus on global warming', is from *The Australian*, 10 July 2007, www.theAustralian.com.au/opinion/bob-carter-its-good-sense-to-avoid-consensus-on-global-warming/news-story/69b1f1 500ada0a1169ad7f5960f1ccc6.

Page 88: 'the greatest hoax ever perpetrated . . .' James Inhofe to the US Congress, 28 July 2003, *Congressional Record*, vol. 149, no. 113, US Government Publishing Office, Washington DC, 2003. www.inhofe. senate.gov/imo/media/doc/The Facts and Science of Climate Change1.pdf.

Page 88: 'the drum major of the denial parade'. The quote on Marc Morano is from Leslie Kaufman, 'Dissenter on warming expands his campaign', *New York Times*, 9 April 2009, www.nytimes.com/2009/04/10/us/politics/10morano.html.

Page 88: 'In his characteristically exuberant style . . .' Professor Carter's quotes are from his original testimony to the US Senate Committee on Environment and Public Works, 6 December 2006, www.govinfo.gov/content/pkg/CHRG-109shrg52324/html/CHRG-109shrg52324.htm.

Page 89: 'Many studies incontrovertibly link . . .' The quote on coral reefs and bleaching is from 'Climate change 2007: Impacts, adaptation and vulnerability', Working Group II, *Fourth Assessment Report of the IPCC*, p. 235, www.ipcc.ch/report/ar4/wg2.

Page 90: 'for their efforts to build up and disseminate . . .' 'The Nobel Peace Prize 2007', The Nobel Prize, www.nobelprize.org/prizes/peace/2007/summary.

Page 90: 'Will those responsible for decisions . . .' The Nobel Lecture of Dr Rajendra Pachauri is at www.nobelprize.org/prizes/peace/2007/ipcc/lecture.

Chapter 6: The defining challenge

Page 91: 'Some 10,000 ministers, officials . . .' Details of the 2007 UNFCCC Bali conference were gathered by the author, who reported on it for the *Sydney Morning Herald* and *The Age*. Delegate numbers and countries can be found at 'Bali Climate Change Conference—December 2007', United Nations Climate Change, unfccc.int/process-and-meetings/conferences/past-conferences/bali-climate-change-conference-december-2007/bali-climate-change-conference-december-2007-0.

Page 91: 'A short time before, Rudd . . .' Kevin Rudd's handover of the Kyoto Protocol to the secretary-general was reported by the Earth Negotiations Bulletin at the Bali conference, 13 December 2007, enb.iisd.org/vol12/enb12352e.html.

Page 91: 'I did so, and my government has done so . . .' The quotes from Rudd are from his speech to the Bali UN Climate Conference, 12 December 2007. See also Marian Wilkinson, 'PM's message to US: world wants rich to fight warming', *Sydney Morning Herald*, 13 December 2007, www.smh.com.au/environment/pms-message-to-us-world-wants-rich-to-fight-warming-20071213-gdrsu9.html.

Page 92: 'He had been prime minister . . .' The reference to writing the speech on the flight to Bali is from Kevin Rudd, *The PM Years*, Sydney: Pan Macmillan, 2018, p. 14.

Page 92: 'I was not conscious that there was . . .' Kevin Rudd's comments on his reception at Bali are from an author interview, 5 April 2019, and from Wilkinson, 'PM's message to US'.

Page 92: 'Bonnie went straight . . .' Al Gore's quote is from Pooley, *The Climate War*, p. 119.

Page 92: 'Developing countries had to be . . .' Howard Bamsey's comments are from the author's interview, 27 November 2018.

Page 93: 'It was a dangerous game of chicken.' On the division in Bali, see Marian Wilkinson and Mark Forbes, 'Rudd to fly into climate battle as US resists greenhouse plan', *Sydney Morning Herald*, 11 December 2007, www.theage.com.au/environment/rudd-to-fly-into-climate-battle-as-us-resists-greenhouse-plan-20071211-ge6hr9.html; and Marian Wilkinson and Mark Forbes, 'Gore blasts US obstruction', *Sydney Morning Herald*, 14 December 2007, www.theage.com.au/national/gore-blasts-us-obstruction-20071214-ge6icn.html.

Page 93: 'By 2007 China had overtaken . . .' China's 2007 emissions figure is from Rudd, *The PM Years*, p. 14.

Page 93: 'The year before, a choking smoke haze . . .' Indonesia as the world's third-largest emitter in 2007 is referenced in Pooley, *The Climate War*, p. 12. For details of the Borneo and Sumatran fires and smoke haze, see John Aglionby, 'Forest fire haze brings misery to Indonesia and beyond', *The Guardian*, 6 October 2006, www.theguardian.com/environment/2006/oct/06/indonesia.pollution.

Page 93: 'All countries, all governments . . .' The quote from Yvo de Boer, then executive secretary of the UNFCC, is from Marian Wilkinson and Stephanie Peatling, 'Just warming up, but already Rudd feels the Bali heat', *Sydney Morning Herald*, 7 December 2007, www.smh.com.au/environment/just-warming-up-but-rudd-already-feels-bali-heat-20071207-gdrrlb.html.

Page 94: 'Rudd tried to sidestep the issue . . .' On Rudd avoiding a 2020 commitment in Bali, see Michelle Grattan, Mark Forbes and Marian Wilkinson, 'Rudd walks tightrope on climate in Bali', *The Age*, 12 December 2007, www.theage.com.au/environment/rudd-walks-tight-rope-on-climate-in-bali-20071212-ge6hyz.html.

Page 94: 'Climate scientists warned that global greenhouse gas . . .' On the IPCC advice on 2020 targets for the developed countries, see Chad Carpenter, *The Bali Action Plan: Key Issues in the climate negotiations, summary for policymakers*, United Nations Development Programme, September 2008, p. 3, www.undp.org/content/dam/undp/library/Environment and Energy/ClimateChange/Bali_Road_Map_Key_Issues_Under_Negotiation. pdf. See also Narelle Towie, 'Scientists issue declaration at Bali', *Nature*, 2007, vol. 450, no. 7171, doi:10.1038/news.2007.361.

Page 94: 'We must make the leap forward . . .' De Boer's quote on the *Planet of the Apes* was reported in Wilkinson, 'PM's message to US'.

Page 94: 'Anger boiled over at one point . . .' For the account of the dispute with the US, the booing of the US delegation and Australia's role in the compromise, see Marian Wilkinson, 'Bali climate change victory', *Sydney Morning Herald*, 16 December 2007, www.smh.com.au/environment/bali-climate-change-victory-20071216-gdrtb6.html.

Page 95: 'It means we know this road map . . .' Don Henry's reaction is reported in Marian Wilkinson, 'Australia comes in from the cold', *Sydney*

Morning Herald, 17 December 2007, www.smh.com.au/environment/ Australia-comes-in-from-the-cold-20071217-gdrtgq.html, and his quote was reported by *ABC News*, 15 December 2007.

Page 96: 'Howard had warned Labor's . . .' The News Ltd coverage of Peter Garrett during the 2007 election campaign and Rudd's reaction come from comments from Garrett to the author, 7 March 2019. The comment on the 'Garrett recession' is from Peter Garrett, *Big Blue Sky: A memoir*, Sydney: Allen & Unwin, 2015, p. 319.

Page 97: 'I had formed a deep appreciation . . .' Kevin Rudd's quote on Penny Wong is from the author's interview, 5 April 2019.

Page 98: 'I'm not used to giving in . . .' Ross Garnaut's comments are from the author's interview, 19 March 2019.

Page 98: 'Its Hong Kong parent company . . .' For the structure of TRUenergy and CLP, see Marian Wilkinson and Ben Cubby, 'Pollution pays off for billionaire', *Sydney Morning Herald*, 12 August 2009.

Page 98: 'By April, McIndoe had written . . .' Richard McIndoe's letter to Rudd, Wong and Victorian premier John Brumby was reported in 'Climate proposal "would cost jobs"', *The Age*, 30 April 2008, www.theage.com. au/environment/climate-proposal-would-cost-jobs-20080430-ge70sq. html.

Page 99: 'Only eight other countries . . .' The emissions intensity comparison of Australia's energy system with the rest of the world is quoted in Professor Ross Garnaut, *Climate Change Review: Final report*, Melbourne: Cambridge University Press, 2008, Chapter 7 at 7.1.4.

Page 99: 'Hazelwood had long been targeted . . .' For the campaigns against Hazelwood Power Station by Environment Victoria, WWF and others, plus its emissions record, see Jordan Ward and Mick Power, 'Cleaning up Victoria's power sector: The full social cost of Hazelwood power station', Harvard Kennedy School of Government, Boston, 24 February 2015, environmentvictoria.org.au/2015/02/24/cleaning-victorias-power-sector-full-social-cost-hazelwood-power-station. See also Philip Chubb, *Power Failure: The inside story of climate politics under Rudd and Gillard*, Melbourne: Black Inc., 2014, pp. 37–40.

Page 99: 'He had been hand-picked by John Howard . . .' For Tony Concannon's role, see media release, Prime Ministerial Task Group on Emissions Trading, 10 December 2006, pmtranscripts.pmc.gov.au/release/ transcript-22624.

Page 99: 'Nethercote had also been a trusted source . . .' Ian Nethercote's opposition to an emissions trading scheme comes from the author's interview with Roger Beale, 20 February 2019.

Page 99: 'Nethercote also sat on the board . . .' Nethercote's membership of the IPA board is documented in its Financial Report for June 2008, ipa. org.au/wp-content/uploads/archive/IPA_2008_Financials.pdf.

Page 100: 'They were the state's economic lifeblood . . .' The figure of 90 per cent of Victoria's power supply coming from the La Trobe Valley generators comes from Melissa Fyfe, 'Victoria: It's time to come clean', *The Age*, 6 July 2008, www.smh.com.au/environment/victoria-its-time-to-come-clean-20080705-329r.html. By 2012 this was around 84 per cent, according to Ward and Power, 'Cleaning up Victoria's power sector'.

Page 100: 'Economic analysis showed . . .' Ross Garnaut's comments are from the author's interview, 16 August 2019.

Page 101: 'I flew to Adelaide the other day . . .' The lobbying campaign by Woodside is outlined in Marian Wilkinson and Ben Cubby, 'Wong's dose of shock therapy', *The Age* and *Sydney Morning Herald*, 10 May 2008, www.smh.com.au/environment/weather/wongs-dose-of-shock-therapy-20080510-gdsd1x.html.

Page 102: 'He always liked to be the cleverest . . .' The comment on Malcolm Turnbull is from a confidential interview with a former senior government official.

Page 102: 'The bottom line is if you implement . . .' Michael Costa's attack on the Garnaut draft report is from his interview with ABC on *The World Today*, 7 July 2008, www.abc.net.au/worldtoday/content/2008/s2296454.htm.

Page 102: 'the generators will be effectively bankrupt . . .' Richard McIndoe's response to the Garnaut draft report is from, Fyfe, 'Victoria: It's time to come clean'.

Page 103: 'The generators were particularly concerned . . .' Martin Parkinson's comment on Garnaut is from the author's interview, 16 August 2019.

Page 103: 'It will be an act of environmental suicide . . .' Brendan Nelson's comments on the emissions trading scheme are from ABC's *The World Today* program, 7 July 2008, www.abc.net.au/worldtoday/content/2008/s2297676.htm.

Page 104: 'If we delay action any longer . . .' The statement on the costs to 'our children and grandchildren' is from the Carbon Pollution Reduction Scheme Green Paper, July 2008, published by the Department of Climate Change. See this paper for the emissions per unit of revenue for the various industries—aluminium, cattle production, etc., p. 313.

Page 105: 'Industries like aluminium were supposed . . .' For details on the assistance, see the CPRS green paper, p. 219.

Page 105: 'Thanks to both Labor and Liberal . . .' On the secret subsidies on electricity prices for the Victorian aluminium industry, see Royce Millar, 'Smelters costing us $4.5 billion', *The Age*, 17 October 2009, www.smh.com.au/national/smelters-costing-us-45-billion-20091016-h17r.html.

Page 107: 'He began relentlessly stalking . . .' For details on Turnbull's campaign against Brendan Nelson, see Paddy Manning, *Born to Rule: The unauthorised biography of Malcolm Turnbull*, Melbourne: Melbourne University Press, 2015, p. 310.

Page 107: 'It changed everything.' The Parkinson quote comes from the author's interview, 16 August 2019.

Page 108: 'The GFC was my greatest ally.' The comment from Paul Howes on the GFC is from Marian Wilkinson, Ben Cubby and Flint Duxfield, 'Come in spinner', *Sydney Morning Herald*, 7 November 2009, www.smh.com. au/environment/come-in-spinner-20091106-i261.html.

Page 108: 'It included his advice ...' Garnaut's advice on the targets was first flagged before the GFC at the National Press Club in Canberra on 5 September 2008 in a draft report, and then in the final report on 30 September. On the targets and costs, see Garnaut, *Climate Change Review*, Chapter 11. For an analysis of the changing targets, see Peter Christoff, 'Aiming high: on Australia's emissions reduction targets', *UNSW Law Journal*, 2008, vol. 31, no. 3, pp. 861–79, www.austlii.edu. au/au/journals/UNSWLawJl/2008/46.pdf.

Page 109: 'Failure of the world to act now ...' The scientists' letter to Rudd was reported in Marian Wilkinson, 'Don't go soft on climate, PM warned', *Sydney Morning Herald*, 29 September 2008, www.smh.com. au/environment/weather/dont-go-soft-on-climate-pm-warned-20080929-gdsws7.html.

Page 109: 'In mid-December, Rudd and Wong ...' For Rudd's announcement on the targets, see 'Australia's low pollution future: Launch of Australian government's white paper on the Carbon Pollution Reduction Scheme', National Press Club, Canberra, 15 December 2008, pmtranscripts.pmc. gov.au/release/transcript-16318.

Page 110: 'And the scheme now included a large pot ...' Details of the assistance to generators were given in Rudd, 'Australia's low pollution future', and also Marian Wilkinson, 'Coal-fired generators escape the blacklist', *Sydney Morning Herald*, 18 December 2008, www.smh. com.au/environment/weather/coal-fired-generators-escape-the-blacklist-20081219-gdt730.html. For the assistance offered to the trade-exposed industries, see the CPRS executive summary archived at pandora.nla.gov. au/pan/99543/20090515-1610/www.climatechange.gov.au/whitepaper/report/pubs/pdf/V100eExecutiveSummary.pdf, p. 22.

Page 110: 'Never in the history of public finance ...' For the critical statement from Ross Garnaut, see Chubb, *Power Failure*, p. 51.

Page 110: 'Garnaut was operating ...' The comment from Rudd on Garnaut was to the author, 16 February 2020.

Page 111: 'I knew that we just needed ...' Penny Wong's response is from the author's interview, 26 February 2019.

Page 111: 'Mr Rudd has betrayed the science ...' Greenpeace's John Hepburn's response to the white paper quoted in 'Protesters heckle Rudd', *Sydney Morning Herald*, 16 December 2008, www.smh.com.au/environment/weather/protesters-heckle-rudd-20081216-gdt6ox.html.

Page 111: 'A very disappointing day.' Don Henry's quote is from Tom Arup, 'Angry Greens accuse PM of going easy on polluters', *The Age*, 16 December 2019, www.smh.com.au/national/angry-greens-accuse-pm-of-going-easy-on-polluters-20081215-6z27.html.

Page 112: 'My conversations with Murdoch ...' Rudd's comment about Murdoch comes from the author's interview, 5 April 2019.

Page 112: 'We had report after report . . .' Penny Wong's quote on the campaign against the government's CPRS is from the author's interview, 26 February 2019.

Chapter 7: The heat goes on

Page 113: 'Senator Cory Bernardi was perspiring . . .' The details of the launch of *Thank God for Carbon* are from the author's interview with Cory Bernardi, 24 October 2018.

Page 113: 'It was January 2009 . . .' Details of the heatwave across southern Australia at that time are from Marian Wilkinson and Ben Cubby, 'Living in a climate of fear', *Sydney Morning Herald*, 14 February 2009, www.smh.com.au/national/living-in-a-climate-of-fear-20090213-8779.html.

Page 113: 'The ABC was there . . .' Bernardi's comment on the heat is from the author's interview, 24 October 2018.

Page 113: '. . . even his wife, Sinéad, famously . . .' Sinéad Bernardi's quip about her husband Cory is from Sally Neighbour, 'All about Cory', *The Monthly*, December 2011–January 2012, www.themonthly.com.au/cory-bernardi-conservative-warrior-all-about-cory-sally-neighbour-4327.

Page 114: 'A bit too much like socialism . . .' Bernardi's quote is from his speech at the Lavoisier Group launch for *Thank God for Carbon*, 27 January 2009, www.lavoisier.com.au/articles/climate-policy/science-and-policy/TGFC-LaunchBernardi.php.

Page 114: 'He'd shot to national fame . . .' On the Senate inquiry into swearing, see Phil Coorey, 'Chef's foul speciality sparks swearing inquiry', *Sydney Morning Herald*, 21 March 2008, www.smh.com.au/entertainment/chefs-foul-specialty-sparks-swearing-inquiry-20080321-gds674.html.

Page 115: 'Under the heading "Cool Heads Needed" . . .' Bernardi's opinion piece, 'Cool heads needed on global warming', was published in the Adelaide *Advertiser*, 25 April 2007. It was cut down from a longer essay written on 20 April 2007, www.adelaidenow.com.au/news/cool-heads-needed-on-global-warming/news-story/654c06d965f97b9347dac8d06353ff57. Both quoted the Oregon Petition.

Page 115: 'The petition was supposedly signed . . .' For the critique of the Oregon Petition, including its false signatures, see H. Josef Hebert, 'Jokers add fake names to warming petition', *Seattle Times*, 1 May 1998, archive.seattletimes.com/archive/?slug=2748308&date=19980501; Pooley, *The Climate War*, pp. 41–2; and H. Josef Hebert, 'Odd names added to greenhouse plea', Associated Press, 1 May 1998, apnews.com/aec8beea85d7fe76fc9cc77b8392d79e.

Page 115: 'You're absolutely right, but . . .' Bernardi's quote on it not being helpful is from the author's interview, 24 October 2018.

Page 115: 'I can't pretend that I knew . . .' Bernardi's quotes on his involvement with Morgan and Evans, and the US right activists group, come from the author's interview, 24 October 2018, and his training with Morton Blackwell is from the author's interview with Myron Ebell, 25 March 2019.

Page 116: 'Bernardi wrapped up the launch . . .' Bernardi's *Thank God for Carbon* launch speech at Lavoisier Group, 27 January 2009.

Page 117: 'The Liberal Party doesn't know . . .' The quote is from Ray Evans' letter to Andrew Robb, 3 December 2008, www.lavoisier.com.au/articles/climate-policy/science-and-policy/evans2009-1.php.

Page 117: '"The first issue," he told the crowd . . .' For the quotes from Ray Evans' speech at the Adelaide launch, see www.lavoisier.com.au/articles/climate-policy/science-and-policy/TGFC-LaunchREvans.php.

Page 117: 'Such an event appears to be . . .' For details on the Adelaide temperatures and extreme weather event, see Wilkinson and Cubby, 'Living in a climate of fear', 14 February 2009. For hospital data, see Monika Nitschke and Graeme Tucker, 'The unfolding story of heat waves in metropolitan Adelaide', SA Health, Government of South Australia, May 2017.

Page 118: 'By the end, the shocking toll . . .' The casualties on the Black Saturday fires are from the National Museum of Australia, www.nma.gov.au/defining-moments/resources/black-saturday-bushfires.

Page 118: 'The projections are based . . .' The quote from Kevin Hennessy is from Wilkinson and Cubby, 'Living in a climate of fear', 14 February 2009.

Page 119: 'Heartland had got funding . . .' For details of ExxonMobil funding of the Heartland Institute, see www.documentcloud.org/documents/1019878-2004-exxon-giving-report.html#document/p4/a260559.

Page 119: 'It gave its first $1 million grant . . .' The funding details of the Mercer Foundation to Heartland are from the Mercer Foundation's tax returns supplied to the author by the Climate Investigations Center.

Page 120: 'We only learned of their years . . .' The quote is from the author's interview with Kert Davies, founder and director of the Climate Investigations Center in the US, 28 March 2019.

Page 120: 'If that happened it could be a game changer . . .' The importance of the Waxman–Markey bill to Copenhagen is from the author's interview with Bruce Wolpe, a former staff member for Representative Waxman, 21 August 2019. For a full account of the issue, see Bryan W. Marshall and Bruce C. Wolpe, *The Committee: A study of policy, power, politics and Obama's historic legislative agenda on Capitol Hill*, Ann Arbor: University of Michigan Press, 2018, Chapter 2.

Page 121: 'If you want to fight this stuff . . .' Ebell's quote from the 2008 Heartland Conference is from Pooley, *The Climate War*, p. 51 (emphasis in original).

Page 121: 'They didn't understand it was going to raise . . .' Ebell's quote on the Waxman–Markey bill is from the author's interview, 25 March 2019.

Page 122: 'Arrogant and patronising . . .' Greg Combet's account of his meeting with the Australian Coal Association is from the author's interview, 14 January 2019.

Page 122: '. . . as a young engineering student . . .' Greg Combet's background as an economist and engineer are from the author's interview, 14 January

2019, and also his memoir, Greg Combet with Mark Davis, *The Fights of My Life*, Melbourne: Melbourne University Press, 2014.

Page 123: 'Contemptuous, he was . . .' Combet's quotes on the coal lobby message and Xstrata boss Mick Davis are from the author's interview, 14 January 2019.

Page 123: 'A carbon price in Australia . . .' Combet's quotes on the impact of the emissions trading scheme on coal come from the author's interview, 14 January 2019.

Page 124: 'Coal exports were worth $54.8 billion . . .' For Australia's coal exports, both thermal and metallurgical, see Frank Bingham and Brent Perkins, 'Australia's coal and iron ore exports, 2001 to 2011', Department of Foreign Affairs and Trade, dfat.gov.au/about-us/publications/Documents/Australias-coal-and-iron-ore-exports-2001-to-2011.pdf.

Page 124: 'Two powerful billionaires were interested . . .' For the opening up of the Galilee Basin, including Gina Rinehart's Hancock Prospecting interest and Clive Palmer's Waratah Coal interest, see 'Joint statement by Premier and Minister for the Arts the Honourable Anna Bligh, Treasurer and Minister for Employment and Economic Development, and the Honourable Andrew Fraser: Premier drops in on new developments set to unlock Galilee Basin's potential', 1 November 2009, statements.qld.gov.au/Statement/Id/67206.

Page 124: 'He announced a new target . . .' On Rudd's backflip on the 2020 target, see Prime Minister, Treasurer, Minister for Climate Change and Water, press release, 'A new target for reducing Australia's carbon pollution', 4 May 2009, parlinfo.aph.gov.au/parlInfo/download/media/pressrel/HFGT6/upload_binary/hfgt60.pdf.

Page 125: 'consistent with Australia having the prospect . . .' For Rudd's quote on the Great Barrier Reef, see Marian Wilkinson, 'Call Rudd and Wong's bluff on climate policy', *Sydney Morning Herald*, 7 May 2009, www.smh.com.au/environment/climate-change/call-rudd-and-wongs-bluff-on-climate-policy-20090506-ava4.html .

Page 125: 'Did I see it as pie in the sky?' Rudd's quote on Obama is from the author's interview, 5 April 2019.

Page 126: 'To reassure the coal industry . . .' For details of the package, see Martin Ferguson, Peter Garrett, Penny Wong and Kim Carr, press release, '$4.5 billion Clean Energy Initiative', 12 May 2009, parlinfo.aph.gov.au/parlInfo/download/media/pressrel/CQJT6/upload_binary/cqjt62.pdf. For RET funding figures, see https://archive.budget.gov.au/2009-10/ministerial_statements/ms_climate_change.pdf.

Page 126: 'In this round the generators' assistance . . .' On the cut to assistance for the generators, see Chubb, *Power Failure*, p. 57.

Page 126: 'Within weeks, their peak lobby . . .' On the letter from the National Generators Forum, see Wilkinson, Cubby and Duxfield, 'Come in spinner'.

Page 126: 'You're going to shut us down . . .' Combet's quote on International Power is from the author's interview, 14 January 2019.

Page 127: 'John Pierce took a very aggressive stance . . .' Combet's quote on Pierce is from the author's interview, 14 January 2019.

Page 127: 'They agreed to hire merchant bank . . .' Details of the Morgan Stanley negotiations are from confidential government sources but are also referenced in Chubb, *Power Failure*, p. 60.

Page 127: 'Kevin Rudd made it clear he did not want . . .' On Rudd's reluctance to work with the Greens, see Christine Milne, *An Activist Life*, Brisbane: UQP, 2017, pp. 185–8.

Page 127: 'We have to make it about him . . .' Wong's strategy with Turnbull is from the author's interview, 26 February 2019.

Page 128: 'Rudd tore Turnbull to shreds . . .' Rudd's speech on Turnbull over the Godwin Grech affair is quoted in Manning, *Born to Rule*, p. 328.

Page 128: 'There was a group of us . . .' Bernardi's quotes on MPs' opposition to the emissions trading scheme is from the author's interview, 24 October 2018.

Page 129: 'a job-destroying rabid dog . . .' Warren Truss's quote is reported in Paul Kelly, *Triumph and Demise: The broken promise of a Labor generation*, Melbourne: Melbourne University Press, 2014, p. 243.

Page 129: 'We explained that the policy . . .' Bernardi's quote is from a speech to the Heartland International Climate Change Conference, Chicago, May 2010, www.heartland.org/multimedia/videos/cory-bernardi-iccc4.

Page 129: 'If you live on a diet . . .' Barnaby Joyce's roast lamb speech, 'Carbon trading and dinner: A note from Barnaby Joyce', was run on IPA member Jennifer Marohasy's blog, 31 July 2009, jennifermarohasy.com/2009/07/carbon-trading-and-dinner-a-note-from-barnaby-joyce.

Page 130: 'Its senior scientist, Jay Lehr . . .' Dr Lehr wrote a letter to the *Courier-Mail* promoting his speech at the Brisbane Institute, which was run under Jay Lehr, 'Real heat generated by vested interests', *Courier-Mail*, 12 August 2009, www.couriermail.com.au/news/real-heat-generated-by-vested-interests/news-story/31ec511797b13238369f38edbca591f1?sv=a3865a977fefc361dc6b7db723972e5f. The tour was originally reported by Greenpeace's Cindy Baxter in *Doubting Australia: The roots of Australia's climate denial*, 2010, author copy (no longer available at www.cana.net.au/sites/default/files/DoubtingAustralia.pdf).

Page 130: 'Many in Labor and the Greens dismissed . . .' Bernardi's description of himself and Joyce as 'mad uncles' by their opponents is from his Heartland speech, May 2010.

Page 130: '. . . in the hierarchy I was the most senior member . . .' Nick Minchin's quote on his role in the Coalition is from the author's interview, 28 February 2019, as are his quotes on Turnbull wanting to avoid a fight and on the Renewable Energy Target as 'madness'.

Page 131: 'Robb was still a climate sceptic . . .' Andrew Robb's quotes on his opposition to the emissions trading scheme are from his memoir, *Black Dog Daze: Public life, private demons*, Melbourne: Melbourne University Press, 2011, p. 169.

Page 132: 'Every time there is a problem . . .' Barnaby Joyce's Senate speech on the emissions trading scheme as a tax is from *Hansard*, Senate, 13 August 2009.

Page 132: 'John who worked for me . . .' Penny Wong's quote is from the author's interview, 26 February 2019.

Chapter 8: Triumph of the sceptics

Page 133: 'One of the great achievements . . .' The two quotes from Malcolm Turnbull's London meeting with David Cameron are from Christian Kerr, 'Turnbull lost his footing down the Tory green path', *The Australian*, 10 April 2010. For David Cameron's support for the UK Climate Change Act, see David Nussbaum, 'David Cameron has gone cold on climate change', *The Guardian*, 21 November 2012, www.theguardian.com/commentisfree/2012/nov/20/david-cameron-gone-cold-on-climate-change; and UK Institute for Government, 'The Climate Change Act (2008)', www.instituteforgovernment.org.uk/sites/default/files/climate_change_act.pdf.

Page 134: 'I have no doubt as to which . . .' Cory Bernardi's quote on his advice in Washington and his meetings with Senator Inhofe and Representative Sensenbrenner are from his speech to the Heartland Institute's Fourth International Conference on Climate Change, Chicago, May 2010, www.heartland.org/multimedia/videos/cory-bernardi-iccc4.

Page 134: 'Bernardi left convinced . . .' Details of the Republican campaign against the Waxman–Markey bill and the campaign by Americans for Prosperity are from Pooley, *The Climate War*, pp. 399–403. See also Marianne Lavelle, 'A case of lowered expectations in the US', International Consortium of Investigative Journalists (ICIJ), 19 March 2012, www.icij.org/investigations/global-climate-change-lobby/case-lowered-expectations; and Jeff Goodell, 'As the world burns: How Big Oil and Big Coal mounted one of the most aggressive lobbying campaigns in history to block progress on global warming', *Rolling Stone*, 7 January 2010, www.rollingstone.com/politics/politics-news/as-the-world-burns-2-199797.

Page 135: 'Greenpeace was now calling . . .' The Greenpeace 'kingpin' quote on the Koch brothers' empire is cited in Jane Mayer, 'Covert operations: The billionaire brothers who are waging a war against Obama', *New Yorker*, 23 August 2010, www.newyorker.com/magazine/2010/08/30/covert-operations.

Page 135: 'Andrews was a protégé . . .' Tim Andrews' background in the Liberal Party as a student, his relationship with Ray Evans and his training as a Koch Associate come from the author's interview with Andrews in Washington, 28 March 2019.

Page 136: 'Soon after Bernardi left Washington . . .' *The Australian*'s survey of Liberal backbenchers is cited in Manning, *Born to Rule*, pp. 336–7.

Page 136: 'Cory was the one who arranged it . . .' Andrews' comments on Bernardi's role in *The Australian* story is from the author's interview, 28 March 2019.

Page 136: 'If this is half true . . .' Julie Bishop's comment on the survey is cited in Kelly, *Triumph and Demise*, p. 246.

Page 137: 'I will not lead a party . . .' Turnbull's comment on leading the party on climate change was made on ABC Radio National, 1 October 2009.

Page 137: 'Indeed, that July, Abbott had written . . .' Abbott's opinion piece on the emissions trading scheme, 'Turnbull is right, the Coalition can't win this fight', was run in *The Australian*, 24 July 2009, www.news.com.au/news/turnbull-is-right-the-coalition-cant-win-this-fight/news-story/7ae46 3ec25a924d34d8c0c2056191683.

Page 137: 'What the hell is this, Tony?' Minchin's response to Abbott is from the author's interview with Minchin, 28 February 2019.

Page 138: 'I gave a speech then there were questions . . .' Abbott's account of the night in Beaufort is from the author's interview, 1 March 2019, and also from Tony Abbott, 'Afterword: A week can be a really long time in politics', *Battlelines*, updated edn, Melbourne: Melbourne University Press, 2009, p. 183.

Page 138: 'The editor of the local newspaper . . .' Abbott's quote in the *Pyrenees Advocate* was reported in Stuart Rintoul, 'Town of Beaufort changed Tony Abbott's view on climate change', *The Australian*, 12 December 2019, www.theAustralian.com.au/archive/politics/the-town-that-turned-up-the-temperature/news-story/6fe0d32a32e42341a12b999f6da82ec5.

Page 138: 'Abbott credited Minchin with turning him . . .' The influence of Nick Minchin on Abbott comes from the author's interview with Abbott, 1 March 2019.

Page 139: 'Hockey and his leader were forced . . .' The Hockey–Turnbull joint press conference on 8 October 2009 was reported in Sarah Ferguson, 'Malcolm and the malcontents', *Four Corners*, ABC TV, 9 November 2009, www.abc.net.au/4corners/malcolm-and-the-malcontents/10166562.

Page 139: 'I saw the opportunity to negotiate . . .' Bernardi's quotes on the negotiations between Macfarlane and Wong as an opportunity and on the mobilisation strategy come from his speech to the Heartland conference, May 2010.

Page 140: 'The coal lobby hired some of the best . . .' Details of the Australian Coal Association campaign are from Wilkinson, Cubby and Duxfield, 'Come in spinner'.

Page 140: 'If International Power's balance sheet . . .' Macfarlane's quote on International Power is from Ferguson, 'Malcolm and the malcontents', *Four Corners*, 9 November 2009.

Page 141: 'I believe Australia has drawn a line . . .' Quotes from Barnaby Joyce and Bob Carter's Roma meeting, from Ferguson 'Malcolm and the malcontents', *Four Corners*, 9 November 2009.

Page 141: 'When Abbott was driving home . . .' The account of Tony Abbott listening to Turnbull's interview with Alan Jones is from David Marr, 'Political animal: The making of Tony Abbott', *Quarterly Essay*, no. 47, Melbourne: Black Inc., 2012, p. 73.

Page 142: 'She revealed that she was simply . . .' Julia Gillard's recollection of her meeting with Penny Wong is from Julia Gillard, *My Story*, Sydney: Random House, 2014, p. 369.

Page 142: 'Why shit on someone . . .' Greg Combet's account of Rudd's staff tactics is from Combet, *The Fights of My Life*, p. 222.

Page 143: 'We had a core conversation . . .' Rudd's comments on the negotiations and Turnbull's future are from the author's interview, 5 April 2019.

Page 143: 'You've got to know when . . .' Rudd's Lowy speech, 6 November 2009, pmtranscripts.pmc.gov.au/release/transcript-16904.

Page 143: 'Three nights later, in a move . . .' Minchin, Abbott and Joyce's comments are from Ferguson, 'Malcolm and the malcontents', *Four Corners*, 9 November 2009.

Page 144: 'The Australian Conservation Foundation's chief . . .' Don Henry was quoted in an Australian Conservation Foundation media release, 'Hard slog ahead on climate with a weak CPRS', 25 November 2009.

Page 144: 'The Rudd government had agreed . . .' The full changes to the CPRS were summed up in a media release from the prime minister and others, 'A Carbon Pollution Reduction Scheme in the national interest', 24 November 2009, parlinfo.aph.gov.au/parlInfo/search/display/display. w3p;query=Id:"media/pressrel/8HAV6.

Page 145: 'It was a total sell-out.' Andrew Robb's role in the critique of the deal between Macfarlane and Wong and his quotes are detailed in his memoir, *Black Dog Daze*, pp. 175–7.

Page 146: 'If I was in that position . . .' Ian Macfarlane's response to Robb is from the author's interview, 17 May 2019.

Page 146: 'I got the most uncomplimentary . . .' Andrew Robb's comment is from *Black Dog Daze*, p. 177.

Page 147: 'Actually yes, I spoke to him . . .' Cory Bernardi's comment on supporting Abbott is from the author's interview, 24 October 2018.

Page 147: 'While all three senators were careful to say . . .' The reasons for the resignations of Cormann, Fifield and Mason were in their press release, 25 November 2009, parlinfo.aph.gov.au/parlInfo/search/display/display. w3p;query=Id:"media/pressrel/I7SV6.

Page 147: 'It was obvious in the party room . . .' Abbott's quote on Turnbull's suicide mission is from the author's interview, 1 March 2019.

Page 147: 'I had people from the partyroom meetings . . .' Tim Andrews' comments on his blog, leaks and lobbying MPs are from the author's interview, 28 March 2019.

Page 148: 'There was hundreds and hundreds . . .' Bernardi's outline of his strategy is from his speech to the Heartland conference, May 2010.

Page 148: 'We must retain our credibility . . .' The quotes from Turnbull's press conference are from the *PM* program, ABC Radio, 26 November 2009, www.abc.net.au/pm/content/2009/s2754757.htm.

Page 149: 'To the surprise of the climate sceptics . . .' Hockey's quote on not wanting to appear a sceptic are from the author's interview with Bernardi, 24 October 2018.

Page 149: 'For you to make your first act . . .' Minchin's response to Hockey's conscience vote proposal is from the author's interview, 28 February 2019.

Page 149: 'Joe couldn't make up his mind . . .' Abbott's comments on Hockey are from the author's interview, 1 March 2019.

Page 150: 'If that's what they want . . .' Macfarlane's conversation with Turnbull before the ballot is from the author's interview with Macfarlane, 17 May 2019.

Page 151: 'This revolt by the party rank and file . . .' Hugh Morgan's account of the sceptics' role in the fight against 'green despotism' and the Liberal Party rank and file revolt are from his president's report to the Lavoisier Group, 9 November 2010, www.lavoisier.com.au/articles/climate-policy/science-and-policy/morgan2010-presidents-report.php.

Page 151: 'Any policy that is announced . . .' Malcolm Turnbull's analysis of Abbott and the sceptics is from his blog, cited in Emma Rodgers, 'Turnbull savages Abbott over climate "bullshit"', *ABC News*, 7 December 2007, www.abc.net.au/news/2009-12-07/turnbull-savages-abbott-over-climate-bullshit/1171972.

Page 152: 'From this time on . . .' For Malcolm Turnbull's full account of his post-leadership depression, see his memoir, *A Bigger Picture*, Hardie Grant Books, Melbourne, 2020, pp. 169–72.

Page 152: 'If the Greens saw Conan the Barbarian . . .' Rudd's quote on the Greens and Abbott is from the author's interview, 5 April 2019.

Chapter 9: Cop-out

Page 153: 'The room was named after . . .' Kevin Rudd's account of conditions in the small meeting room are from the author's interview, 5 April 2019, and from his memoir, *The PM Years*, p. 222. The atmospherics and weather in Copenhagen are from the author's observations from covering the Copenhagen climate conference.

Page 153: 'More disconcerting for Rudd . . .' The absence of the Danish prime minister from the talks at a vital point and the quote about Rudd's decision to take over are from his memoir, *The PM Years*, p. 226.

Page 154: 'The Chinese delegates had . . .' The Chinese delegation refusing to include the reference to the 2050 emissions cuts is from Per Meilstrup, 'The runaway summit: The background story of the Danish Presidency of COP15, the UN Climate Change Conference', in Nanna Hvidt and Hans Mouritzen (eds), *Danish Foreign Policy Yearbook 2010*, Copenhagen: Danish Institute for International Studies, 2010, pp. 113–35, www.fao.org/fileadmin/user_upload/rome2007/docs/What really happen in COP15.pdf.

Page 154: 'But now the conference venue . . .' The conditions in the Bella Center are from the author's observations.

Page 154: 'I want to see Wen.' The quote from Barack Obama is from Meilstrup, 'The runaway summit', p. 132.

Page 155: 'Jaws dropped when they saw us . . .' The details of the meeting between Obama and Wen Jiabao and the quote from Hillary Clinton are from Hillary Rodham Clinton, *Hard Choices*, New York: Simon & Schuster, 2014, p. 416.

Page 155: 'That left him promising a weak . . .' The US target was reported in Darren Samuelsohn and Lisa Friedman, 'Obama announces 2020 emissions target, Dec. 9, Copenhagen visit', *New York Times*, 25 November 2009, archive.nytimes.com/www.nytimes.com/cwire/2009/11/25/25climate wire-obama-announces-2020-emissions-target-dec-9-22088.html?page wanted=3.

Page 155: 'China in turn did not want . . .' The Chinese negotiating position was reported in, 'Chinese climate negotiator provides candid take on what happened in Copenhagen', *Inside Climate News*, 29 August 2010, insideclimatenews.org/news/20100829/chinese-climate-negotiator-provides-candid-take-what-happened-copenhagen; David Adam and James Randerson, 'Secret Copenhagen recording reveals resistance from China and India', *The Guardian*, 8 May 2010, www.theguardian.com/environment/2010/may/07/secret-copenhagen-talks-climate-recording; and Marian Wilkinson, 'Turning up the heat in Copenhagen', *Sydney Morning Herald*, 12 December 2009, www.smh.com.au/environment/climate-change/turning-up-the-heat-in-copenhagen-20091211-kokr.html.

Page 155: 'That was resisted down to . . .' The quote from Todd Stern is from the author's interview, 4 April 2019.

Page 156: 'Even Clinton admitted it was . . .' Clinton's quote on the deal being far from perfect and the meeting with the European leaders is from Clinton, *Hard Choices*, pp. 422, 500.

Page 156: 'They wanted a legal treaty . . .' Clinton's quote is from *Hard Choices*, p. 500.

Page 156: 'For the first time in history all major . . .' The quotes from President Obama are from his Copenhagen press conference at the Bella Center, 10.30 pm, 18 December 2009, obamawhitehouse.archives.gov/the-press-office/remarks-president-during-press-availability-copenhagen.

Page 157: 'Obama cut the deal . . .' The quotes from Alden Meyer are from the author's interview, 30 March 2019.

Page 157: 'because I'm still in the room . . .' The quotes from Kevin Rudd are from the author's interview, 5 April 2019.

Page 157: 'This was really, really hard . . .' Kevin Rudd and Penny Wong's reactions to the final Copenhagen deal are from Marian Wilkinson, 'The Copenhagen Accord: A deal far from perfect', *Sydney Morning Herald*, 20 December 2009, www.smh.com.au/environment/climate-change/the-copenhagen-accord-a-deal-far-from-perfect-20091219-l6oi.html.

Page 157: 'This represents a significant global agreement . . .' Rudd's quotes on the accord are from his press conference at the Bella Center, 19 December 2009, pmtranscripts.pmc.gov.au/release/transcript-16986.

Page 158: 'I remember seeing him in the room . . .' The quote from Penny Wong about Rudd in Copenhagen is from the author's interview, 26 February 2019.

Page 158: Rudd's explanation of the 'ratfucked' comment is from the author's interview, 5 April 2019. The leak of the briefing was reported in David Marr, 'Power trip: The political journey of Kevin Rudd', *Quarterly Essay*, no. 38, Melbourne: Black Inc., 2010. According to Marr, Rudd said, 'Those Chinese fuckers are trying to rat-fuck us.' Rudd insisted he was describing the talks, not the Chinese and objected to the account.

Page 158: 'They handed out a leaked UN report . . .' The details of the UN report on the targets leaked to Greenpeace were reported in Marian Wilkinson, 'Blessed or blamed? A little of both', *Sydney Morning Herald*, 19 December 2009, www.smh.com.au/environment/climate-change/blessed-or-blamed-a-little-of-both-20091218-l5ol.html.

Page 159: 'After leaving the world leaders . . .' Details of the call Rudd took at 1 am are from the author's interviews with Rudd, 5 April 2019; Karl Bitar, 14 May 2019; Bruce Hawker, 9 October 2019; and Mark Arbib, 4 October 2019. There was genuine confusion about who was on the call, but Arbib and Bitar both confirm Arbib was not present. Arbib said, 'I was never on a call with Rudd during Copenhagen or after about the CPRS,' Bitar confirms Arbib was not in the room in Hurstville with him and Hawker.

Page 160: 'If Copenhagen is not going well . . .' The quote from Bitar about ditching the CPRS is from the author's interview with Bitar, 14 May 2019.

Page 160: 'A few days later, Rudd arrived . . .' The descriptions of the meeting at the PM's Sydney office are from Julia Gillard, *My Story*, Sydney: Random House, 2014, p. 372; Wayne Swan, *The Good Fight*, Sydney: Allen & Unwin, 2014, p. 186; and the author's interview with Karl Bitar, 14 May 2019.

Page 160: 'If you stick with it . . .' Bitar's quote is from the author's interview, 14 May 2019.

Page 160: 'My view was that if we were going . . .' Wayne Swan's quote is from the author's interview, 12 March 2019.

Page 161: 'Karl's report told us that carbon pricing . . .' Julia Gillard's comment is from *My Story*, pp. 372–3.

Page 161: 'I was pleading with him . . .' John Faulkner's quote is from the author's interview, 13 December 2018.

Page 161: 'In early January, Rudd and Gillard met again.' Rudd's and Gillard's descriptions of the Kirribilli meeting on 4 January 2010 come from their respective memoirs, *The PM Years*, p. 233; and *My Story*, p. 373.

Page 161: 'She has a definitive conversation . . .' Rudd's quotes on Gillard's view on the CPRS are from the author's interview, 5 April 2019.

Page 162: 'I was left with the clearest . . .' Gillard's quote is from *My Story*, p. 373.

Page 162: 'I said at the time . . .' Faulkner's quote is from Kelly, *Triumph and Demise*, p. 277.

Page 163: 'There is absolutely no question . . .' Swan's quote is from the author's interview, 12 March 2019.

Page 163: 'The more we moved away . . .' Faulkner's quote on climate science is from the author's interview, 13 December 2018.

Page 163: 'Now not everyone agrees . . .' Abbott's quote on Ian Plimer is from Ferguson, 'Malcolm and the malcontents', *Four Corners*, 9 November 2009.

Page 164: 'The hypothesis that human activity . . .' Plimer's quote on global warming is from James Delingpole, 'Meet the man who has exposed the great climate change con trick', *The Spectator*, 8 July 2009, www.spectator.co.uk/2009/07/meet-the-man-who-has-exposed-the-great-climate-change-con-trick.

Page 164: 'It is not "merely" atmospheric scientists . . .' Professor Michael Ashley's review appeared in *The Australian*, 9 May 2009, www.theAustralian.com.au/arts/books/ian-plimer-heaven-and-earth/news-story/211cf8e7a30d14ff69e97f5bb3ba5cf1.

Page 165: 'Within two weeks, the "Climategate" story . . .' The revelation that the Climategate emails were uploaded to a Russian server was reported by Patrick Courrielche, *Breitbart News* online, in Part II of a three-part series, 'Peer-to-peer review (Part II): How "Climategate" marks the maturing of a new science movement', 10 January 2010, www.breitbart.com/the-media/2010/01/10/peer-to-peer-review-part-ii-how-climategate-marks-the-maturing-of-a-new-science-movement. Part I of the series on 8 January describes the number of web hits. The spread of the story was described in Morgan Goodwin, 'Climategate: An autopsy', DeSmog Blog, 30 March 2010, www.desmogblog.com/climatgate-autopsy.

Page 165: 'As Mann put it . . .' For Mann's version of Climategate, see Michael E. Mann, *The Hockey Stick and the Climate Wars: Dispatches from the Front Lines*, New York: Columbia University Press, 2012.

Page 166: 'Tell me why Kevin Rudd . . .' Andrew Bolt's column 'Climategate's most damning emails' was published in the *Herald Sun*, 25 November 2009, www.dailytelegraph.com.au/blogs/andrew-bolt/column--climategates-most-damning-emails/news-story/7c44e73fa414ea180333344c38f133c3.

Page 166: 'A UK parliamentary inquiry . . .' The clearing of Phil Jones of scientific misconduct was reported in George Monbiot, 'The "climategate" inquiry at last vindicates Phil Jones—and so must I', *The Guardian*, 8 July 2010, www.theguardian.com/commentisfree/cif-green/2010/jul/07/russell-inquiry-i-was-wrong. Michael Mann's clearing was reported in Suzanne Goldenberg, '"Hockey stick" graph creator Michael Mann cleared of academic misconduct', *The Guardian*, 4 February 2010, www.theguardian.com/environment/2010/feb/03/climate-scientist-michael-mann.

Page 166: 'But it was a howler . . .' The IPCC mistake and the quotes from Will Steffen are from Marian Wilkinson, 'Crisis of climate-change confidence', *Sydney Morning Herald*, 13 February 2010, www.smh.com.au/environment/crisis-of-climatechange-confidence-20100212-nxmb.html.

Page 167: 'By 2010 a Lowy Institute poll . . .' For the change in the poll, see 'Media release: 2019 Lowy Institute poll—Australian attitudes to climate change', 8 May 2019, www.lowyinstitute.org/publications/media-release-2019-lowy-institute-poll-Australian-attitudes-climate-change.

Page 167: 'essentially a baseless scientific scam'. Lord Monckton's comment on climate change was reported in Daniel Flitton, 'Lord of climate change sceptics hot under collar', *The Age*, 11 November 2009, www.smh.com. au/environment/climate-change/lord-of-climate-change-sceptics-hot-under-collar-20091110-i7ks.html.

Page 167: 'When their tour hit Perth . . .' The tour for Plimer and Monckton in Perth was posted on the website of high-profile sceptic Jo Nova, jonova. s3.amazonaws.com/monckton/perth.pdf.

Page 168: 'With the support of Queensland's Labor government . . .' On Hancock Prospecting and its coal licences in the Galilee Basin, see joint statement, Premier and Minister for the Arts Anna Bligh and Treasurer and Minister for Employment and Economic Development Andrew Fraser, 'Premier drops in on new developments set to unlock Galilee Basin's potential', 1 November 2009, statements.qld.gov.au/Statement/Id/67206.

Page 168: 'If this, as the Prime Minister says . . .' The quote from Tony Abbott is from *Hansard*, House of Representatives, 2 February 2010.

Page 169: 'Instead, he promised a new . . .' The opposition's Direct Action plan is from Tony Abbott MP, media release, 'Direct Action on the environment and climate change', 2 February 2010, parlinfo.aph.gov.au/ parlInfo/search/display/display.w3p;query=Id:media/pressrel/GMSV6.

Page 169: 'Rudd's chief of staff, Alister Jordan . . .' Details of the call to Alister Jordan and the polling of focus groups on the CPRS are from the author's interview with Karl Bitar, 14 May 2019.

Page 170: 'Given how hard KR has supported . . .' The quote of the email from Bitar to Jordan is from the original supplied by Bitar. This exchange was also reported in Kelly, *Triumph and Demise*, p. 282.

Page 170: 'A few weeks later, the prime minister's staff . . .' The examination of options by the PM's staff and Gillard and Swan is detailed in Gillard, *My Story*, p. 376.

Page 170: 'Instead of just polling on the CPRS . . .' Penny Wong's response to Bitar's briefing is from the author's interview with Wong, 26 February 2019.

Page 170: 'We can't afford it politically.' Gillard's quote to Rudd is from Rudd, *The PM Years*, p. 258.

Page 171: 'the immediate conclusion of the public . . .' The speakerphone conversation with Penny Wong and the other ministers is from Rudd, *The PM Years*, p. 260.

Page 171: 'I was perfectly poleaxed.' Rudd's comment is from the author's interview, 5 April 2019.

Page 171: 'The *Sydney Morning Herald*'s . . .' Lenore Taylor's story of the leak was published as 'ETS off the agenda until late next term', *Sydney Morning Herald*, 27 April 2010, www.smh.com.au/politics/federal/ets-off-the-agenda-until-late-next-term-20100426-tnbc.html.

Page 171: 'You can't tell someone for years . . .' Penny Wong's reaction to the leak and the impact on Rudd is from the author's interview, 26 February 2019.

Page 171: 'It, to me, was the final nail ...' Martin Parkinson's reaction is from the author's interview, 16 August 2019.

Page 172: 'Rudd's approval rating plummeted ...' The Newspoll data was supplied by Karl Bitar.

Page 173: 'There will be no carbon tax ...' Gillard's quote is from her campaign interview with Deborah Knight and Bill Woods, *Ten News*, Channel Ten, 16 August 2010. Her comment on a carbon emissions trading scheme in the future was in her interview with Paul Kelly and Dennis Shanahan, 'Julia Gillard's carbon price promise', *The Australian*, 20 August 2010, www.theAustralian.com.au/national-affairs/julia-gillards-carbon-price-promise/news-story/48f865d2c27454ff3fc082815abaaedf.

Chapter 10: Battlelines

Page 175: 'That's what she knew was going ...' Christine Milne's comment on Julia Gillard and the price of power is from the author's interview, 14 October 2019.

Page 176: 'I always found her personable ...' Milne's comment on dealing with Gillard is from the author's interview, 14 October 2019.

Page 176: 'I needed a scientist there ...' Milne's comment on wanting Will Steffen on the multi-party committee is from the author's interview, 14 October 2019.

Page 177: 'After months of wrangling ...' On the fight over the 2020 emissions target and the decision to use the Climate Change Authority to deliver the target by 2014, see Milne, *An Activist Life*, pp. 195–6; and Combet, *The Fights of My Life*, pp. 249–50.

Page 178: 'Gillard war-gamed a strategy.' On announcing the deal with the Greens on an emissions trading scheme, see Gillard, *My Story*, p. 393.

Page 178: 'I'm determined to price carbon ...' On the Gillard press conference with Bob Brown, see Tim Leslie, 'Gillard unveils carbon price details', *ABC News*, 24 February 2011, www.abc.net.au/news/2011-02-24/gillard-unveils-carbon-price-details/1955968.

Page 178: 'Well it's pretty clear ...' Tony Abbott's quote on Bob Brown leading the government is from Barrie Cassidy, 'Setting a fixed price on carbon', *Insiders*, ABC, 2 March 2011, www.abc.net.au/insiders/setting-a-fixed-price-on-carbon/1959064.

Page 178: 'In parliament that afternoon ...' Tony Abbott's censure motion and the Lady Macbeth comments are from *Hansard*, House of Representatives, 24 February 2011.

Page 179: 'Combet and Behm argued ...' Greg Combet and Allan Behm's advice to Gillard is from the author's interview with Allan Behm, 23 October 2018.

Page 179: 'Gillard went into the ABC studio ...' Julia Gillard's comments to ABC TV's *7.30 Report* were on 24 February 2011, www.abc.net.au/7.30/gillard-explains-carbon-scheme/2670250.

Page 179: 'The most profound mistake we made . . .' Behm's comment is from the author's interview, 23 October 2018.

Page 180: 'If you stand back and look . . .' Wayne Swan's comment is from the author's interview, 12 March 2019.

Page 180: 'I don't believe it's going to happen . . .' Tony Abbott's quotes on the 'people's revolt' and fighting the carbon tax every inch are from 'Carbon tax will inspire people's revolt: Abbott', *ABC News*, 24 February 2011, www.abc.net.au/news/2011-02-24/carbon-tax-will-inspire-peoples-revolt-abbott/1956028.

Page 180: 'That summer the Millennium Drought . . .' The flood details are from 'Queensland Floods, Commission of Inquiry, Final Report, March 2012', http://www.floodcommission.qld.gov.au/publications/final-report/. The La Niña details are from www.bom.gov.au/climate/enso/lnlist/.

Page 181: 'That was a unique thing . . .' Tim Andrews' comments on the 'unique campaign', 'the death knell' and 'Koch people' are all from the author's interview, 28 March 2019.

Page 182: 'I'm not in favor of abolishing . . .' Norquist's bathtub quote was reported in Karen Rothmyer, 'The original vast right-wing conspirator: What the left could learn from the late Richard Mellon Scaife', *Politico Magazine*, 9 July 2014, www.politico.com/magazine/story/2014/07/richard-mellon-scaife-the-original-koch-brother-108729.

Page 182: 'This was very much a Grover Norquist-style campaign.' Tim Andrews' comment on Grover Norquist is from the author's interview, 28 March 2019, as is Andrews' work with Americans for Tax Reform.

Page 182: 'Norquist had a take-no-prisoners style . . .' Grover Norquist's use of the phrase 'political throat slitters' is from Mayer, 'Covert operations'.

Page 182: 'Throughout 2011, Andrews and Bernardi . . .' The document listing the groups united in the stopgillardstax campaign was obtained by ABC TV's *Four Corners* for 'The carbon war', reporter Marian Wilkinson with producer Deborah Masters, 19 September 2011, www.abc.net.au/4corners/interview-with-prime-minister-julia-gillard/2907738.

Page 183: 'Bernardi helped fund . . .' Bernardi's assistance in funding Andrews' blog Menzies House is from the author's interview with Cory Bernardi, 24 October 2018, as is the work of CANdo.

Page 183: 'They set up stopgillardscarbontax.com . . .' The links between the website campaign and the Big Lie were reported in 'The carbon war', *Four Corners*, 19 September 2011.

Page 183: 'The conference was part of a talkfest . . .' The co-sponsors of the Pacific Rim exchange are from its website, www.americansfortaxreformfoundation.org/userfiles/PAC_Rim_final_agenda2010.pdf. For Heartland's Sydney Conference, see its website, http://climateconferences.heartland.org/iccc5/.

Page 184: 'I was cannon fodder.' Cory Bernardi's comments are from the author's interview, 24 October 2018.

Page 185: 'In a rousing address to the Lavoisier faithful . . .' Hugh Morgan's comments on Tea Party politics and his warning to 'rent-seekers' is from his president's report to the Lavoisier Group, November 2010, www.lavoisier.com.au/articles/climate-policy/science-and-policy/morgan2010-presidents-report.php.

Page 185: 'Hugh Morgan signed up as an early member . . .' For the original members of ANDEV, see Adele Ferguson, *Gina Rinehart: The untold story of the richest woman in the world*, Sydney: Pan Macmillan, 2012, p. 333.

Page 186: 'There is no problem with global warming . . .' Ian Plimer's quote is from Delingpole, 'Meet the man who has exposed the great climate change con trick', 8 July 2009.

Page 186: 'Mrs Rinehart suggests . . .' Gina Rinehart's views on Plimer and climate sceptics were given in response to written questions from ABC's *Four Corners* program, 'The power of one', reporter Marian Wilkinson, producer Janine Cohen, 20 June 2012, see www.scribd.com/document/98218171/Gina-Rinehart-responds-to-Four-Corners-questions.

Page 186: 'The sooner we can, as a majority . . .' Rinehart's quote is from Gina Rinehart, 'Australian business leaders—Where are you?', *Australian Resources and Investment Magazine*, May 2011, posted by joannenova.com.au.

Chapter 11: The carbon war

Page 187: 'It was absolutely surreal.' The description of Gina Rinehart's arrival at Swan's office and the comments about the meeting are from the author's interview with Wayne Swan, 12 March 2019.

Page 188: 'The National Party's Senate leader . . .' For Barnaby Joyce and Julie Bishop's trip to the wedding with Gina Rinehart, see James Robertson and Jonathan Swan, 'Coalition MPs in "overseas study" claim for wedding', *Sydney Morning Herald*, 6 October 2013, www.smh.com.au/politics/federal/coalition-mps-in-overseas-study-claim-for-wedding-20131005-2v13w.html.

Page 188: 'Despite this, the Indian conglomerate . . .' The staggered payment to Hancock prospecting with $500 million up-front was reported in Tony Grant-Taylor, 'Dusty Alpha becomes billionaires' row', *Courier-Mail*, 17 October 2011, www.couriermail.com.au/business/dusty-alpha-becomes-billionaires-row/news-story/77a038765c226f67feafedf1565a34e7?sv=828192ca37671dc83e1e7768d03e8cdf; and Barry FitzGerald, 'Riches flow for Rinehart with $1.2b Galilee sale', *Sydney Morning Herald*, 12 September 2011, www.smh.com.au/business/riches-flow-for-rinehart-with-1-2b-galilee-sale-20110918-1kg23.html.

Page 189: 'Australia will need to look beyond . . .' The quotes from Marius Kloppers are from his speech to the Australian British Chamber of Commerce at the Hilton Hotel, Sydney, 15 September 2010, reported widely in the media.

Page 189: 'He came out much to everybody's surprise . . .' Richard McIndoe's comments are from the author's interview, 18 October 2019.

Page 190: 'For at least a decade, BHP . . .' BHP's Paul Anderson's attempts to get a carbon price were discussed with the author for 'Delayed reaction', *Sydney Morning Herald*, 24 March 2007. Anderson was responsible for the position statement on climate change released by BHP on 2 August 2000, cited in *The Heat is On: Australia's Greenhouse Future*, Canberra: Senate Environment, Communications, Information Technology and the Arts References Committee, 2000, www.aph.gov.au/parliamentary_business/committees/senate/environment_and_communications/completed_inquiries/1999-02/gobalwarm/report/c09.

Page 190: 'While BHP Billiton had big coalmines . . .' BHP Billiton's gas buys were reported in Michael J. de la Merced, 'BHP Billiton to buy Petrohawk for $12.1 billion', *New York Times*, 14 July 2011, dealbook.nytimes.com/2011/07/14/bhp-billiton-to-buy-petrohawk-for-12-1-billion.

Page 190: 'The Golden Age of Gas had arrived.' 'Are we entering the Golden Age of Gas?', flagging the gas boom was part of the International Energy Agency's annual *World Energy Outlook* report in 2011.

Page 190: 'We took the view that . . .' The comment from Richard McIndoe is from the author's interview, 18 October 2019.

Page 191: 'It was going to happen . . .' Greg Combet's comments on the mood for change and on Gérard Mestrallet are from the author's interview, 14 January 2019.

Page 191: 'They were all in favour of a carbon price . . .' Swan's comment on the carbon price is from the author's interview, 12 March 2019.

Page 191: 'I used to comment to my staff . . .' Combet's comment on rent-seeking is from the author's interview, 14 January 2019.

Page 191: 'In the middle of 2011, a brawl . . .' The description of the meeting with Gillard, Combet, Milne, Pierce and other officials is from the author's interview with Christine Milne and is also mentioned in her book *An Activist Life*, p. 199.

Page 192: 'After hours of wrangling . . .' Milne's account of the conversation with Gillard is from the author's interview, as is the discussion with Garnaut and the request to Pierce for the advice in writing, 14 October 2019.

Page 192: 'He also recommended paying . . .' The comment from John Pierce on the necessity of the compensation is from the letter he supplied to Christine Milne and Julia Gillard, which was given to the author.

Page 193: 'In the end the Greens leader . . .' The debate over the carbon price of $23 is covered in detail in Milne, *An Activist Life*, p. 197; and Combet, *The Fights of My Life*, p. 258. Both Labor and the Greens take credit for the Clean Energy Finance Corporation.

Page 194: '. . . the most sweeping reform . . .' The details of the Clean Energy Package were set out on 11 July and are available in a report by Frank Jotzo, 'Australia's clean energy future', *Environmental Finance*, December 2011–January 2012, pp. 14–15.

Page 194: 'The fact that the politics . . .' The quote from Combet on his condition at the end of the process is from *The Fights of My Life*, p. 258.

Page 194: 'We will see ordinary people . . .' The quotes from the Coal Association's Ralph Hillman, and from Tony Abbott and Andrew Bolt, are from the ABC wrap, 'Carbon Sunday: Who said what', 11 July 2011, www.abc. net.au/news/2011-07-11/carbon-sunday-who-said-what/2789628.

Page 195: 'Tony Abbott jumped up onto the makeshift stage . . .' The details of the anti-carbon tax rally, including the Abbott quote are from 'The carbon war', *Four Corners*, 19 September 2011.

Page 195: 'The woman's off her tree . . .' The quote from Alan Jones was reported in 'The carbon war', *Four Corners*, 19 September 2011.

Page 196: 'They are stealing your money . . .' The quote from Malcolm Roberts comes from 'The carbon war', *Four Corners*, 19 September 2011.

Page 196: '. . . a long list of climate-sceptic scientists . . .' The list of the Galileo Movement's 'independent advisers' is given on its website, www. galileomovement.com.au/who_we_are.php, as is Alan Jones's patronage of the organisation. Alan Jones's patron position was also confirmed by former advisers to the Galileo Movement.

Page 197: 'We were constantly being pilloried . . .' The comments from Tim Flannery on the media experience and the Parramatta meeting are from the author's interview, 19 April 2019.

Page 197: 'The brunt of the attacks . . .' The report of Flannery's remarks on Perth were retrieved by Andrew Bolt in his column on 17 February 2017, 'Flannery washed out again', *Herald Sun,* www.heraldsun.com. au/blogs/andrew-bolt/flannery-washed-out-in-perth/news-story/dce43bec 13c5610492502d78c226c50e. They were, however, originally made in 2004—see 'Sydney's future eaten: The Flannery prophecy', *Sydney Morning Herald*, 19 May 2004, www.smh.com.au/environment/sydneys-future-eaten-the-flannery-prophecy-20040519-gdiyd2.html.

Page 197: 'After one climate presentation . . .' Professor Will Steffen's comments are from the author's interview, 5 November 2018.

Page 198: 'Prominent climate-sceptic scientists . . .' The abusive letters to Bernardi were reported by Sally Neighbour in 'All about Cory'.

Page 198: 'Every time I think it's reached a low . . .' The comment from Professor Ian Chubb is from 'The carbon war', *Four Corners*, 19 September 2011.

Page 198: 'Sydney's *Daily Telegraph* managed . . .' The *Telegraph* headline on the 'death tax' is referenced by Combet in *The Fights of My Life*, p. 262.

Page 198: 'The coal lobby warned . . .' The figure of 'potentially' 25 mines closing is from the former CEO of the Australian Coal Association, Dr Nikki Williams, interviewed by the author in 'The carbon war', *Four Corners*, 19 September 2011.

Page 198: 'the longest suicide note . . .' Abbott's comment was reported in 'The carbon war', *Four Corners*, 19 September 2011.

Page 198: 'We govern in a world of change . . .' Gillard's statement is from *Hansard*, House of Representatives, 13 September 2011.

Page 199: 'It was elation really . . .' Milne's comment is from the author's interview, 14 October 2019.

Page 200: 'In the 1990s we were warned . . .' George Pell's lecture to the Global Warming Policy Foundation, 'One Christian perspective on climate change', was on 26 October 2011, www.thegwpf.org/images/stories/gwpf-reports/pell-2011_annual_gwpf_lecture_new.pdf.

Page 200: 'I think Professor Plimer has done. . .' John Howard's quote at the launch of Ian Plimer's book is from Mitchell Nadin, 'New book takes climate debate back to school', *The Australian*, 13 December 2011, www.theAustralian.com.au/national-affairs/climate/new-book-takes-climate-debate-back-to-school/news-story/b3902c87e9d63f8de0a28e364d9b0da0.

Page 200: 'Rinehart had just been awarded . . .' Gina Rinehart's award is noted in the *IPA Annual Report 2013*, p. 6, ipa.org.au/wp-content/uploads/2016/11/2013-Annual-Report.pdf. Details of the IPA's 70th anniversary dinner, including the high-profile guests and Pell's attendance, were reported in Ben Butler, 'Protests greet Murdoch', *The Age*, 4 April 2013, www.theage.com.au/national/victoria/protests-greet-murdoch-20130404-2h9sw.html.

Page 200: 'The Coalition will indeed repeal . . .' Abbott's speech is on the IPA YouTube channel, www.youtube.com/watch?v=j4pA5nTr8i0.

Page 201: 'A few weeks later a report . . .' The reduction in emissions from electricity generation was reported in *How Australia's Carbon Price is Working One Year On*, released by the Gillard government, June 2013, archived at webarchive.nla.gov.au/awa/20130904221045/http://www.cleanenergy future.gov.au/wp-content/uploads/2013/08/carbon-price-one-year-on.pdf.

Page 201: 'Thanks to both state and federal subsidies . . .' Figures on the solar panel increase are from Mark Butler, *Climate Wars*, Melbourne: Melbourne University Press, 2017, p. 73.

Page 201: 'Abbott was supreme.' Greg Combet's quote is from the author's interview, 14 January 2019.

Page 201: 'Industry had just got into . . .' Mark Butler's quotes are from the author's interview, 22 October 2018.

Page 202: 'The EU's carbon price had temporarily . . .' The collapse of the carbon price is from the author's interview with Greg Combet, 14 January 2019.

Page 202: 'The government of Australia has changed.' Tony Abbott's victory speech was run in the *Sydney Morning Herald*, 7 September 2013, www.smh.com.au/politics/federal/tony-abbotts-victory-speech-20130907-2tcxc.html.

Page 202: '. . . Rinehart, Australia's richest woman, celebrated . . .' Gina Rinehart's celebration with Barnaby Joyce and his then wife, Natalie, was reported by *ABC News*, 7 September 2013, www.abc.net.au/news/2013-09-07/4943706?nw=0.

Chapter 12: The state of the Reef II

Page 203: 'That was the *wow* moment.' The quotes from Jon Day are from the author's interview, 9 November 2019.

Page 204: 'Not only had the Gillard government . . .' The failure to inform the World Heritage Committee of the Curtis Island plans is detailed in the document 'Decisions adopted by the World Heritage Committee at its 35th Session', UNESCO, Paris, 2011, p. 55, whc.unesco.org/archive/2011/whc11-35com-20e.pdf. The plans had been flagged to the committee but not the approval.

Page 204: 'The World Heritage Committee formally noted . . .' The 'extreme concern' quote is from 'Decisions adopted by the World Heritage Committee', p. 55.

Page 204: '"Extreme concern" is very tough language . . .' The comment from Jon Day is from the author's interview, 9 November 2019.

Page 205: 'If you look at everything individually . . .' The comment from Tim Badman was made in 'Great Barrier Grief', *Four Corners*, ABC TV, reporter Marian Wilkinson, producer Clay Hichens, 3 November 2011, www.abc.net.au/4corners/great-barrier-grief/3652070.

Page 205: 'But Garrett "hit the pause button" . . .' Peter Garrett's comments on the Curtis Island project are from his memoir, *Big Blue Sky*, p. 397.

Page 205: 'The Gladstone Ports Corporation . . .' On the scale of the Gladstone Harbour and Curtis Island dredging project and the construction of the retaining wall, see 'Great Barrier Grief', *Four Corners*, 3 November 2011. See also 'Western Basin dredging and disposal project: Coordinator-general's report for an environmental impact statement', Queensland government, Brisbane, July 2010, wwwstatedevelopment.qld.gov.au/resources/project/port-of-gladstone-western-basin-dredging-disposal-project/wbdp-cg-report-22-july-10.pdf.

Page 206: 'We raised extreme concerns . . .' For Russell Reichelt's comment, see 'Great Barrier Grief', *Four Corners*, 3 November 2011.

Page 206: 'Ferguson had already touted . . .' Martin Ferguson's interview on the British Gas contract for Curtis Island was reported by ABC Radio's *PM* program, 24 March 2010, www.abc.net.au/pm/content/2010/s2855159.htm.

Page 206: 'His replacement, Tony Burke . . .' The details of Tony Burke's approval of the project are from 'Great Barrier Grief', *Four Corners*, 3 November 2011.

Page 206: 'Swan and Ferguson joined Cath Tanna . . .' For the press conference, see James McCullough, 'British Gas approves LNG plant at Curtis Island in Gladstone', *Courier-Mail*, 31 October 2010, www.couriermail.com.au/business/british-gas-approves-lng-plant-at-curtis-island-in-gladstone/news-story/f681dbc1d078837dbdac47159d54171e.

Page 207: 'We made sure that each . . .' For Tony Burke's comments on the approval conditions for the project, see 'Great Barrier Grief', *Four Corners*, 3 November 2011.

Page 207: 'It's going to mean 6000 jobs.' For Julia Gillard's comment on Curtis Island, 'Great Barrier Grief', *Four Corners*, 3 November 2011.

Page 207: 'Bligh was even more enthusiastic . . .' For Anna Bligh's comment on the Curtis Island project, see Anna Bligh, Premier and Minister for

Reconstruction, 'Premier heralds new "gas age" for Queensland', media statement, 27 May 2011, statements.qld.gov.au/Statement/Id/74946.

Page 207: 'Switching from coal to gas . . .' On the emissions reductions from switching to gas from coal, see Heinz Schandl, Tim Baynes, Nawshad Haque et al., *Whole of Life Greenhouse Gas Emissions Assessment of a Coal Seam Gas to Liquefied Natural Gas Project in the Surat Basin, Queensland, Australia: Final Report for GISERA Project G2*, Canberra: CSIRO Energy, 2019, gisera.csiro.au/wp-content/uploads/2019/07/ GISERA_G2_Final_Report-whole-of-life-GHG-assessment.pdf.

Page 208: 'Also, importantly, producing LNG . . .' On the emissions created by the LNG process, see the report by Origin and ConocoPhillips, *The Australia Pacific LNG Project*, vol. 4, *LNG Facility*, Chapter 14, 'Greenhouse Gases', p. 22, www.aplng.com.au/content/dam/aplng/compliance/ eis/Volume_4/Vol_4_Chapter14_GreenhouseGas.pdf.

Page 208: 'Just months after dredging started . . .' On the fish disease problem in Gladstone Harbour in 2011 and the response of the Gladstone Ports Corporation and the Queensland government, see 'Great Barrier Grief', *Four Corners*, 3 November 2011.

Page 209: 'They say it's naturally occurring . . .' For Mark McMillan's quote, see 'Great Barrier Grief', *Four Corners*, 3 November 2011.

Page 209: 'The result of the mission was a damning report . . .' Fanny Douvere and Tim Badman, 'Reactive monitoring mission to Great Barrier Reef (Australia) 6th to 14th March 2012', UNESCO, Paris, 2012, whc.unesco. org/en/documents/117104.

Page 209: 'The committee concluded . . .' On the findings of the World Heritage Committee meeting in St Petersburg, see 'Decisions adopted by the World Heritage Committee at its thirty-sixth session (Saint-Petersburg, 2012)', 24 June–6 July 2012, pp. 57–8, whc.unesco.org/archive/2012/ whc12-36com-19e.pdf.

Page 210: 'The Great Barrier Reef has been . . .' For Greenpeace comment and Tony Burke's response to the World Heritage Committee warning it would consider listing the Reef as in danger, see 'UN report scathing of Barrier Reef plan', *ABC News*, 5 June 2012, www.abc.net.au/news/2012-06-02/ un-report-scathing-of-barrier-reef-plan/4048498. Burke's comment, 'you may as well give up', was cited in, 'Cooking the climate, wrecking the Reef: The global impact of coal exports from Australia's Galilee Basin', Greenpeace Pacific, September 2012, secured-static.greenpeace.org/Australia/ Global/Australia/images/2012/Climate/Galilee Report(4.2MB).pdf.

Page 210: 'We are in the coal business.' For Campbell Newman's response to the World Heritage Committee, see 'UN report scathing of Barrier Reef plan'.

Page 210: 'The previous year, local and global investors . . .' For the investments in energy projects in Australia in 2011, see Robert New, Allison Ball, Alan Copeland et al., *Minerals and Energy: Major Development Projects—April 2011 Listing*, Canberra: ABARES, 2011, daff.ent.sir sidynix.net.au/client/en_AU/search/asset/1027512/0.

Page 211: 'These are some of the biggest mines . . .' Michael Roche's comments come from 'Battle for the Reef', *Four Corners*, ABC TV, reporter Marian Wilkinson, producer Ali Russell, 18 August 2014, www. abc.net.au/4corners/battle-for-the-reef/5680580.

Page 212: 'The Galilee projects, if they got up . . .' Bob Burton's quote is from the author's interview, 11 November 2019.

Page 212: 'The anti-coal strategy called for . . .' The details of the anti-coal strategy are from John Hepburn, Bob Burton and Sam Hardy, 'Stopping the Australian coal export boom', November 2011, archived at www.abc. net.au/mediawatch/transcripts/1206_greenpeace.pdf. Graeme Wood later said he had no knowledge of the report. It was funded by a grant from the US Rockefeller Family Fund.

Page 212: 'Change the story of coal . . .' The quote is from, 'Stopping the Australian coal export boom'.

Page 212: '. . . the strategy document leaked . . .' The *Australian Financial Review* story on the anti-coal strategy is by Matthew Stevens, Gemma Daley and Marcus Priest, 'Revealed: Coal under green attack', 6 March 2012, see www.abc.net.au/mediawatch/transcripts/1206_afr5.pdf.

Page 212: 'I think many people confuse . . .' The reactions to the anti-coal strategy from Wayne Swan, Craig Emerson, Martin Ferguson, Rio Tinto and BHP Billiton are from 'Revealed: Coal under green attack'.

Page 213: 'destroying one country's coal industry . . .' Peter Freyberg's comment is from Adam Morton and David Wroe, 'Greenpeace leads war against coal', *Sydney Morning Herald*, 7 March 2012, www.smh.com.au/ national/greenpeace-leads-war-against-coal-20120306-1uic5.html.

Page 213: 'The reaction reflected their attitude . . .' Bob Burton's comment is from the author's interview, 11 November 2019.

Page 213: 'If our coal is not supplied . . .' Michael Roche's comment is from 'Battle for the Reef', *Four Corners*, 18 August 2014.

Page 214: 'The committee repeated its request . . .' The recommendation of the World Heritage Committee's meeting is from, 'Decisions adopted by the World Heritage Committee at its 37th Session (Phnom Penh, 2013)', pp. 63–4, whc.unesco.org/archive/2013/whc13-37com-20-en.pdf.

Page 214: 'This time it was the coal port . . .' On the plan to expand Abbot Point port, see 'Battle for the Reef', *Four Corners*, 18 August 2014.

Page 215: 'Sure, they may cost more . . .' Jon Day's quote is from 'Battle for the Reef', *Four Corners*, 18 August 2014.

Page 215: 'The likely impact of the dredging . . .' The advice from Adam Smith was contained in an email released from the Great Barrier Reef Marine Park Authority under FOI and cited in 'Battle for the Reef', *Four Corners*, 18 August 2014.

Page 215: 'The former Gold Coast real estate broker . . .' Clive Palmer's Waratah Coal lease details are from, 'Palmer drama', *Four Corners*, ABC TV, reporter Marian Wilkinson, producer Mary Ann Jolley, 25 November 2013, www.abc.net.au/4corners/palmer-drama/5117552.

Page 215: 'A lifetime member of the Queensland LNP . . .' For the background on Clive Palmer's relationship with the Queensland National Party, see Jamie Walker, 'Clive Palmer: Having it all', *The Australian*, 18 August 2012, www.theAustralian.com.au/news/inquirer/clive-palmer-having-it-all/news-story/4b2f55be2981cd1d3e1689879a941a37.

Page 216: 'He's the most successful political leader . . .' The quote from Palmer on Campbell Newman is from, 'Palmer drama', *Four Corners*, 25 November 2013.

Page 216: 'I don't think it was anything against us . . .' The quote from Palmer on GVK is from 'Palmer drama', *Four Corners*, 25 November 2013.

Page 217: 'God bless the backbench . . .' The quote from Palmer on Campbell Newman is from *Lateline*, ABC TV, 12 September 2012, www.abc.net.au/lateline/clive-palmer-attacks-increase-in-royalties/4258354.

Page 217: 'I think a lot of the politics of Australia . . .' The quote from Bob Katter is from 'Palmer drama', *Four Corners*, 25 November 2013.

Page 217: 'He comes across as a sort of . . .' The quote from Barnaby Joyce on Palmer is from 'Palmer drama', *Four Corners*, 25 November 2013.

Page 218: 'By his own calculation, Labor's carbon scheme . . .' The estimate of the carbon price on Queensland Nickel is from the author's interview with Clive Palmer, 4 November 2019.

Page 218: 'Perhaps not surprisingly, the PUP campaign . . .' Details of the campaign funding of the Palmer United Party are from the company's 'Donor to political party disclosure return', 2013–14, AEC, transparency.aec.gov.au/AnnualDonor/ReturnDetail?SubmissionId=55&ClientId=29314. Queensland Nickel Pty Ltd lists donations of $15,216,476 to the Palmer United Party over 2013 and 2014. Some of this was spent on the re-run of the WA Senate race. After first being declared a winner for PUP in Western Australia, Dio Wang lost the Senate seat on a recount. PUP contested the recount and the High Court ordered a new half-Senate election in Western Australia in April 2014. Dio Wang won a Senate seat for PUP with an increased vote.

Page 219: 'Climate change remains the most serious . . .' The Great Barrier Reef Marine Park Authority's draft *Great Barrier Reef Region Strategic Assessment: Strategic assessment report* (Townsville: GBRMPA) was published in August 2013. For the quotes on the risk from climate change and emissions, see p. 11–11, for the risks to biodiversity on the Reef region from warming waters, see p. 10–10.

Chapter 13: The purge

Page 220: 'It was a short and courteous conversation . . .' Flannery's account of the sacking is from the author's interview, 19 April 2019.

Page 220: 'I think we were the first definitive action . . .' Steffen's account of the sacking is from the author's interview, 5 November 2018.

Page 220: 'The website that we'd spent . . .' Flannery's comments on the website are from the author's interview, 19 April 2019.

Page 221: 'help build the consensus required . . .' The brief of the Climate Change Commission is from the media release by Greg Combet, 'Launch of the Climate Commission', 10 February 2011, archived at webarchive. nla.gov.au/awa/20110226204027/http://www.climatechange.gov.au/ minister/greg-combet/2011/media-releases/February/mr20110210.aspx.

Page 221: 'I was almost proud to be sacked . . .' Beale's comment on his sacking was made in comments to the author, 20 February 2019.

Page 221: 'It raised almost $1 million . . .' See 'Climate Council donations "nearing $1 million", Business Spectator, 27 September 2013, www.theAustralian.com.au/business/business-spectator/news-story/ ff1b4fdbb8f099ff2fbf97cf0822511c.

Page 221: 'Hunt argued shutting down . . .' Greg Hunt's comments on the commission's abolition and on duplication of advice were reported in 'Tim Flannery defends Climate Commission after Government scrapping', *ABC News*, 20 September 2013, www.abc.net.au/news/2013-09-19/ federal-government-scraps-climate-commission/4968816.

Page 222: 'I think that the proposition . . .' Tony Abbott's comment on climate change is from the author's interview, 1 March 2019.

Page 222: 'He struck an agreement with Abbott . . .' On Hunt's agreement not to publicly challenge the science of climate change, see Niki Savva, *The Road to Ruin*, Melbourne: Scribe, 2016, p. 135.

Page 222: 'The Saturday after Abbott's election . . .' This account of Martin Parkinson's removal as Treasury secretary, along with Parkinson's quotes, is from the author's interview, 16 August 2019.

Page 224: 'The prime minister's only explanation . . .' Abbott's comments on Parkinson's removal are from Gareth Hutchens and James Massola, 'Criticism mounts over Abbott removal of Treasury mandarin', *Sydney Morning Herald*, 14 March 2014, www.smh.com.au/politics/federal/ criticism-mounts-over-abbott-removal-of-treasury-mandarin-20140313- 34pm9.html.

Page 224: 'Watt also had to remove . . .' On the Comley and Russell sackings, see Paddy Gourley, 'Night of the short knives: Sacking the nose to spite the face', *Sydney Morning Herald*, 5 November 2013, www.smh.com. au/public-service/night-of-the-short-knives-sacking-the-nose-to-spite-the- face-20131103-2wugk.html.

Page 224: 'People were sent hell . . .' The quote from Allan Behm is from the author's interview, 23 October 2018.

Page 225: 'It is my clear intention . . .' Tony Abbott's letter to Dr Ian Watt on drafting the bills was dated 5 August 2013 and was released by Abbott on the same day.

Page 225: 'That same day, Abbott had sent . . .' Tony Abbott's letter to Jillian Broadbent was dated 5 August 2013 and released by Abbott on the same day.

Page 225: Greg Hunt and Andrew Robb's letter to the Clean Energy Finance Board, dated 3 February 2013, after then prime minister, Kevin Rudd, named the election day as 14 September 2013. The letter was made available to the author.

Page 225: 'I said I don't know exactly . . .' Oliver Yates's response to the shadow minister is from the author's interview, 28 February 2019.

Page 225: 'Not Hunt, Hockey or Abbott . . .' Legal advice on the status of the Clean Energy Finance Corporation (CEFC)was given to the Australian Conservation Foundation (ACF) by Stephen Keim SC on 19 September 2013 and released by the ACF.

Page 226: 'We were meeting with our . . .' The comments from Jillian Broadbent on the meeting with Joe Hockey and Mathias Cormann over the CEFC are from the author's interview, 21 May 2019, as is the account of the contact from Treasury officials.

Page 227: 'The numbers were always . . .' Oliver Yates's comment on the Senate numbers are from the author's interview, 28 February 2019.

Page 227: 'Morgan compared them . . .' Hugh Morgan's comments on 'Chicken Little' were quoted in Andrew Burrell, 'IPCC this century's "Chicken Little"', *The Australian*, 4 November 2013, www.theAustralian. com.au/national-affairs/climate/ipcc-this-centurys-chicken-little/news-story/0ac46fd549790176bae8c208aee9df3e.

Page 227: 'Not long before this . . .' The $300,000 donation from the Cormack Foundation to the IPA on 18 September 2013 was listed in Hugh Morgan's witness statement in Federal Court Case VID1270/2017 *Alston v Cormack Foundation*.

Page 227: 'The think tank's latest publication . . .' For the authors of *Climate Change: The Facts 2014*, see ipa.org.au/publications-ipa/books/climate-change-facts-2014.

Page 227: 'The former Liberal prime minister . . .' John Howard's address, 'One religion is enough', was the 2013 Annual GWPF Lecture, delivered on 5 November 2013, www.thegwpf.org/content/uploads/2013/12/Howard-2013-Annual-GWPF-Lecture.pdf.

Page 228: 'A prominent climate change denier . . .' For Maurice Newman's opposition to wind farms, see Tim Elliott, 'Maurice Newman, the million-dollar smiler', *Sydney Morning Herald*, 5 July 2014, www.smh.com.au/national/maurice-newman-the-milliondollar-smiler-20140630-3b2qs. html.

Page 229: 'Newman was furious that Australia . . .' Maurice Newman's quotes on climate change madness and the climate crusade are from Graham Lloyd, 'Climate policies helped kill manufacturing says Maurice Newman', *The Australian*, 31 December 2013, www.theAustralian.com. au/national-affairs/climate/climate-policies-helped-kill-manufacturing-says-maurice-newman/news-story/9e9eab698a0ffb979c079b80c13e 50a6.

Page 229: 'I am not a denier of climate change . . .' Dick Warburton's quote is from Sid Maher, 'RET reviewer Dick Warburton: I'm not a climate sceptic', *The Australian*, 18 February 2014, www.theAustralian.com.au/national-affairs/

climate/ret-reviewer-dick-warburton-im-not-a-climate-sceptic/news-story/5
b02d71333655e924085a89fc07376f0.

Page 229: 'I would've liked to have . . .' Tony Abbott's comment on the
Renewable Energy Target (RET) is from the author's interview, 1 March
2019.

Page 230: 'It absolutely froze it . . .' Oliver Yates's comment on the investment
collapse is from the author's interview, 28 February 2019.

Page 230: '"Good," he said later . . .' Tony Abbott's comment on the damage
from the RET and loss of jobs is from the author's interview, 1 March
2019.

Page 230: 'Put simply, the Renewable Energy Target . . .' The details about
the RET are from the *Renewable Energy Target Review* by the Climate
Change Authority, Canberra, 2014, climatechangeauthority.gov.au/sites/
prod.climatechangeauthority.gov.au/files/files/reviews/ret/2014/review.
pdf. For how the RET works, see p. 16; for details on solar rooftop panels,
see p. 18; for large-scale sources of renewables, including wind farms, see
p. 18; for costs to the household bills in 2014, see p. 27.

Page 231: 'I drive to Canberra to go to parliament . . .' For Joe Hockey's
comments to Alan Jones, see Peter Hannan, 'Joe Hockey warns clean
energy and "utterly offensive" windfarms are in his budget crosshairs',
Sydney Morning Herald, 2 May 2014, www.smh.com.au/politics/federal/
joe-hockey-warns-clean-energy-and-utterly-offensive-windfarms-are-in-
his-budget-crosshairs-20140502-zr3cc.html.

Page 231: 'The new MP for Hume . . .' For Angus Taylor's opposition to
wind farms for the ACT located in his electorate, see Fleta Page, 'Put
windfarms on Red Hill, not in NSW, say politicians', *Canberra Times*,
1 April 2014, www.canberratimes.com.au/story/6143736/put-windfarms-
on-red-hill-not-in-nsw-say-politicians.

Page 231: 'Not surprisingly, when the Warburton review . . .' For the
analysis of the review recommendations and the cost of the RET, see
Grant Anderson, 'The Warburton report', Allens Linklaters, Melbourne,
8 September 2014, www.allens.com.au/insights-news/insights/2014/09/
the-warburton-report.

Page 232: 'It was a jobs issue.' The comment from Jay Weatherill is from the
author's interview, 27 February 2019.

Page 232: 'After a protracted fight . . .' For the compromise deal with Labor
on the RET, see Butler, *Climate Wars*, pp. 37–9.

Page 232: 'It's a well-kept secret . . .' Maurice Newman's column on the UN's
'new world order', see 'The UN is using climate change as a tool not an issue',
The Australian, 8 May 2013, www.theAustralian.com.au/commentary/
opinion/the-un-is-using-climate-change-as-a-tool-not-an-issue/news-story
/1e6586191def1fb6d970ba052a4faf08.

Page 233: 'It's a sensible package . . .' The quote from Greg Hunt is from
his doorstop press conference with Ian Macfarlane in Melbourne, 8 May
2015, on the RET.

Page 233: 'In February 2014, the authority . . .' The Climate Change Authority's report on the 2020 target is *'Reducing Australia's Greenhouse Gas Emissions—Targets and Progress Review, Final Report'*, Canberra, 2014, www.climatechangeauthority.gov.au/files/files/Target-Progress-Review/Targets and Progress Review Final Report.pdf. This report was followed up with more detailed statements by the authority on the 2030 targets (see Chapter 14). The CCA's proposed 2020 target was a 15 per cent cut plus 4 per cent from the 'carryover credits' Australia had in surplus from overachieving its first Kyoto target, making a total target of a 19 per cent cut in emissions for 2020.

Page 234: 'He wasn't interested in what . . .' The comments from Bernie Fraser on Greg Hunt are from the author's interview, 22 November 2019.

Page 234: 'So in March 2014, Fraser gave . . .' For Bernie Fraser's Press Club speech on 13 March 2014, see www.climatechangeauthority.gov.au/files/files/Target-Progress-Review/Transcript - FRASER, BERNIE THURS 13 MARCH 2014.pdf.

Page 235: 'I was trying to make a case . . .' Bernie Fraser's quote on not scaring people is from the author's interview, 22 November 2019.

Page 236: 'The Climate Change Authority in 2014 . . .' Oliver Yates's comments on the authority's report and not defeating science are from the author's interview, 28 February 2019.

Page 236: 'No cuts to education.' Abbott's quote on no cuts is from *SBS News*, 6 September 2013, cited in 'House of cards', *Four Corners*, ABC TV, reporter Marian Wilkinson, producer Karen Michelmore, 16 March 2015, www.abc.net.au/4corners/house-of-cards---promo/6313942.

Page 237: 'Gore was coming to Australia . . .' The quote from Don Henry is from the author's interview, 18 April 2019.

Page 238: 'We came to the agreement, then had dinner . . .' Clive Palmer's quote on Al Gore is from the author's interview, 4 November 2019.

Page 238: 'His credibility in Australia . . .' Christine Milne's quote on Al Gore is from her memoir, *An Activist Life*, p. 266.

Page 238: 'As the Climate Change Authority later reported . . .' See Climate Change Authority, *Renewable Energy Target Review*, p. 28.

Page 239: 'So we effectively got two years . . .' Richard McIndoe's comment is from the author's interview, 18 October 2019.

Page 239: 'His Emissions Reduction Fund . . .' For the critique of the fund, see Tim Baxter, 'The emissions reduction fund was deeply flawed—and no rebranding will change this', *The Guardian*, 25 February 2019, www.theguardian.com/commentisfree/2019/feb/25/the-emissions-reduction-fund-was-deeply-flawed-and-no-rebranding-will-change-this.

Page 240: 'It was a lifeline from Palmer . . .' Bernie Fraser's quote on Palmer's vote for the Emissions Reduction Fund is from the author's interview, 22 November 2019.

Page 240: 'With the help of Greg Hunt . . .' Abbott's quote on Hunt's policy is from the author's interview, 1 March 2019.

Page 240: 'In November 2014, the IPCC ...' The IPCC finding on human-induced greenhouse gas emissions being the highest in history is from 'Headline statements from the summary for policymakers', *Climate Change 2014: Synthesis report*, 5th Assessment Report, 2014, www.ipcc.ch/site/assets/uploads/2018/02/ar5_syr_headlines_en.pdf.

Page 240: 'Australia's climate, they reported ...' The climate projections are from the CSIRO and Bureau of Meteorology's *State of the Climate Report 2014*, Canberra: Commonwealth of Australia, 2014, www.bom.gov.au/state-of-the-climate/2014.

Page 241: '... these demands came from high-profile climate sceptics ...' For the attacks on the Bureau of Meteorology, see Graham Lloyd, 'Bureau of Meteorology "altering climate figures"', *The Australian*, 23 August 2014, www.theAustralian.com.au/nation/climate/bureau-of-meteorology-altering-climate-figures/news-story/5bccf49433ae80f23b332d3493437106. On the push to investigate the bureau, see Jake Sturmer, 'Tony Abbott's department discussed investigation into Bureau of Meteorology over global warming exaggeration claims, FOI documents reveal', *ABC News*, 24 September 2015, www.abc.net.au/news/2015-09-24/government-discussed-bom-investigation-over-climate-change/6799628.

Page 241: 'Newman savaged the bureau ...' Maurice Newman's original column was replaced on the website and an updated column, 'Climate is right for a probe into the Bureau of Meteorology', appeared in *The Australian* on 24 June 2015, www.theAustralian.com.au/commentary/opinion/climate-is-right-for-a-probe-into-the-bureau-of-meteorology/news-story/84b7eeeb1c087a38ac3b9de624d1eb68.

Chapter 14: The target

Page 242: 'The sumptuous state banquet ...' The account of the banquet and meeting between President Obama and President Xi are drawn from Emily Rauhala, 'APEC closes with a "historic" climate deal between the U.S. and China', Associated Press, 12 November 2014, see time.com/3577820/apec-climate-change-barack-obama-xi-jinping-greenhouse-gas; and Mark Landler, 'U.S. and China reach climate accord after months of talks', *New York Times*, 11 November 2014, www.nytimes.com/2014/11/12/world/asia/china-us-xi-obama-apec.html.

Page 242: 'By making this announcement ...' President Obama's quote is from 'Remarks by President Obama and President Xi Jinping in joint press conference', White House press release, 12 November 2014, obamawhitehouse.archives.gov/the-press-office/2014/11/12/remarks-president-obama-and-president-xi-jinping-joint-press-conference.

Page 243: 'The US had agreed to cut ...' The details of the emissions goals by the US and China are from 'U.S.–China joint announcement on climate change', White House press release, 12 November 2014. obamawhitehouse.archives.gov/the-press-office/2014/11/11/us-china-joint-announcement-climate-change.

Page 243: 'I think it was a pivotal event.' The quote from Todd Stern is from the author's interview, 4 April 2019.

Page 244: 'He said, "I need someone . . ."' The quote from John Podesta is from the author's interview, 26 March 2019. .

Page 244: 'I think they had to be . . .' Podesta's quote on the surprise announcement is from the author's interview, 26 March 2019.

Page 244: 'We needed to make that announcement.' Podesta's quote on the one–two punch, from the author's interview, 26 March 2019.

Page 245: 'I put aside diplomatic niceties.' Podesta's comments on Abbott and his government are from the author's interview, 26 March 2019.

Page 245: 'For him it was "a Bob Brown bank . . ."' Tony Abbott's description of the UN Climate Fund is from Peter Hartcher, 'Abbott sniffs the wind on climate change', *Sydney Morning Herald*, 13 December 2014, www.smh.com.au/opinion/abbott-sniffs-the-wind-on-climate-change-20141212-12660c.html.

Page 245: 'The wrangling between Washington . . .' For an account of the dispute between Australia and the US over the G20 and climate change, see Peter Hartcher, 'G20 Brisbane Summit: Australia's adolescence on show', *The Interpreter*, 8 December 2014, www.lowyinstitute.org/the-interpreter/g20-brisbane-summit-Australias-adolescence-show.

Page 245: 'Obama insisted that climate . . .' The comments from Ben Rhodes were given to the author, 9 May 2019.

Page 245: 'Abbott gave a low-key speech . . .' For Abbott's opening speech to the G20, see 'G20: Prime minister's remarks at G20 leaders' retreat about domestic issues "weird and graceless"', *ABC News*, 16 November 2014, www.abc.net.au/news/2014-11-15/shorten-slams-abbott-remarks-at-g20-leaders-retreat/5894142.

Page 245: 'Obviously I would like this . . .' Abbott's quote on the G20 agenda is from Lenore Taylor and Daniel Hurst, 'Tony Abbott "whingeing" about domestic agenda on world stage', *The Guardian*, 15 November 2014, www.theguardian.com/world/2014/nov/15/g20-tony-abbott-whingeing-about-domestic-agenda-on-world-stage.

Page 246: 'It allows us to help . . .' The quotes from Obama's speech are from 'Remarks by President Obama at the University of Queensland', White House press release, 15 November 2014, obamawhitehouse.archives.gov/the-press-office/2014/11/15/remarks-president-obama-university-queensland.

Page 246: 'The incredible natural glory . . .' Obama's quote on the Great Barrier Reef is from 'Remarks by President Obama at the University of Queensland'.

Page 246: 'The speech was a public stake . . .' John Podesta's quote is from the author's interview, 26 March 2019.

Page 246: 'A very deliberate smash . . .' The quote on payback is from a confidential interview with a senior government figure.

Page 247: 'What was interesting is that . . .' Ben Rhodes's comments were given to the author, 9 May 2019.

Page 247: 'Tony asked that I make . . .' Julie Bishop's comments on Abbott's reaction were given to the author, 11 December 2019.

Page 247: 'In the end, despite all this aggravation . . .' The quotes on the leaders' decision are from 'G20 leaders' communique Brisbane Summit', 15–16 November 2014, www.oecd.org/g20/summits/brisbane/brisbane_g20_leaders_summit_communique1.pdf.

Page 247: 'All of us support strong . . .' Abbott's comments on climate change and on coal are from his closing press conference at the G20, 17 November 2014, https://pmtranscripts.pmc.gov.au/release/transcript-23969.

Page 248: 'There was regular consultation . . .' Bishop's comments on the decision to contribute to the UN Green Climate Fund were given to the author, 11 December 2019.

Page 248: 'Earlier that year, Peabody's . . .' On Peabody Energy's views on the Abbott government's support of the coal industry, see Amanda Saunders, 'Coal always wins and will stay No. 1, says carbon king Boyce', *Australian Financial Review*, 12 August 2014, www.afr.com/companies/mining/coal-always-wins-and-will-stay-no-1-says-carbon-king-boyce-20140812-jekle.

Page 248: 'The company's big message . . .' The details of the campaign come from Peabody Energy's own publication, 'Let's brighten the many faces of global energy poverty', 2014, www.coalblog.org/wp-content/uploads/2014/09/Lets-Brighten-the-Many-Faces-of-Global-Energy-Poverty_0.pdf. See also Kate Sheppard, 'World's biggest coal company, world's biggest PR firm pair up to promote coal for poor people', Huffpost, 28 March 2014, www.huffingtonpost.com.au/2014/03/27/peabody-burson-marstellar-coal_n_5044962.html?ri18n=true; and Graham Readfearn, 'How big coal is lobbying G20 leaders and trying to capture the global poverty debate', *The Guardian*, 14 October 2014, www.theguardian.com/environment/planet-oz/2014/oct/14/how-big-coal-is-lobbying-g20-leaders-and-trying-to-capture-the-global-poverty-debate.

Page 249: 'Despite this, by the time . . .' For a critique of Peabody's 500,000 social media supporters in its energy poverty campaign, see Greg McNevin, 'Has Peabody been fibbing about its G20 coal supporters?', *Business Spectator*, 25 November 2014, www.theAustralian.com.au/business/business-spectator/news-story/has-peabody-been-fibbing-about-its-g20-coal-supporters/8513a47b593f11baa6a2579402cd142c.

Page 249: 'Peabody was also under . . .' As an indicator of Peabody's financial difficulties, the US parent would file for bankruptcy protection in April 2016, see Sue Lannin, 'Coal miner Peabody Energy files for bankruptcy protection; Australian operations to continue as usual', *ABC News*, 13 April 2016, www.abc.net.au/news/2016-04-13/peabody-energy-files-for-bankruptcy-protection/7324534.

Page 249: 'In the past, the company had . . .' For Peabody's promotion of climate scepticism, see former CEO Greg Boyce's testimony to the Select Committee for Energy Independence and Global Warming, US

House of Representatives, 14 April 2010, www.markey.senate.gov/
GlobalWarming/files/HRG/041410futureCoal/Boyce.pdf. See also Suzanne
Goldenberg and Helena Bengtsson, 'Biggest US coal company funded
dozens of groups questioning climate change', *The Guardian*, 13 June 2016,
www.theguardian.com/environment/2016/jun/13/peabody-energy-coal-
mining-climate-change-denial-funding.

Page 249: 'The campaign to end energy poverty . . .' For Peabody's lobbying
on energy poverty back in 2011, see Frank Clemente on behalf of Peabody
Energy, 'Australia's coal in context', 7 January 2011.

Page 250: 'The council's new chief executive . . .' On Brendan Pearson's
appointment as CEO of the Minerals Council of Australia and his background with Peabody, see Vicky Validakis, 'Minerals Council of Australia
appoints new CEO', *Australian Mining*, 3 December 2013, www.Australian
mining.com.au/news/minerals-council-of-australia-appoints-new-ceo.

Page 250: 'In May 2014 the prime minister . . .' Abbott's quotes on coal
from his Address to the Minerals Week 2014 Annual Minerals Industry
Parliamentary Dinner, Canberra, 28 May 2014, pmtranscripts.pmc.gov.
au/release/transcript-23528.

Page 250: 'Abbott's mantra became . . .' See 'Coal "good for humanity",
Prime Minister Tony Abbott says at $3.9b Queensland mine opening',
ABC News, 13 October 2014, www.abc.net.au/news/2014-10-13/
coal-is-good-for-humanity-pm-tony-abbott-says/5810244.

Page 251: 'The environment minister, Greg Hunt . . .' For Hunt's approval of
the Adani mine and his quotes on energy poverty, see 'Battle for the Reef',
Four Corners, 18 August 2014.

Page 251: 'Hunt had gone ahead and approved . . .' For the details of the
Abbot Point dredging and Hunt's comments on the approval, 'Battle for
the Reef', *Four Corners*, 18 August 2014.

Page 252: 'The committee's critical view . . .' For the World Heritage Committee's findings at the Qatar meeting, see 'Decisions adopted by the World
Heritage Committee at its 38th session (Doha, 2014)', 15–25 June 2014,
whc.unesco.org/archive/2014/whc14-38com-16en.pdf.

Page 252: 'It just didn't go well . . .' Richard Leck's quote is from the author's
interview, 5 November 2019.

Page 252: 'If you've got an endangered listing . . .' The implications for the
coal industry of an 'in Danger' listing on the Reef comes from a confidential interview. For a history of the crisis over the Reef governance,
see Tiffany H. Morrison (James Cook University), 'Evolving polycentric
governance of the Great Barrier Reef', *PNAS*, 8 March 2017, vol. 114,
no. 15, pp. E3013–21, doi:10.1073/pnas.1620830114.

Page 253: 'The biggest developer they were . . .' The quote from Professor
Morrison is from the author's interview, 13 February 2019.

Page 253: 'If all goes to plan . . .' Abbott's quote on the largest ever mine is
from *Hansard*, House of Representatives, 18 November 2014.

Page 253: 'He cheered on Australia . . .' Prime Minister Modi's quote on the
glaciers is from *Hansard*, House of Representatives, 18 November 2014.

Page 254: 'You talk about a figurative . . .' Peter Costello's quote on the barbecue-stopper is from, 'Dethroning Tony Abbott', reporters John Lyons and Marian Wilkinson, *Four Corners*, ABC TV, 21 September 2015, www.abc.net.au/4corners/dethroning-tony-abbott-promo/6786660.

Page 255: 'We have decided that we are not . . .' Abbott's quote on dysfunctional government is from 'House of cards, *Four Corners*, 16 March 2015.

Page 255: 'They will stick . . .' Amanda Vanstone's quote is from 'House of cards, *Four Corners*, 16 March 2015.

Page 255: 'I lost the prime ministership . . .' Abbott's quote on why he lost the prime ministership is from the author's interview, 1 March 2019.

Page 256: 'The authority urged . . .' On the Climate Change Authority's advice and the 2025 and 2030 targets, see 'Final report on Australia's future emissions reduction targets', Climate Change Authority, Canberra, 2 July 2015, climatechangeauthority.gov.au/sites/prod.climatechangeauthority.gov.au/files/Final-report-Australias-future-emissions-reduction-targets.pdf.

Page 257: 'He, along with Bishop . . .' On keeping the Reef off the endangered list, see 'Decisions adopted by the World Heritage Committee at its 39th session (Bonn, 2015)', 28 June–8 July 2015, whc.unesco.org/archive/2015/whc15-39com-19-en.pdf.

Page 257: 'In the lead-up to the committee's annual meeting . . .' For Hunt's announcement, see Mark Ludlow, 'Greg Hunt bans dumping in Great Barrier Reef marine park', *Australian Financial Review*, 16 March 2015, www.afr.com/politics/greg-hunt-bans-dumping-in-great-barrier-reef-marine-park-20150316-1m02z.

Page 257: 'It looked like a copy . . .' For the comparison of Australia's 2030 targets with those of the US and the EU, see Bernie Fraser, 'Some observations on Australia's post-2020 emissions reduction target', Climate Change Authority, Canberra, 14 August 2015, climatechangeauthority.gov.au/sites/prod.climatechangeauthority.gov.au/files/files/CFI/CCA-statement-on-Australias-2030-target.pdf. See also 'Paris Climate Summit: Catalyst for further action?' Climate Institute, November 2015, www.climate institute.org.au/articles/publications/paris-climate-summit-brief.html.

Page 258: 'The advice I had with Greg Hunt . . .' The quotes on Hunt from Abbott are from the author's interview, 1 March 2019.

Page 259: 'Environment groups scorned these . . .' On the 'hot air loophole' and other countries cancelling their Kyoto credits, see Lenore Taylor, 'Australia isolated as developed nations cancel carryover credits from Kyoto', *The Guardian*, 5 December 2015, www.theguardian.com/Australia-news/2015/dec/05/Australia-climate-talks-developed-nations-cancel-carryover-emissions-reduction-credits-kyoto.

Page 259: 'Abbott unveiled the government's . . .' For the final Paris target, see 'Australia's intended nationally determined contribution to a new climate change agreement', August 2015, www4.unfccc.int/sites/ndcstaging/PublishedDocuments/Australia First/Australias Intended Nationally Determined Contribution to a new Climate Change Agreement – August 2015.pdf.

Page 259: 'We are committed to tackling . . .' The quote on power prices is from the prime minister, minister for foreign affairs and minister for the environment, 'Australia's 2030 Emissions Reduction Target', media release 11 August 2015, parlinfo.aph.gov.au/parlInfo/search/display/display.w3p;query=Id%3A"media%2Fpressrel%2F4008134";src1=sm1.

Page 259: 'Conveniently, the day before . . .' The report on the cost of Labor's target is from Simon Benson, 'ALP's $600b carbon bill', *Daily Telegraph*, 10 August 2015.

Page 260: 'He issued a statement . . .' Fraser's criticisms of the modelling interpretation used by the *Daily Telegraph* are in 'Some observations on Australia's post-2020 emissions reduction target', 14 August 2015. Hunt's defence of the modelling was in a media release posted by his office, 'Australia's post-2020 targets' on 14 August 2015, www.greghunt.com.au/Australias-post-2020-targets.

Page 260: 'In a phone call with Hunt . . .' Fraser's comment during the call witnessed by his daughter is from the author's interview, 22 November 2019, as is his description of relations with the government.

Page 260: 'The Climate Institute warned . . .' The Climate Institute's quote is from 'Paris Climate Summit: Catalyst for further action?', 20 November 2015, www.climateinstitute.org.au/articles/publications/paris-climate-summit-brief.html.

Page 261: 'I always regarded this as hocus pocus . . .' Abbott's comment about 1.5 versus 2 degrees Celsius is from the author's interview, 1 March 2019.

Page 261: 'The polls meant the bush . . .' Costello's quote on the polls is from 'Dethroning Tony Abbott', *Four Corners*, 21 September 2015.

Page 262: 'Any policy that is announced . . .' For Turnbull's earlier quotes on the Coalition policy, see Chapter 8.

Page 262: 'It was a very, very good . . .' Turnbull's quote on the policy when he returned as leader is from his press conference, 14 September 2015, www.malcolmturnbull.com.au/media/transcript-vote-on-the-liberal-party-leadership.

Page 262: 'There will be no wrecking . . .' Abbott's quote was reported in Lisa Cox, '"There will be no wrecking, no undermining, and no sniping": Tony Abbott's final statement as prime minister', *Sydney Morning Herald*, 15 September 2015, www.smh.com.au/politics/federal/there-will-be-no-wrecking-no-undermining-and-no-sniping-tony-abbotts-final-statement-as-prime-minister-20150915-gjmxzv.html.

Chapter 15: Hostage

Page 263: 'We are all together.' For Malcolm Turnbull's comments outside the Bataclan Theatre, see Tom Arup, 'Paris UN climate conference 2015: Turnbull pays respect to terror victims', *Sydney Morning Herald*, 30 November 2015, www.smh.com.au/environment/climate-change/paris-un-climate-conference-2015-turnbull-pays-respect-to-terror-victims-20151130-glb18d.html.

Page 263: 'We are not daunted . . .' For Turnbull's speech to the Paris summit on 1 December 2015, see '2015 United Nations Climate Change Conference speech' on his own website, www.malcolmturnbull.com.au/media/2015-united-nations-climate-change-conference-speech.

Page 264: 'That summer, said Obama . . .' For Barack Obama's speech to the Paris summit, see 'Remarks by President Obama at the First Session of COP21', 30 November 2015, obamawhitehouse.archives.gov/the-press-office/2015/11/30/remarks-president-obama-first-session-cop21.

Page 264: 'At the last minute, Foreign Minister Julie Bishop . . .' On Australia's support for the High Ambition Coalition, see Lenore Taylor, 'Australia belatedly joins "coalition of ambition" at Paris climate talks', *The Guardian*, 12 December 2015, www.theguardian.com/environment/2015/dec/12/Australia-belatedly-joins-coalition-of-ambition-at-paris-climate-talks.

Page 264: 'In reality, the pledges in Paris . . .' On the Paris commitment pledges, see 'Paris Climate Summit: Catalyst for further action?' Policy Brief, Carbon Institute, November 2015, www.climateinstitute.org.au/articles/publications/paris-climate-summit-brief.html.

Page 264: 'It is a historic agreement . . .' Erwin Jackson's comments appeared in 'Paris Agreement—it is done', Climate Institute's Paris blog, 13 December 2015, www.climateinstitute.org.au/news/paris-climate-summit-daily-updates.html.

Page 265: 'Unless "clean coal" could work . . .' For the phase-out times for coal-fired power, see Marcia Rocha, Bill Hare, Paota Yanguas Parra et al., *Implications of the Paris Agreement for Coal Use in the Power Sector*, Berlin: Climate Analytics, November 2015, climateanalytics.org/media/climateanalytics-coalreport_nov2016_1.pdf.

Page 265: 'The time for piecemeal . . .' For John Connor's quote, see 'Paris climate agreement: Now it's time for Australia's real work to start', Climate Institute media statement, 13 December 2015, www.climateinstitute.org.au/articles/media-releases/paris-agreement.html.

Page 266: 'As part of that deal . . .' The National Party agreement with Turnbull for 'no carbon tax' was confirmed by former leader Barnaby Joyce, 18 December 2019.

Page 266: 'It sounds about right . . .' Turnbull's quote on the climate sceptics in the Coalition during his government is from the author's interview, 15 March 2019.

Page 266: 'Former western Sydney Liberal MP . . .' Craig Laundy's breakdown of the climate sceptics in the Coalition is from the author's interview, 22 October 2018.

Page 267: 'There is a particular constituency . . .' Turnbull's comments on Maurice Newman and Hugh Morgan's climate scepticism is from the author's interview, 15 March 2019.

Page 268: 'Despite their differences, in March 2016 . . .' The lunch at the Athenaeum Club was reported in Pamela Williams, 'Federal election 2016: How Labor's Mediscare plot was hatched', *The Australian*, 4 July 2016,

www.theAustralian.com.au/nation/health/federal-election-2016-inside-story-of-how-labors-mediscare-plot-was-hatched/news-story/51ee0223
bfe1f0922cd6d15f447e692f.

Page 268: 'I spoke to Hugh and Charles.' Turnbull's comment on the Cormack Foundation's donation is from the author's interview, 15 March 2019.

Page 268: 'In the end the Cormack Foundation . . .' For the donations from the Cormack Foundation to the federal Liberal Party in 2016, see Hugh Morgan's Witness Statement in *Alston v Cormack Foundation and others*, [2018] FCA 895. This also lists the donations to Bob Day and David Leyonhjelm. The donations to the latter helped spark the legal dispute between the Victorian Liberal Party with its then president, Michael Kroger, and the Cormack Foundation.

Page 268: 'With the Cormack Foundation keeping . . .' For Turnbull's personal donations to the Liberal Party in 2016, see the Liberal Party's 2016–17 'Party political disclosure form', AEC, transparency.aec.gov.au/Download/ReturnImageByMoniker?moniker=64-XJTK3.

Page 269: 'Malcolm made it absolutely crystal clear . . .' Abbott's account of the meeting with Turnbull is from the author's interview, 1 March 2019.

Page 269: 'It was very obvious to me . . .' Turnbull's quote on Abbott and cabinet is from the author's interview, 15 March 2019.

Page 269: 'Shortly before 4 pm . . .' Details of the start of the South Australian blackout are from a personal account written by Jay Weatherill and made available to the author.

Page 270: 'That afternoon an extraordinary burst . . .' The details on the weather event that hit South Australia are from Gary Burns, Leanne Adams and Guy Buckley, 'Independent review of the extreme weather event, South Australia, 28 September to 5 October 2016: Report presented to the premier of South Australia', www.dpc.sa.gov.au/__data/assets/pdf_file/0003/15195/Independent-Review-of-Extreme-Weather-complete.pdf; and 'Severe thunderstorm and tornado outbreak South Australia 28 September, 2016' by the Bureau of Meteorology, www.dpc.sa.gov.au/__data/assets/pdf_file/0007/15199/Attachment-3-BoM-Severe-Thunderstorm-and-Tornado-Outbreak-28-September-2016.pdf.

Page 270: 'Three major transmission lines . . .' How the blackout happened is detailed in 'Black system South Australia, 28 September 2016—final report', Australian Energy Market Operator (AEMO), March 2017, p. 6, www.aemo.com.au/-/media/Files/Electricity/NEM/Market_Notices_and_Events/Power_System_Incident_Reports/2017/Integrated-Final-Report-SA-Black-System-28-September-2016.pdf.

Page 270: 'With chaos engulfing his state . . .' Weatherill's account of the emergency cabinet meeting and the quote on public safety and stranded commuters are from his written account of the crisis given to the author.

Page 271: 'Forty per cent of South Australia's . . .' For Chris Uhlmann's original story and the controversy it generated, see Giles Parkinson, 'Is ABC's Chris Uhlmann the new face of the anti-wind lobby?, reneweconomy.com.

au/is-abcs-chris-uhlmann-the-new-face-of-the-anti-wind-lobby-90343. Uhlmann would defend his analysis but also refine it in another ABC story, 'SA storms: 'Rushing to Renewable Energy Targets puts sector's reputation at risk', *ABC News*, 30 September 2016, www.abc.net.au/news/2016-09-29/ rushing-to-renewables-risks-sectors-reputation:-uhlmann/7888290.

Page 271: 'Not long after, independent . . .' For Nick Xenophon's comments, see Rachel Baxendale, 'SA blackout: Heads will roll, Nick Xenophon says', *Weekend Australian*, 28 September 2016, www.theAustralian. com.au/nation/nation/sa-blackout-heads-will-roll-nick-xenophon-says/ news-story/1a67198f0c3c7901f288301723440e6b.

Page 271: 'It was Frydenberg.' Weatherill's comment on Frydenberg, Turnbull et al. is from the author's interview, 27 February 2019.

Page 271: 'Turnbull had put Frydenberg.' For the critical comments on South Australia from Frydenberg and Turnbull, see Gareth Hutchins, 'Malcolm Turnbull says South Australia blackout a wake-up call on renewables', *The Guardian*, 29 September 2016, www.theguardian.com/ Australia-news/2016/sep/29/jay-weatherill-accuses-barnaby-joyce-of-pushing-anti-windfarm-agenda-over-blackouts.

Page 272: 'Not one offer of assistance . . .' Weatherill's quote on Turnbull is from the author's interview, 27 February 2019.

Page 273: 'At that point, as one . . .' The quote from a senior government figure on owning the electricity issue is from a confidential interview with the author.

Page 273: 'Trump promised he would pull the US . . .' For Donald Trump's first promise to pull out of Paris, see Benjy Sarlin, 'Donald Trump pledges to rip up Paris climate agreement in energy speech', *NBC News*, 26 May 2016, www.msnbc.com/msnbc/donald-trump-pledges-rip-paris-climate-agreement-energy-speech.

Page 273: 'The adviser asked Ebell . . .' Myron Ebell's account of the phone call from the Trump adviser is from the author's interview, 25 March 2019.

Page 274: 'Within weeks of Trump's victory . . .' On the Scott Pruitt appointment, see Oliver Milman, 'Donald Trump picks climate change sceptic Scott Pruitt to lead EPA', *The Guardian*, 8 December 2016, www. theguardian.com/us-news/2016/dec/07/trump-scott-pruitt-environmental-protection-agency.

Page 275: 'Shortly before the federal parliament opened . . .' For the weather events of the 2016–17 summer, see Andrew Stock, Will Steffen, David Alexander and Martin Rice, 'Angry summer 2016/17: Climate change super-charging extreme weather', Climate Council, www.climatecouncil. org.au/resources/angry-summer-report.

Page 275: 'But he did it by launching . . .' For the quotes from Turnbull's Press Club speech on 1 February 2017, see www.malcolmturnbull.com. au/media/address-at-the-national-press-club-and-qa-canberra.

Page 275: 'Turnbull insisted he wanted . . .' Turnbull's comments on being 'technology agnostic' are from the author's interview, 15 March 2019.

Page 276: 'When firming power was needed . . .' For a simple explanation of how pumped hydro works, see Roger Dargaville, 'Five gifs that explain how pumped hydro actually works', The Conversation, 7 March 2019, https://theconversation.com/five-gifs-that-explain-how-pumped-hydro-actually-works-112610.

Page 276: 'nobody was politically more at risk . . .' Turnbull's comment about energy policy as a toxic time bomb that put him at political risk is from his memoir, *A Bigger Picture*, p. 597.

Page 277: 'That evening the state suffered . . .' For the numbers affected by the second South Australian blackout, see Giles Parkinson, 'Frydenberg on blackouts: No mention of failing network, gas, software', *Renew Economy*, 21 February 2017, reneweconomy.com.au/frydenberg-on-blackouts-no-mention-of-failing-network-gas-software-19688.

Page 277: 'The outage was caused . . .' On the causes of the 8 February blackout, see Australian Energy Regulator (AER), 'Electricity spot prices above $5000/MWh: South Australia, 8 February 2017', 27 April 2017, www.aer.gov.au/wholesale-markets/market-performance/prices-above-5000-mwh-8-february-2017-sa. For details of the AER lawsuit, see 'Pelican Point in court for alleged breaches of National Electricity Rules', AER, 27 August 2019, www.aer.gov.au/news-release/pelican-point-in-court-for-alleged-breaches-of-national-electricity-rules.

Page 277: 'But before the causes of the blackout . . .' The comments from Turnbull, Frydenberg and Morrison are from *Hansard*, House of Representatives, 9 February 2017.

Page 278: 'That was a crazy thing to do.' The comment from Turnbull on Morrison holding up the lump of coal is from the author's interview, 15 March 2019.

Page 278: 'The day after the bizarre theatrics . . .' For the power problems in New South Wales in February, see AEMO, 'System event report New South Wales—10 February 2017', 22 February 2017, www.aemo.com.au/-/media/Files/Electricity/NEM/Market_Notices_and_Events/Power_System_Incident_Reports/2017/Incident-report-NSW-10-February-2017.pdf.

Page 279: 'This was New South Wales . . .' Weatherill's quote on the NSW power crisis is from his written account of the crisis given to the author.

Page 279: 'By then the right had . . .' Weatherill's comments on Turnbull and the Liberal Party are from the author's interview, 27 February 2019.

Page 279: 'Not to be outdone, just days later . . .' For Turnbull's comments on Snowy 2.0, see 'Securing Australia's energy future with Snowy Mountains 2.0', 15 March 2017, on his own website, www.malcolmturnbull.com.au/media/securing-Australias-energy-future-with-snowy-mountains-2.0.

Page 280: 'Scathingly, Morrison dismissed South Australia's . . .' For Scott Morrison's comments on the Tesla battery, see Phillip Coorey, 'Scott Morrison mocks SA's big battery as like the "big banana"', *Australian Financial Review*, 27 July 2017, www.afr.com/politics/scott-morrison-mocks-sas-big-battery-as-like-the-big-banana-20170727-gxjqbz.

Page 280: 'We are at a critical turning point.' The comment on the power market is from Alan Finkel, chair, *Independent Review into the Future Security of the National Electricity Market: Blueprint for the future*, Canberra: Commonwealth of Australia, 2017, p. 5 www.energy.gov.au/sites/default/files/independent-review-future-nem-blueprint-for-the-future-2017.pdf.

Page 281: 'Modelling for the Finkel review . . .' For details of the modelling on renewable energy in the Finkel review, see *Independent Review into the Future Security of the National Electricity Market*, p. 13.

Page 281: 'There were a lot of speakers.' Abbott's comments on Finkel are from the author's interview, 1 March 2019.

Page 282: 'The company would later describe . . .' Glencore's descriptions of Project Caesar were in a statement to *The Guardian*'s Christopher Knaus, see 'Revealed: Glencore bankrolled covert campaign to prop up coal', 7 March 2019, www.theguardian.com/business/2019/mar/07/revealed-glencore-bank rolled-covert-campaign-to-prop-up-coal. Glencore's description is from their statement given to *The Guardian* and confirmed to the author by Glencore. Caesar was dropped by Glencore in February 2019.

Page 282: 'As Sir Lynton Crosby reportedly . . .' Lynton Crosby's comment on Crosby Textor's clients were reported in Nick Tabakoff, 'Media diary', *The Australian*, 24 November 2019, www.theAustralian.com.au/business/media/boris-johnsons-aussie-strategy-boss-sir-lynton-crosby-in-flying-sydney-visit/news-story/5769d141bf9d87184d04134a0aef0a32.

Page 283: 'But even on the industry's best estimates . . .' On the emissions cuts from HELE plants, see the NSW Minerals Council submission to the 'Inquiry Into Electricity Supply, Demand and Prices in NSW', 19 June 2018, www.parliament.nsw.gov.au/lcdocs/submissions/60936/0244 NSW Mineral Council.pdf. See also Geoff Chambers, 'Clean-coal power station cheaper option than renewables', *The Australian*, 3 July 2017, www.theAustralian.com.au/nation/politics/cleancoal-power-station-cheaper-option-than-renewables-bill/news-story/b42432238b46c17f 53f57db08fbb207c.

Page 283: 'Glencore refused to confirm . . .' is from Glencore's communication with the author, 24 January 2020.

Page 283: 'George Christensen, the federal LNP member . . .' The details of Project Caesar were published in Christopher Knaus, 'Revealed: Glencore bankrolled covert campaign'.

Page 283: 'Crosby Textor polled voters . . .' For the NSW Minerals Council poll, see Geoff Chambers, 'Malcolm Turnbull's home state of NSW backs clean coal plant', *The Australian*, 7 September 2017, www.theAustralian.com.au/business/mining-energy/malcolm-turnbulls-home-state-of-nsw-backs-clean-coal-plant/news-story/4c45aa515451ca4047b06c86cef6 24f9.

Page 284: 'BHP's senior executive . . .' On the background to the MCA–BHP split over Brendan Pearson, see Ben Potter, '"Clean coal" crusade claims Minerals Council CEO Brendan Pearson', *Australian Financial Review*,

22 September 2017, www.afr.com/politics/clean-coal-crusade-claims-minerals-council-ceo-brendan-pearson-20170922-gyn01f.

Page 284: 'As well, BHP was increasingly worried . . .' On the Minerals Council support for an inquiry into the charity status of environment groups, see Liz Hobday and Gus Goswell, 'Charity crackdown would be a "torpedo" to environmental groups, Bob Brown says', *ABC News*, 31 August 2017, www.abc.net.au/news/2017-08-31/charities-crackdown-threatens-status-of-environmental-groups/8859792. Also, confidential industry sources confirmed the concerns of BHP over this issue.

Page 285: 'Lamenting Pearson's departure . . .' Janet Albrechtsen's quote on BHP is from her column 'Coal loses as Pearson leaves Minerals Council', *The Australian*, 3 October 2017, www.theAustralian.com.au/nation/politics/bhp-wins-coal-loses-as-brendan-pearson-leaves-minerals-council/news-story/4a4ddf5646136196386dbaf9e6a1850c.

Page 285: 'This was despite an expert panel . . .' For the chair's communiqué of the findings of the expert panel, see Dr Robert Sandland, 'Technical Advisory Forum meets for the final time . . .', 15 May 2017, www.bom.gov.au/climate/data/acorn-sat/documents/2017_TAF_communique.pdf.

Page 285: 'Governments don't want to know . . .' Alan Jones remarks are from 'Why there needs to be a public inquiry into the BOM', Sky News, 2 August 2017, www.youtube.com/watch?v=mmg2NTzqL74.

Chapter 16: Sacrificing goats

Page 286: 'Primitive people once killed goats . . .' For the quotes from Abbott's speech to the Global Warming Policy Foundation, 'Daring to doubt', given 9 October 2017, see www.thegwpf.org/tony-abbott-daring-to-doubt.

Page 287: 'Among the dignitaries in the crowd . . .' Turnbull's speech at the opening of the Sir John Monash Centre at Villers-Bretonneux on 25 April 2018 recognised Tony Abbott's role, see www.malcolmturnbull.com.au/media/speech-at-the-opening-of-the-sir-john-monash-centre-villers-bretonneux-fran.

Page 287: 'The flyer that called them to arms . . .' The flyer on the launch of the Monash Forum was reported in David Crowe, 'Monash Forum trick leaves conservatives divided', *Sydney Morning Herald*, 5 April 2018, www.smh.com.au/politics/federal/monash-forum-trick-leaves-conservatives-divided-20180405-p4z7yi.html.

Page 288: 'It was about ensuring . . .' Tony Abbott's comment on the Monash Forum is from the author's interview, 1 March 2019.

Page 288: 'Coal became a kind of leitmotif . . .' Turnbull's comment on coal-fired power stations is from the author's interview, 15 March 2019.

Page 289: 'But not everyone in business . . .' On Energy in Australia endorsing the Monash Forum, see Christopher Knaus, 'Revealed: Glencore bank-rolled covert campaign to prop up coal'.

Page 289: 'The NEG, as it was dubbed . . .' The outline of the NEG is from 'Retailer reliability and emissions guarantee', 7 November 2017, www.

coagenergycouncil.gov.au/sites/prod.energycouncil/files/publications/
documents/Energy Security Board Overview of the National Energy Guar-
antee.pdf, and the 'National Energy Guarantee, final detailed design',
1 August 2018, coagenergycouncil.gov.au/publications/energy-security-
board—final-detailed-design-national-energy-guarantee, both published
by the Energy Security Board.

Page 290: 'Innes Willox, the AIG chief . . .' The industry leaders' briefing
of the Coalition backbenchers and their quotes are detailed in Katharine Murphy, 'Turnbull quashes Abbott's bid to give party room a say on
energy guarantee', *The Guardian*, 26 June 2018, www.theguardian.com/
Australia-news/2018/jun/26/turnbull-quashes-abbotts-bid-to-give-party-
room-a-say-on-energy-guarantee.

Page 290: 'The party room must've been . . .' Abbott's comment is from the
author's interview, 1 March 2019.

Page 291: 'We can't possibly get there . . .' Shane Rattenbury's comment on
the NEG is from the author's interview, 2 December 2019.

Page 291: 'They thought the Paris targets . . .' The quote from the ACF was
posted on its Twitter account the day after the protest on 26 June 2018,
twitter.com/ausconservation/status/1011798256970448896.

Page 291: 'In July, Abbott turned up the pressure.' Abbott urged pulling
out of Paris in his Bob Carter commemorative lecture on 3 July 2018
(http://tonyabbott.com.au/2018/07/2018-bob-carter-commemorative-
lecture-australian-environment-foundation-melbourne/), but he retracted
his call in the 2019 election campaign. Hugh Morgan and Gina Rinehart
publicly backed exiting the Paris Agreement, adding their names to 'The
Saltbush Club', a lobby group advocating this set up in November 2018.

Page 291: 'At this point, Turnbull lost . . .' For a full account of Dutton's
growing disaffection with Turnbull, see Niki Savva, *Plots and Prayers:
Malcolm Turnbull's demise and Scott Morrison's ascension*, Melbourne:
Scribe, 2019, p. 22.

Page 292: 'When you add all the Liberal Party members . . .' Laundy's quote
on Turnbull losing support is from the author's interview, 22 October 2018.

Page 292: 'Abbott was trenchantly opposed . . .' Turnbull's comment on the
opposition in the party room to the NEG is from the author's interview,
15 March 2019.

Page 292: 'In a series of humiliating backdowns . . .' For the business reaction
to Turnbull dropping the NEG emissions target, see Gareth Hutchins,
'Business condemns Turnbull's shift on emissions and "extreme intervention"', *The Guardian*, 20 August 2018, www.theguardian.com/
Australia-news/2018/aug/20/business-condemns-turnbulls-shift-on-
emissions-and-extreme-intervention.

Page 293: 'Everyone has been dancing . . .' Weatherill's comments are from
the author's interview, 27 February 2019.

Page 293: 'I said to Dutts . . .' Abbott's comments on Dutton are from the
author's interview, 1 March 2019.

Page 293: 'Rupert, this insurgency . . .' Turnbull's account of his conversation with Rupert Murdoch is from his memoir, *A Bigger Picture*. See pp. 626–7 for the full version, some of which is disputed by Murdoch. Turnbull's remark to Mathias Cormann is on p. 628.

Page 294: 'Mathias was strongly in favour . . .' Turnbull's comments on Mathias Cormann are from the author's interview, 15 March 2019.

Page 294: 'So, insofar as there has been chaos . . .' Turnbull's quote is from his final prime ministerial press conference, 24 August 2018, www.malcolmturn bull.com.au/media/press-statement-palriament-house-24-august-2018.

Page 295: 'The country had already warmed . . .' The 'State of the Climate' report projections are at 'State of the climate 2018: Report at a glance', Bureau of Meteorology, www.bom.gov.au/state-of-the-climate.

Chapter 17: The state of the Reef III

Page 296: 'We just got savaged.' Ove Hoegh-Guldberg's account of the Coalition backbench committee and his quote come from the author's interview, 11 March 2019.

Page 296: 'The atmosphere and ocean . . .' For the role of Howden and Church in the IPCC Fifth Assessment Report, see www.ipcc.ch/site/assets/uploads/2018/02/SYR_AR5_FINAL_full.pdf. The quote from the IPCC report is on p. 2. For Hoegh-Guldberg's role in the IPCC report on oceans, see gci.uq.edu.au/ipcc-fifth-assessment-report-chapter-30-ocean.

Page 297: 'Rounding out the IPA team . . .' For Brett Hogan's report, 'The life saving potential of coal', 1 August 2015, see ipa.org.au/ipa-review-articles/the-life-saving-potential-of-coal.

Page 297: 'We were all surprised at being treated . . .' Mark Howden's quote on the Coalition backbench committee meeting comes from the author's interview, 29 November 2018.

Page 297: 'You have got to look at this . . .' Hoegh-Guldberg's quote is from the author's interview, 11 March 2019.

Page 298: 'What amused Bob and I . . .' John Roskam's amusement over *The Guardian*'s account of the meeting was from his Carter tribute, 1 April 2016, ipa.org.au/ipa-review-articles/in-memoriam-professor-robert-m-carter. For Graham Readfearn's article on the backbench committee meeting, see 'Coalition committee tries to balance climate science briefings by inviting denialists from think tank', *The Guardian*, 23 October 2015, www.theguardian.com/environment/planet-oz/2015/oct/23/coalition-committee-tries-to-balance-climate-science-briefings-by-inviting-denialists-from-think-tank.

Page 299: 'A few months earlier, in June 2015 . . .' For the NOAA warnings on coral bleaching on 2 June 2015, see '2014–16 bleaching event continues: June 2015 update', coralreefwatch.noaa.gov/satellite/analyses_guidance/global_bleaching_update_20150602.php.

Page 299: 'NOAA's forecasts were particularly extreme.' Terry Hughes's quote on the NOAA is from the author's interview, 12 February 2019.

Page 299: 'The National Coral Bleaching Taskforce put together ...' For the membership of the taskforce, see 'Coral Bleaching Taskforce documents most severe bleaching on record', 29 March 2016, www. coralcoe.org.au/media-releases/coral-bleaching-taskforce-documents-most-severe-bleaching-on-record.

Page 300: 'While the surveying was supposed ...' The first reports of bleaching in January 2016 reported in Morgan Pratchett and Janice Lough, 'Coral Bleaching Taskforce: More than 1000 km of the Great Barrier Reef has bleached', The Conversation, 7 April 2016, theconversation.com/coral-bleaching-taskforce-more-than-1-000-km-of-the-great-barrier-reef-has-bleached-57282.

Page 300: 'One of the first people ...' For Joe Bast's tribute comments on Professor Bob Carter, Carter's role as a policy adviser to Heartland and his Lifetime Achievement Award, see 'Robert M. Carter (1942–2016)', www. heartland.org/about-us/who-we-are/robert-m-carter-1942-2016.

Page 300: 'The right-wing free-market think tank ...' The Mercer Foundation's funding of the Heartland Institute comes from its IRS statements supplied to the author by Kert Davies of the Climate Investigations Center.

Page 300: 'That year Mercer also became ...' Robert Mercer's donations to Donald Trump's election are cited in Jane Mayer, 'The reclusive hedge-fund tycoon behind the Trump presidency', *New Yorker*, 17 March 2017, www.newyorker.com/magazine/2017/03/27/the-reclusive-hedge-fund-tycoon-behind-the-trump-presidency.

Page 301: 'I and many, many people ...' Myron Ebell's tribute to Bob Carter is from Heartland's obituary page, 'Robert M. Carter (1942–2016)'.

Page 301: 'They had not forgotten Carter's role ...' On Carter campaigning with Barnaby Joyce, see Sarah Ferguson, 'Malcolm and the malcontents', *Four Corners*, 9 November 2009.

Page 301: 'Murdoch columnist Andrew Bolt ...' Bolt's tribute to Carter is from Heartland's obituary page, 'Robert M. Carter (1942–2016)'.

Page 301: 'Bob was an immensely valued colleague ...' For John Roskam's tribute to Carter, see 'In memoriam Professor Robert M. Carter'.

Page 301: 'A few weeks after Carter's death ...' On the sea-surface temperature rises for the Reef for February to June 2016, see GBRMPA, *Final Report: 2016 coral bleaching event on the Great Barrier Reef*, Canberra: Commonwealth of Australia, pp. 5–6, elibrary.gbrmpa.gov.au/jspui/bitstream/11017/3206/1/Final-report-2016-coral-bleaching-GBR.pdf. The extreme temperature rise at Lizard Island was reported by the Australian Museum's Dr Lyle Vail in his blog post, 'Coral bleaching on the doorstep of Lizard Island', Australian Museum, 11 March 2016, Australianmuseum. net.au/blog/amri-news/coral-bleaching-on-the-doorstep-of-lizard-island.

Page 302: 'You can tell if the coral ...' Morgan Pratchett's account of his surveying on the *James Kirby* and his quote are from the author's interview, 13 February 2019.

Page 302: 'Corals usually tolerate only ...' For the description of heat stress in corals, bleaching and mortality, see *Final Report: 2016 coral bleaching event on the Great Barrier Reef*, p. 1.

Page 302: 'When we started to measure . . .' Hughes's quote about the corals being cooked is from the author's interview, 12 February 2019.

Page 302: 'You became almost immune . . .' Pratchett's quote on the death and destruction is from the author's interview, 13 February 2019.

Page 303: 'The biggest cyclone ever to make landfall . . .' On the role of Cyclones Winston and Tatiana, see *Final Report: 2016 coral bleaching event on the Great Barrier Reef*, p. 6.

Page 303: 'That's the only way you can . . .' Hughes's quotes on the aerial surveys are from the author's interview, 12 February 2019.

Page 303: 'GBRMPA, AIMS and other outfits . . .' On the role of the GBRMPA vessels in the surveying, see *Final report: 2016 Coral Bleaching Event on the Great Barrier Reef*, Appendix A, p. 34.

Page 304: 'On reefs and sites around Lizard Island . . .' The quote from Dr Anne Hoggett is from the media release by the National Coral Bleaching Taskforce, 'National Coral Taskforce unleashes an armada of experts', 5 April 2016, www.jcu.edu.au/news/releases/2016/april/national-coral-taskforce-unleashes-an-armada-of-experts.

Page 304: 'For me, personally, it was devastating . . .' Hughes's interview on ABC TV's *7.30* program was broadcast on 28 March 2016, www.abc.net.au/news/2016-03-28/great-barrier-reef-coral-bleaching-95-per-cent-north-section/7279338.

Page 305: 'As Hunt put it in the approval decision . . .' For Hunt's statement on Adani and greenhouse gas emissions, see 'Statement of reasons for approval of a proposed action under the *Environment Protection and Biodiversity Conservation Act 1999* (Cth)', 14 October 2015, www.environment.gov.au/system/files/pages/cb8a9e41-eba5-47a4-8b72154d0a5a6956/files/carmichael-statement-reasons.pdf. See par 53. For an analysis of the final Adani decision, see Samantha Hepburn, Adam Lucas, Craig Froome et al., 'Greg Hunt approves Adani's Carmichael coal mine, again: Experts respond', The Conversation, 16 October 2015, theconversation.com/greg-hunt-approves-adanis-carmichael-coal-mine-again-experts-respond-49227.

Page 305: 'The Great Barrier Reef is at a crisis . . .' The 21 April 2016 advertisement in the *Courier-Mail* was funded by supporters of the Climate Council, see 'Climate change is destroying our reefs. We must phase out coal', 26 April 2016, www.climatecouncil.org.au/reefstatement.

Page 305: '. . . virtually all reefs . . .' Russell Reichelt's testimony is from *Hansard*, Senate Estimates, Environment and Communications, 5 May 2016.

Page 306: 'At the end of May, Hughes released . . .' For Hughes's 30 May 2016 statement on the bleaching, see 'Coral death toll climbs on Great Barrier Reef', www.jcu.edu.au/news/releases/2016/may/coral-death-toll-climbs-on-great-barrier-reef.

Page 306: 'There was a definite effort . . .' Hughes's comment on the government hosing down the bleaching is from the author's interview, 12 February 2019.

Page 306: 'The strains in the taskforce burst . . .' For the original GBRMPA and AIMS statement of 3 June 2016, see 'The facts on Great Barrier Reef coral mortality', www.gbrmpa.gov.au/news-room/latest-news/latest-news/coral-bleaching/2016/the-facts-on-great-barrier-reef-coral-mortality.

Page 306: '*The Australian* headlined its story . . .' On the split in the task-force and for Reichelt's comments, see Graham Lloyd, 'Great Barrier Reef: Scientists "exaggerated" coral bleaching', *The Australian*, 4 June 2016, www.theAustralian.com.au/nation/nation/great-barrier-reef-scientists-exaggerated-coral-bleaching/news-story/99810c83f5a420727b12ab2552 56774b.

Page 307: 'The authority's final report . . .' See *Final report: 2016 coral bleaching event on the Great Barrier Reef*, p. v. The figure for bleaching was raised to an average of 30 per cent of shallow-water corals (at depths between 2 and 10 metres) across the whole Reef in GBRMPA's 'Position statement: Climate change', 25 June 2019 elibrary.gbrmpa.gov.au/jspui/bitstream/11017/3460/5/v1-Climate-Change-Posistion-Statement-for-eLibrary.pdf.

Page 308: 'When you looked at the pattern . . .' Will Steffen's comment, see the author's interview, 5 November 2018.

Page 308: 'That year Ridd wrote an essay . . .' Peter Ridd, 'The extraordinary resilience of Great Barrier Reef corals, and problems with policy science', in Jennifer Marohasy (ed.), *Climate Change: The Facts 2017*, Redland Bay, Queensland: Connor Court, 2017, pp. 9–23. Criticism by Ridd of other marine scientists over their research eventually led to Ridd's sacking by JCU. Ridd fought this in court with the backing of the IPA. He won his unfair dismissal case against JCU in September 2019 and was awarded $1.2 million in compensation. At the time of writing, JCU had lodged an appeal against the decision. For the IPA's account of the case, see 'Peter Ridd's fight for freedom of speech on climate change', IPA, ipa.org.au/peterridd.

Page 308: 'But both GBRMPA and AIMS took . . .' On the complexity of coral recovery, see GBRMPA, *Great Barrier Reef Outlook Report 2019*, Canberra: Commonwealth of Australia, 2019, p. 245, elibrary.gbrmpa.gov.au/jspui/bitstream/11017/3474/10/Outlook-Report-2019-FINAL.pdf.

Page 309: 'Much of the damage was surveyed . . .' On the survival of corals in the 2017 bleaching, see Terry P. Hughes, James T. Kerry, Sean R. Connolly et al., 'Ecological memory modifies the cumulative impact of recurrent climate extremes', *Nature Climate Change*, 2019, vol. 9, pp. 40–3, doi:10.1038/s41558-018-0351-2.

Page 309: 'But GBRMPA later found . . .' See GBRMPA, 'Position statement: climate change', 25 June 2019.

Page 309: 'The second bleaching was another . . .' Hughes's estimates on the 2017 bleaching are from the author's interview, 12 February 2019, as is the comment on marine heatwaves.

Page 309: 'Some reefs in the south . . .' On coral resistance to bleaching and coral recovery, see Hughes et al., 'Ecological memory modifies the cumulative impact of recurrent climate extremes'.

Page 309: 'On the negative side . . .' On the crash of coral recruitment after the bleachings, see Terry P. Hughes, James T. Kerry, Andrew H. Baird et al., 'Global warming impairs stock–recruitment dynamics of corals', *Nature*, 2019, vol. 568, pp. 387–90, www.nature.com/articles/s41586-019-1081-y. See also GBRMPA, *Great Barrier Reef Outlook Report 2019*, p. 25.

Page 310: 'JCU's Morgan Pratchett had surveyed . . .' The information on Pratchett's surveys of butterflyfish is from the author's interview, 13 February 2019.

Page 310: 'Over a decade before . . .' The studies on the feminisation of turtles are cited in GBRMPA, *Great Barrier Reef Outlook Report 2019*, pp. 65–6. See also GBRMPA, 'Position statement: Climate change'.

Page 310: 'The latest work is saying . . .' David Wachenfeld's comment on the turtle study is from the author's interview, 13 March 2019.

Page 311: 'The Great Barrier Reef is a massive plus . . .' On the economic contribution to the Reef, see GBRMPA, 'Position statement: Climate change'.

Page 311: 'Our best estimate is that . . .' Wachenfeld's comment on climate change and the reef is from the author's interview, 13 March 2019.

Page 311: 'It was called *Juliana v United States* . . .' For details on the case, see 'Juliana v. United States: Youth climate lawsuit', Our Children's Trust, www.ourchildrenstrust.org/juliana-v-us.

Page 312: 'It's basically on the denial . . .' The quote from Ove Hoegh-Guldberg on denying kids a future is from the author's interview, 11 March 2019.

Page 312: 'The fact that the U.S. Government . . .' The quote is from Hoegh-Guldberg's affidavit, 2 August 2018, *Juliana v United States*, see www.ourchildrenstrust.org/court-orders-and-pleadings.

Page 312: 'We have to say something . . .' Hoegh-Guldberg's comment is from the author's interview, 11 March 2019.

Page 312: 'The children's case would later be thrown out . . .' For the 17 January 2020 decision of the US Court of Appeals for the Ninth Circuit, see http://cdn.ca9.uscourts.gov/datastore/opinions/2020/01/17/18-36082.pdf.

Chapter 18: No Regrets

Page 314: 'The government, he assured them . . .' For the account of Scott Morrison's visit to the Tully farm, see Anna Henderson, 'Prime minister leaves climate change debate for "another day" during drought trip', ABC News online, 27 August 2018, www.abc.net.au/news/rural/2018-08-27/pm-leaves-climate-change-debate-for-another-day-on-drought-tour/10168860. For the amount of Queensland in drought in August 2018, see longpaddock.qld.gov.au/drought/archive.

Page 315: 'Morrison batted the question away . . .' For Morrison's remarks at the Tully farm, including the exchange on climate change, see 'Press conference, Quilpie Qld: transcript', 27 August 2018, Prime Minister of Australia website, www.pm.gov.au/media/press-conference-quilpie-qld.

Page 316: 'The year 2018 was going to be . . .' For the Bureau of Meteorology report on the early bushfire season in New South Wales in 2018 and it being the warmest on record, see 'New South Wales in 2018: warmest year on record, very dry', 10 January 2019, www.bom.gov.au/climate/current/annual/nsw/archive/2018.summary.shtml. On the drought, see BOM's Annual climate statement 2018, www.bom.gov.au/climate/current/annual/aus/2018. On the NSW drought, see 'NSW 100 per cent in drought: minister', *Sydney Morning Herald*, 8 August 2018, www.smh.com.au/environment/weather/nsw-100-per-cent-in-drought-minister-20180808-p4zw62.html.

Page 316: 'I'm interested in getting . . .' For Morrison's remark on electricity prices, see 'Press conference, Quilpie Qld'.

Page 316: 'But Morrison brushed off . . .' For Morrison's remarks on the IPCC report, see Paul Karp, 'Australian government backs coal in defiance of IPCC climate warning', *The Guardian*, 9 October 2018, www.theguardian.com/Australia-news/2018/oct/09/Australian-government-backs-coal-defiance-ipcc-climate-warning.

Page 316: 'At 1.5 degrees of warming . . .' For the IPCC's Special Report on a rise of 1.5 degrees Celsius and the potential impacts on drought, etc. compared with 2.0 degrees warming, see 'Projected climate change, potential impacts and associated risks', in IPCC, 'Summary for policy makers', *Global Warming of 1.5°C*, 2018, pp. 7–11, www.ipcc.ch/sr15/chapter/spm. On the investment in low-carbon energy needed, see the 'Executive summary' of Chapter 2 in IPCC, *Global Warming of 1.5°C*, www.ipcc.ch/sr15/chapter/chapter-2.

Page 317: 'Interviewed by the ABC about the special IPCC report . . .' Melissa Price was interviewed by the *AM* program, 9 October 2018, ABC, www.abc.net.au/radio/programs/am/melissa-price-paris-commitment,-ipcc-and-the-opera-house/10354540.

Page 318: 'When the prime minister, Scott Morrison . . .' The quotes from Angus Taylor at the AFR summit are from 'Keynote address', 10 October 2018, Department of the Environment and Energy, minister.environment.gov.au/taylor/news/2018/keynote-address-afr-national-energy-summit.

Page 319: 'He's always been close to these guys . . .' Turnbull's remarks on Taylor are from the author's interview, 15 March 2019.

Page 319: 'I'm a great fan of Angus.' Tony Abbott's remarks on Taylor are from the author's interview, 1 March 2019.

Page 319: 'Soon after the *Financial Review* summit . . .' Morrison and Taylor's 'A Fair Deal on Energy', was released on 23 October 2018, see Prime Minister of Australia website, www.pm.gov.au/media/fair-deal-energy.

Page 320: 'A month later they doubled down . . .' For Labor's energy policy, see Nicole Hasham, 'Labor announces $15 billion to "turbo charge" renewables sector', *Sydney Morning Herald*, 22 November 2018, www.smh.com.au/politics/federal/labor-announces-15-billion-to-turbo-charge-renewables-sector-20181122-p50hm2.html.

Page 321: 'The truth is these are reckless policies . . .' For Taylor's remarks at the Tomago smelter, see 'Doorstop with Matt Howell, CEO of

Tomago Aluminium', 22 November 2018, Department of the Environment and Energy, minister.environment.gov.au/taylor/news/2018/doorstop-matt-howell-ceo-tomago-aluminium.

Page 321: 'That was inevitable when, two days before the election . . .' For the pressure on Melissa Price over the approval of the Adani groundwater plans, see Dan Conifer, 'Adani coal mine a step closer with environment minister endorsing groundwater approvals', *ABC News*, 9 April 2019, www.abc.net.au/news/2019-04-09/adani-gains-commonwealth-groundwater-approval/10984134.

Page 321: 'How dare this government . . .' For Tim Flannery's reaction to the Adani approval, see 'Adani mine must be stopped', 9 April 2019, Climate Council, www.climatecouncil.org.au/resources/adani-mine-must-be-stopped.

Page 322: 'We lost credibility with coal industry workers'. Jay Weatherill's remarks are from the author's follow-up interview, 5 January 2020.

Page 322: '. . . borrowing language from the Business Council's CEO . . .' For the history of the dispute between Labor and the Business Council of Australia's Jennifer Westacott over her 'economy wrecking' remarks, see Phillip Coorey, 'Labor attacks BCA for "baseless" claim on emissions target', *Australian Financial Review*, 6 March 2019, www.afr.com/business-summit/labor-at-daggers-drawn-with-the-bca-over-climate-20190306-h1c23z.

Page 322: 'Morrison claimed that an affordable electric car . . .' For Morrison's remarks on Labor's electric car policy, see Ben Packham, 'Shorten to end 4WD fun: claims PM Morrison', *The Australian*, 8 April 2019, www.theAustralian.com.au/nation/shorten-to-kill-4wd-fun-claims-pm-morrison/news-story/77480ce996dbfb2c906787f97bbbd5c3.

Page 322: 'Or, as Murdoch's Sydney tabloid . . .' For the *Daily Telegraph* story on vehicle emissions standards, see Anna Caldwell, 'Bill's $5K CAR-BON TAX', *Daily Telegraph*, 9 April 2019.

Page 323: 'Headlines like "Labor's Carbon plan . . ."' For the cost to miners story, see Perry Williams, 'Labor's carbon plan "to cost miners $2bn"', *The Australian*, 7 May 2019, www.theAustralian.com.au/business/mining-energy/labors-carbon-plan-to-cost-miners-2bn-wood-mackenzie/news-story/568e6b561055fc4bd0c7b19cec686de1.

Page 323: 'More importantly, Labor could not resolve . . .' For Labor's conflicted position on the Adani mine, see 'Review of Labor's 2019 federal election campaign', chaired by Dr Craig Emerson and Jay Weatherill, p. 35, https://alp.org.au/media/2043/alp-campaign-review-2019.pdf.

Page 323: '. . . the CFMEU, a big donor . . .' Details of donations were released by the AEC on 3 February 2020, https://transparency.aec.gov.au/AnnualPoliticalParty. Note in March 2018 the CFMEU was given approval to merge with the Maritime Union to become the CFMMEU.

Page 323: 'Adding to Labor's problems . . .' For details on the route of the anti-Adani convoy, see 'Stop Adani convoy—on the road to action', Bob Brown Foundation, www.bobbrown.org.au/stopadani_co.

Page 324: 'I want to thank Bob Brown . . .' For Matt Canavan's quote on Bob Brown, see Jemima Burt and Rachel McGhee, 'Anti-Adani protesters felt threatened by locals in Queensland coal mining town of Clermont', *ABC News*, 28 April 2019, www.abc.net.au/news/2019-04-27/adani-carmichael-mine-greens-clermont-convoy-qld/11051390. This report also carried the images of Pauline Hanson holding the CFMEU placard, and Clive Palmer's remarks.

Page 324: 'Just days earlier, Palmer had struck a deal . . .' For Clive Palmer's preference deal with the Liberals, see Simon Benson and Ben Packham, 'Palmer sides with Liberals on preferences over economy fears', *The Australian*, 26 April 2019, www.theAustralian.com.au/nation/politics/palmer-sides-with-liberals-on-preferences-over-economy-fears/news-story/aa260ff1317e4213defc3c3c7952f60e.

Page 325: 'His old North Queensland nickel refinery . . .' On the history of the liquidators' battle with Palmer over his nickel refinery, see Mark Ludlow, 'Clive Palmer "pretty happy" with QNI settlement', *Australian Financial Review*, 5 August 2019, www.afr.com/companies/mining/clive-palmer-pretty-happy-with-qni-settlement-20190805-p52dwp.

Page 326: 'Palmer's companies donated . . .' Details of donations were released by the AEC on 3 February 2020, https://transparency.aec.gov.au/AnnualPoliticalParty.

Page 326: 'Labor's post-mortem analysis . . .' For the impact of Palmer and climate change on Labor's campaign, see 'Review of Labor's 2019 federal election campaign'.

Page 327: 'I said that I was going to burn for you . . .' Morrison's quote on election night is from Nicole Hasham, '"I always believed in miracles": Scott Morrison celebrates as Bill Shorten concedes defeat', *Sydney Morning Herald*, 19 May 2019, www.smh.com.au/federal-election-2019/i-always-believed-in-miracles-scott-morrison-celebrates-as-bill-shorten-concedes-defeat-20190518-p51osw.html.

Page 327: 'By September 2019 it was clear . . .' For the bushfire conditions in Spring 2019, the record dry and the early fires, see 'Special climate statement 72—dangerous bushfire weather in spring 2019', Bureau of Meteorology, 18 December 2019, www.bom.gov.au/climate/current/statements/scs72.pdf.

Page 327: 'When the Australian Bureau of Meteorology warns . . .' For a more detailed explanation of the Forest Fire Danger Index, see www.bom.gov.au/nsw/forecasts/fire-map.shtml.

Page 327: 'Almost 60 per cent of Australia . . .' For the spring FFDI statistics and 'catastrophic' NSW fire conditions, see 'Special climate statement 72'.

Page 327: 'We've never seen fire danger indices . . .' For Andrew Sturgess's quote, see www.abc.net.au/news/2019-11-14/fact-check-did-high-fire-danger-indices-come-earlier-this-year/11506082.

Page 327: 'They can expect the same support . . .' Morrison's quote is from Richard Ferguson, 'PM Scott Morrison heads for fire zone', *The*

Australian, 13 September 2019, www.theAustralian.com.au/nation/pm-scott-morrison-heads-for-fire-zones/news-story/c7dcfc0f4950bac256dc25ca4f82bc66.

Page 328: 'Between them, they had 600 years' experience . . .' For the 10 April statement by the former emergency leaders, see Greg Mullins, 'Emergency leaders: Australia unprepared for worsening extremes', Climate Council, 10 April 2019, www.climatecouncil.org.au/emergency-leaders-climate-action.

Page 328: 'Mullins believed Morrison's attitude . . .' Greg Mullins' quote is from the author's interview, 13 January 2020.

Page 329: 'We're coming into what I think . . .' Mullins' interview was broadcast on ABC online, 6 November 2019, https://mobile.abc.net.au/news/2019-11-06/former-fire-chief-worried-about-firefighting-resources/.

Page 329: 'Days later, three people died . . .' On the first fire deaths in NSW, see Caitlin Fitzsimmons and Laura Chung, '"She was in an absolute panic": Family pays tribute to grandmother killed by bushfire', *Sydney Morning Herald*, 9 November 2019, www.smh.com.au/national/nsw/she-was-in-an-absolute-panic-family-pays-tribute-to-grandmother-killed-by-bushfire-20191109-p53926.html.

Page 329: 'Four hours north of Sydney, in the coastal resort town . . .' On the Port Macquarie air quality, see Luisa Rubbo, Kirstie Wellauer et al., 'Waves of ash wash up on NSW beaches, Port Macquarie records world's dirtiest air as bushfires burn', *ABC News*, 16 November 2019, www.abc.net.au/news/2019-11-15/nsw-bushfires-make-port-macquarie-most-polluted-place/11708612.

Page 329: 'Asked why he hadn't met . . .' For Morrison's comments on the fire response and Australia's emissions on 21 November 2019, see 'Radio interview with Sabra Lane: transcript', www.pm.gov.au/media/radio-interview-sabra-lane-abc-am-0.

Page 330: 'The former fire chief wasn't impressed . . .' Greg Mullins' comment to Taylor is from the author's interview, 13 January 2020.

Page 331: 'Do we really want to be remembered . . .' For António Guterres's speech at the opening of the Madrid conference on 2 December 2019, see 'Remarks at opening ceremony of UN Climate Change Conference COP25, United Nations Secretary-General website, www.un.org/sg/en/content/sg/speeches/2019-12-02/remarks-opening-ceremony-of-cop25.

Page 332: 'We can only reduce emissions . . .' For Taylor's speech to the COP25 conference in Madrid on 10 December 2019, see 'National statement COP25, Madrid', Department of the Environment and Energy, minister.environment.gov.au/taylor/news/2019/national-statement-cop25-madrid.

Page 332: 'But many delegates knew those words . . .' For the dispute regarding 'carry-over' credits, see Leslie Hook, 'UN climate talks stymied by carbon markets' "ghost from the past"', *Financial Times*, 16 December 2019, republished at insideclimatenews.org/news/16122019/cop25-carbon-markets-un-climate-talks-fail-madrid-kyoto-protocol. See also James Fernyhough and Elouise Fowler, 'The countries with the biggest hoard of Kyoto credits', *Australian Financial Review*, 16 December 2019, www.afr.com/policy/energy-and-climate/the-countries-with-the-biggest-hoard-of-kyoto-credits-20191216-p53kay.

Page 332: 'We think Australia should be . . .' For Taylor's quote on carry-over credits, see Mark Ludlow, 'Use of carry-over credits is legitimate, says Taylor', *Australian Financial Review*, 9 December 2019, www.afr.com/companies/energy/use-of-carry-over-credits-is-legitimate-says-taylor-20191209-p53i3s.

Page 333: 'All the time it is costing Australia . . .' Frank Jotzo's comment is from the author's interview, 8 January 2020.

Page 333: 'Ominously, the Bureau of Meteorology . . .' On the fire conditions when Morrison left the country, see 'Number of properties burnt set to rise as New South Wales and other states prepare for heatwave', Australian Associated Press, 16 December 2019.

Page 334: 'If the roof was on fire . . .' Greg Mullins' account of the New Year's Eve fires on the south coast are from the author's interview, 13 January 2020.

Page 334: 'The image of Finn Burns . . .' The story of Finn Burns, 'Mallacoota Boy', appeared in many media outlets including Ashleigh McMillan, '"Just worried about getting away from the fire": Mallacoota's boat boy speaks out', *The Age*, 1 January 2020, www.theage.com.au/national/victoria/just-worried-about-getting-away-from-the-fire-mallacoota-s-boat-boy-speaks-out-20200101-p53o5n.html.

Page 334: 'Ultimately, the "Black Summer" fires . . .' The death and destruction toll from the 2019–20 bushfires is from the Royal Commission into National Natural Disasters, transcript, 16 April 2020, p. 11, https://naturaldisaster.royalcommission.gov.au/publications/ceremonial-hearing-transcript. For the estimate of the smoke-related deaths, see Associate Professor Fay Johnston's evidence on 26 May; for the 'ecological disaster' comment, see Threatened Species Commissioner Dr Sally Box's evidence, 27 May.

Page 335: 'Three thousand defence force reservists . . .' On the government response to the bushfires, see 'Press conference—Australian Parliament House: transcript, 5 January 2020, Prime Minister of Australia website, www.pm.gov.au/media/press-conference-Australian-parliament-house-0.

Page 335: 'What our commitment . . .' On Scott Morrison's quote on 'reckless' targets, see 'Press conference—Australian Parliament House: transcript', 12 January 2020, Prime Minister of Australia website, www.pm.gov.au/media/press-conference-Australian-parliament-house-1.

Page 336: 'They've made that quite clear . . .' Frank Jotzo's quote is from the author's interview, 8 January 2020.

Page 337: 'Morrison tried to grapple . . .' For Scott Morrison's speech to the National Press Club, 29 January 2020, see www.pm.gov.au/qa-national-press-club.

Epilogue: The Road Ahead

Page 339: 'an utter tragedy'. Terry Hughes' quote is from his joint article with Morgan Pratchett, 'We just spent two days surveying the Reef . . .', The Conversation, 7 April 2020, https://theconversation.com/we-just-spent-two-weeks-surveying-the-great-barrier-reef-what-we-saw-was-an-utter-tragedy-135197.

Page 339: 'The Reef had been hit with its third mass bleaching . . .' Details of the bleaching, temperature levels and the survey were released by the ARC Centre of Excellence for Coral Reef Studies at James Cook University. See media release, 'Climate change triggers Great Barrier Reef bleaching', www.coralcoe.org.au/media-releases/climate-change-triggers-great-barrier-reef-bleaching.

Page 339: '. . . the royal commission investigating the devastating summer bushfires was visiting . . .' The early visits of the Royal Commission into National Natural Disaster Arrangements to bushfire-affected communities were detailed in the first day of hearings on 16 April 2020, https://naturaldisaster.royalcommission.gov.au/publications/ceremonial-hearing-transcript.

Page 341: 'from the Covid frying pan into the climate fire'. The quote is from Cameron Hepburn, Brian O'Callaghan, Nicholas Stern et al., 'Will COVID-19 fiscal recovery packages accelerate or retard progress on climate change?', Oxford Review of Economic Policy, 8 May 2020, doi:10.1093/oxrep/graa015.

Page 341: '. . . Nev Power, publicly advocated a gas-led recovery'. Nev Power's ideas on the gas-led recovery and his role on the board of Strike Energy were detailed in Phillip Coorey, 'Cheap gas to power recovery', Australian Financial Review, 20 May 2020.

Page 341: 'The commission's special adviser on manufacturing, Andrew Liveris . . .' Andrew Liveris's board positions are listed on the Worley website, www.worley.com/investors/leadership/the-board.

Page 341: 'The first draft report from Liveris's panel . . .' The draft report from the panel headed by Liveris was leaked to ABC News. See Stephen Long, 'Government's COVID Commission manufacturing plan calls for huge public gas subsidies', ABC News, 21 May 2020, www.abc.net.au/news/2020-05-21/leaked-national-covid-commission-gas-manufacturing-report/12269100. A copy of the draft report was made available to the author.

Page 341: 'The gas vision was also backed by the Labor Party's key trade union affiliate . . .' The Australian Workers Union national secretary, Daniel Walton, had previously supported domestic gas expansion for manufacturing and was a member of the manufacturing task force committee. See Angela Macdonald-Smith, '$6b trans-Australia gas pipeline gets fresh legs', Australian Financial Review, 19 May 2020, www.afr.com/companies/energy/trans-australian-6b-gas-pipeline-gets-fresh-legs-20200518-p54u4z.

Page 341: 'As the world's largest exporter of LNG...' The quote on LNG exports is from the *Technology Investment Roadmap Discussion Paper*, Department of Industry, Science, Energy and Resources, May 2020, p. 28, https://consult.industry.gov.au/climate-change/technology-investment-roadmap/supporting_documents/technologyinvestmentroadmapdiscussionpaper.pdf.

Page 342: 'The International Energy Agency sees gas playing a role as a transition fuel...' For the International Energy Agency timetable on gas, see Heymi Bahar, Davide D'Ambrosio, Raimund Malischek and Henri Paillere, *Tracking Power*, May 2019, www.iea.org/reports/tracking-power-2019/natural-gas-fired-power#abstract.

Page 342: 'Taylor touted CCS as a lifeline...' For Taylor's support for CCS in the gas industry, see his interview transcripts with Fran Kelly on ABC's Radio National and Lisa Millar on ABC TV's *News Breakfast*, 21 May 2020, www.minister.industry.gov.au/ministers/taylor/transcripts.

Page 343: '... renewable energy made up well over 20 per cent...' For the estimates of renewable energy as 24 per cent of annual electricity generation in 2019, see the *Clean Energy Council's Clean Energy Australia Report 2020*, p. 11, https://assets.cleanenergycouncil.org.au/documents/resources/reports/clean-energy-australia/clean-energy-australia-report-2020.pdf. The government's Australian Energy Statistics put the figure at 21 per cent, www.energy.gov.au/publications/australian-energy-statistics-table-o-electricity-generation-fuel-type-2018-19-and-2019.

Page 343: 'The vexing problems...' On the problems of the grid being confronted, see the *Renewable Integration Study: Stage One Report*, Australian Energy Market Operator, April 2020.

Page 343: 'While gas generation would still be needed...' For renewable energy generation (without back-up storage) being competitive with gas and coal, see the *Technology Investment Roadmap Discussion Paper*, p. 27.

Page 343: 'The government's own Roadmap paper recognised...' On the use of renewable energy in energy-intensive manufacturing, see the *Technology Investment Roadmap Discussion Paper*, p. 16. The paper does stress this would not work 'if we compromise energy security, reliability or affordability in an effort to reduce emissions'. For the $300m fund for advancing the National Hydrogen Strategy, see Taylor's media release, 4 May 2020, www.minister.industry.gov.au/ministers/taylor/media-releases/backing-australian-hydrogen-industry-grow-jobs-economy-and-exports.

Page 343: 'this is about technology not taxes...' Angus Taylor's quote is from his media release, 21 May 2020, www.minister.industry.gov.au/ministers/taylor/media-releases/harnessing-new-technology-grow-jobs-and-economy-and-lower-emissions.

Page 344: 'Prices for gas, coal and oil plummeted...' On the fall in energy prices caused by the pandemic see the International Energy Agency media release, 30 April 2020, www.iea.org/news/global-energy-demand-to-plunge-this-year-as-a-result-of-the-biggest-shock-since-the-second-world-war.

Page 344: '. . . such dramatic falls in emissions would have to be repeated every year . . .' On the estimated plunge in greenhouse gases needed to reach the 1.5–2 degree target, see Pep Canadell, Corinne Le Quéré, Felix Creutzig et al., 'Coronavirus is a "sliding doors" moment. What we do now could change Earth's trajectory', The Conversation, 20 May 2020, https://theconversation.com/coronavirus-is-a-sliding-doors-moment-what-we-do-now-could-change-earths-trajectory-137838.

SELECTED BIBLIOGRAPHY

Books

Abbott, Tony, *Battlelines*, updated edn, Melbourne: Melbourne University Press, 2009

Butler, Mark, *Climate Wars*, Melbourne: Melbourne University Press, 2017

Chubb, Philip, *Power Failure: The inside story of climate politics under Rudd and Gillard*, Melbourne: Black Inc., 2014

Clinton, Hillary Rodham, *Hard Choices*, New York: Simon & Schuster, 2014

Coll, Steve, *Private Empire: ExxonMobil and American Power*, London: Penguin Books, 2013

Combet, Greg, with Mark Davis, *The Fights of My Life*, Melbourne: Melbourne University Press, 2014

Ferguson, Adele, *Gina Rinehart: The untold story of the richest woman in the world*, Sydney: Pan Macmillan, 2012

Garrett, Peter, *Big Blue Sky: A memoir*, Sydney: Allen & Unwin, 2015

Gelbspan, Ross, *Boiling Point: How politicians, Big Oil and Coal, journalists, and activists are fueling the climate crisis—and what we can do to avert disaster*, New York: Basic Books, 2005

Gillard, Julia, *My Story*, Sydney: Random House, 2014

Howard, John, *Lazarus Rising: A personal and political autobiography*, Sydney: HarperCollins, 2010

Kelly, Dominic, *Political Troglodytes and Economic Lunatics: The hard right in Australia*, Melbourne: La Trobe University Press, 2019

Kelly, Paul, *Triumph and Demise: The broken promise of a Labor generation*, Melbourne: Melbourne University Press, 2014

Kolbert, Elizabeth, *Field Notes from a Catastrophe: Man, nature and climate change*, New York: Bloomsbury Publishing, 2006

Mann, Michael E., *The Hockey Stick and the Climate Wars: Dispatches from the front lines*, New York: Columbia University Press, 2012

Manning, Paddy, *Born to Rule: The unauthorised biography of Malcolm Turnbull*, Melbourne: Melbourne University Press, 2015

Markus, Andrew, *Race: John Howard and the remaking of Australia*, Sydney: Allen & Unwin, 2001

Marr, David, 'Political animal: The making of Tony Abbott', *Quarterly Essay*, no. 47, Melbourne: Black Inc., 2012

Marr, David, 'Power trip: The political journey of Kevin Rudd', *Quarterly Essay*, no. 38, Melbourne: Black Inc., 2010

Marshall, Bryan W. & Bruce C. Wolpe, *The Committee: A study of policy, power, politics and Obama's historic legislative agenda on Capitol Hill*, Ann Arbor: University of Michigan Press, 2018

Milne, Christine, *An Activist Life*, Brisbane: UQP, 2017

Pearse, Guy, *High and Dry: John Howard, climate change and the selling of Australia's future*, Melbourne: Penguin Books, 2007

Pooley, Eric, *The Climate War: True believers, power brokers, and the fight to save the earth*, New York: Hyperion, 2010

Robb, Andrew, *Black Dog Daze: Public life, private demons*, Melbourne: Melbourne University Press, 2011

Rudd, Kevin, *The PM Years*, Sydney: Pan Macmillan, 2018

Sales, Leigh, *Detainee 002: The case of David Hicks*, Melbourne: Melbourne University Press, 2007

Savva, Niki, *Plots and Prayers: Malcolm Turnbull's demise and Scott Morrison's ascension*, Melbourne: Scribe, 2019

Savva, Niki, *The Road to Ruin*, Melbourne: Scribe, 2016

Swan, Wayne, *The Good Fight*, Sydney: Allen & Unwin, 2014

Turnbull, Malcolm, *A Bigger Picture*, Melbourne: Hardie Grant Books, 2020

Watson, R.T., M.C. Zinyowera & R.H. Moss (eds), *The Regional Impacts of Climate Change: An assessment of vulnerability*, Cambridge: Cambridge University Press, 1997

Whitman, Christine Todd, *It's My Party Too*, New York: Penguin Books, 2006

Official reports and documents

Australian Energy Market Operator, 'Black system South Australia, 28 September 2016—final report', March 2017, www.aemo.com.au/-/media/Files/Electricity/NEM/Market_Notices_and_Events/Power_System_Incident_Reports/2017/Integrated-Final-Report-SA-Black-System-28-September-2016.pdf

Burns, Gary, Leanne Adams & Guy Buckley, 'Independent review of the extreme weather event, South Australia, 28 September to 5 October 2016: Report presented to the premier of South Australia', no date, www.dpc.sa.gov.au/__data/assets/pdf_file/0003/15195/Independent-Review-of-Extreme-Weather-complete.pdf

Carpenter, Chad, *The Bali Action Plan: Key issues in the climate negotiations, summary for policymakers*, United Nations Development Programme, September 2008, www.undp.org/content/dam/undp/library/EnvironmentandEnergy/ClimateChange/Bali_Road_Map_Key_Issues_Under_Negotiation.pdf

Climate Institute, 'Paris Climate Summit: Catalyst for further action?' November 2015, www.climateinstitute.org.au/articles/publications/paris-climate-summit-brief.html

Douvere, Fanny & Tim Badman, 'Reactive monitoring mission to Great Barrier Reef (Australia) 6th to 14th March 2012', UNESCO, Paris, 2012, whc.unesco.org/en/documents/117104

Meilstrup, Per, 'The runaway summit: The background story of the Danish Presidency of COP15, the UN Climate Change Conference', in Nanna Hvidt & Hans Mouritzen (eds), *Danish Foreign Policy Yearbook*

2010, Copenhagen: Danish Institute for International Studies, 2010, pp. 113–35, www.fao.org/fileadmin/user_upload/rome2007/docs/What really happen in COP15.pdf

Rocha, Marcia, Bill Hare, Paola Yanguas Parra et al., *Implications of the Paris Agreement for Coal Use in the Power Sector*, 2015, Berlin: Climate Analytics, climateanalytics.org/media/climateanalytics-coalreport_nov2016_1.pdf

Stern, Nicholas, *The Economics of Climate Change: The Stern Review*, Cambridge: Cambridge University Press, 2007, webarchive.national archives.gov.uk/20100407172811tf_/http://www.hm-treasury.gov.uk/stern_review_report.htm

Stock, Andrew, Will Steffen & Martin Rice, 'Angry summer 2016/17: Climate change super-charging extreme weather', Climate Council, 2017, www.climatecouncil.org.au/resources/angry-summer-report

INTERNATIONAL PANEL ON CLIMATE CHANGE

Climate Change 2001: Synthesis Report. A contribution of Working Groups I, II, and III to the Third Assessment Report of the Intergovernmental Panel on Climate Change, Cambridge and New York: Cambridge University Press, www.grida.no/publications/267

Climate Change 2014: Synthesis Report. Contribution of Working Groups I, II and III to the Fifth Assessment Report of the Intergovernmental Panel on Climate Change, Geneva: IPCC, 2014, www.ipcc.ch/site/assets/uploads/2018/02/SYR_AR5_FINAL_full.pdf

Global Warming of 1.5°C: An IPCC Special Report on the impacts of global warming of 1.5°C above pre-industrial levels and related global greenhouse gas emission pathways, in the context of strengthening the global response to the threat of climate change, sustainable development, and efforts to eradicate poverty, Special Report, Geneva: World Meteorological Organization, 2018, www.ipcc.ch/site/assets/uploads/sites/2/2019/05/SR15_SPM_version_report_HR.pdf

'Headline statements from the summary for policymakers', *Climate Change 2014: Synthesis report*, Fifth Assessment Report, 2014, www.ipcc.ch/site/assets/uploads/2018/02/ar5_syr_headlines_en.pdf

'Summary for policymakers', *Climate Change 2007: The physical science basis. Contribution of Working Group I to the Fourth Assessment Report of the Intergovernmental Panel on Climate Change*, Cambridge and New York: Cambridge University Press, 2007, www.ipcc.ch/site/assets/uploads/2018/02/ar4-wg1-spm-1.pdf

AUSTRALIAN GOVERNMENT

Bingham, Frank & Brent Perkins, 'Australia's coal and iron ore exports, 2001 to 2011', Department of Foreign Affairs and Trade, 2012, dfat.gov.au/about-us/publications/Documents/australias-coal-and-iron-ore-exports-2001-to-2011.pdf

Commonwealth of Australia, 'Australia's intended nationally determined contribution to a new climate change agreement', August 2015, www4.unfccc. int/sites/ndcstaging/PublishedDocuments/AustraliaFirst/Australias Intended Nationally Determined Contribution to a new Climate Change Agreement – August 2015.pdf

Finkel, Alan (chair), *Independent Review into the Future Security of the National Electricity Market: Blueprint for the future*, Canberra: Commonwealth of Australia, 2017, www.energy.gov.au/sites/default/ files/independent-review-future-nem-blueprint-for-the-future-2017.pdf

Fraser, Bernie, 'Some observations on Australia's post-2020 emissions reduction target', 14 August 2015, Climate Change Authority, climate changeauthority.gov.au/sites/prod.climatechangeauthority.gov.au/files/ files/CFI/CCA-statement-on-Australias-2030-target.pdf

Garnaut, Ross, *Climate Change Review: Final report*, Melbourne: Cambridge University Press, 2008

Schandl, Heinz, Tim Baynes, Nawshad Haque et al., *Whole of Life Greenhouse Gas Emissions Assessment of a Coal Seam Gas to Liquefied Natural Gas Project in the Surat Basin, Queensland, Australia: Final report for GISERA Project G2*, Canberra: CSIRO Energy, 2019, gisera. csiro.au/wp-content/uploads/2019/07/GISERA_G2_Final_Report-whole-of-life-GHG-assessment.pdf

Senate Environment, Communications, Information Technology and the Arts References Committee, *The Heat is On: Australia's greenhouse future*, Canberra: Commonwealth of Australia, 2000, www.aph.gov.au/ parliamentary_business/committees/senate/environment_and_ communications/completed_inquiries/1999-02/gobalwarm/report/ c09

Shergold, Peter (chair), *Prime Ministerial Task Group on Emissions Trading Report*, Canberra: Commonwealth of Australia, 2007, archived at webarchive.nla.gov.au/awa/20070604000621/http://pandora.nla. gov.au/pan/72614/20070601-0000/www.pmc.gov.au/publications/ emissions/index.html

BUREAU OF METEOROLOGY

'Severe thunderstorm and tornado outbreak South Australia 28 September, 2016', [2016], www.dpc.sa.gov.au/__data/assets/pdf_file/0007/15199/ Attachment-3-BoM-Severe-Thunderstorm-and-Tornado-Outbreak-28-September-2016.pdf

'Special climate statement 72—dangerous bushfire weather in spring 2019', 18 December 2019, www.bom.gov.au/climate/current/statements/scs72. pdf

with CSIRO, *State of the Climate Report 2014*, Canberra: Commonwealth of Australia, 2014, www.bom.gov.au/state-of-the-climate/2014

CLIMATE CHANGE AUTHORITY

'Final report on Australia's future emissions reduction targets', 2 July 2015, climatechangeauthority.gov.au/sites/prod.climatechangeauthority.gov. au/files/Final-report-Australias-future-emissions-reduction-targets.pdf

Reducing Australia's Greenhouse Gas Emissions—Targets and Progress Review, Final Report, Canberra: Commonwealth of Australia, 2014, www.climatechangeauthority.gov.au/files/files/Target-Progress-Review/ Targets and Progress Review Final Report.pdf

Renewable Energy Target Review, Canberra: Commonwealth of Australia, 2014, climatechangeauthority.gov.au/sites/prod.climatechangeauthority. gov.au/files/files/reviews/ret/2014/review.pdf

GREAT BARRIER REEF MARINE PARK AUTHORITY

Final Report: 2016 coral bleaching event on the Great Barrier Reef, Canberra: Commonwealth of Australia, 2017, elibrary.gbrmpa.gov.au/jspui/ bitstream/11017/3206/1/Final-report-2016-coral-bleaching-GBR.pdf

Great Barrier Reef Outlook Report 2019, Canberra: Commonwealth of Australia, 2019, elibrary.gbrmpa.gov.au/jspui/bitstream/11017/3474/10/ Outlook-Report-2019-FINAL.pdf

Great Barrier Reef Region Strategic Assessment: Strategic assessment report, Canberra: Commonwealth of Australia, 2014

Report on the Great Barrier Reef Zoning Plan, 2003, Townsville: GBRMPA, 2005, elibrary.gbrmpa.gov.au/jspui/bitstream/11017/407/1/Report-on-the-Great-Barrier-Reef-Marine-Park-zoning-plan.pdf

with Australian Greenhouse Office, *Great Barrier Reef Climate Change Action Plan 2007–2012*, Canberra: Commonwealth of Australia, 2007, elibrary.gbrmpa.gov.au/jspui/bitstream/11017/198/1/Great-Barrier-Reef-Climate-Change-Action-Plan-2007-2012.pdf

Scholarly reports

Burleson, Elizabeth, 'The Bali Climate Change Conference', *American Society of International Law Insights*, 2008, vol. 12, no. 4, www.asil.org/ insights/volume/12/issue/4/bali-climate-change-conference

Cai, Wenju, Ariaan Purich, Tim Cowan et al., 'Did climate change-induced rainfall trends contribute to the Australian Millennium Drought?', *Journal of Climate*, 2014, vol. 27, no. 9, pp. 3145–68, doi:10.1175/ JCLI-D-13-00322.1

Christoff, Peter, 'Aiming high: On Australia's emissions reduction targets', *UNSW Law Journal*, 2008, vol. 31, no. 3, pp. 861–79, www.austlii.edu. au/au/journals/UNSWLawJl/2008/46.pdf

Hoegh-Guldberg, Ove, *The Implications of Climate Change for Australia's Great Barrier Reef*, WWF, 2004, www.cakex.org/documents/ implications-climate-change-australias-great-barrier-reef

Hoegh-Guldberg, Ove, 'Climate change, coral bleaching and the future of the world's coral reefs', *Marine Freshwater Research*, 1999, vol. 50, no. 8, pp. 839–66, doi:10.1071/MF99078

Hoegh-Guldberg, Ove & G. Jason Smith, 'The effect of sudden changes in temperature, light and salinity on the population density and export of zooxanthellae from the reef corals *Stylophora pistillata* Esper and *Seriatopora hystrix* Dana', *Journal of Experimental Marine Biology and Ecology*, 1989, vol. 129, no. 3, pp. 279–303, doi:10.1016/0022-0981(89)90109-3

Hughes, Terry P., James T. Kerry, Andrew H. Baird et al., 'Global warming impairs stock-recruitment dynamics of corals', *Nature*, 2019, vol. 568, pp. 387–90, www.nature.com/articles/s41586-019-1081-y

Hughes, Terry P., James T. Kerry, Sean R. Connolly et al., 'Ecological memory modifies the cumulative impact of recurrent climate extremes', *Nature Climate Change*, 2019, vol. 9, pp. 40–43, doi:10.1038/s41558-018-0351-2

Jotzo, Frank, 'Australia's clean energy future', *Environmental Finance*, December 2011–January 2012, pp. 14–15

Macintosh, Andrew, 'The Australia clause and REDD: A cautionary tale', *Climate Change*, 2012, vol. 112, no. 2, pp. 169–88, doi:10.1007/s10584-011-0210-xs

Morrison, Tiffany H., 'Evolving polycentric governance of the Great Barrier Reef', *PNAS*, 8 March 2017, vol. 114, no. 15, pp. E3013–21, doi:10.1073/pnas.1620830114

Rootes, Christopher, 'The first climate change election? The Australian general election of 24 November 2007', *Environmental Politics*, 2008, vol. 17, no. 3, pp. 473–80, doi:10.1080/09644010802065815

Ward, Jordan & Mick Power, 'Cleaning up Victoria's power sector: The full social cost of Hazelwood power station', Harvard Kennedy School of Government, Boston, 24 February 2015, environmentvictoria.org.au/2015/02/24/cleaning-victorias-power-sector-full-social-cost-hazelwood-power-station

INDEX